RF and MICROWAVE SEMICONDUCTOR DEVICE HANDBOOK

RF and MICROWAVE SEMICONDUCTOR DEVICE HANDBOOK

Editor-in-Chief
MIKE GOLIO

CRC PRESS
Boca Raton London New York Washington, D.C.

This material was previously published in *The RF and Microwave Handbook.* © CRC Press LLC 2001.

Library of Congress Cataloging-in-Publication Data

RF and microwave semiconductor device handbook / editor-in-chief Mike Golio.
p. cm.
Includes bibliographical references and index.
ISBN 0-8493-1562-X
1. Microwave devices—Handbooks, manuals, etc. 2. Semiconductors—Handbooks, manuals, etc. 3. Very high speed integrated circuits—Handbooks, manuals, etc. I. Golio, John Michael, 1954–

TK7876 .R495 2002
621.384′134—dc21

2002074129
CIP

Direct all inquiries to CRC Press LLC, 2000 N.W. Corporate Blvd., Boca Raton, Florida 33431.

Visit the CRC Press Web site at www.crcpress.com

International Standard Book Number 0-8493-1562-X
Library of Congress Card Number 2002074129
Printed in the United States of America 1 2 3 4 5 6 7 8 9 0
Printed on acid-free paper

Preface

The purpose of the *CRC RF & Microwave Semiconductor Device Handbook* is to provide a single volume comprehensive reference for high frequency semiconductor devices. It is intended to be a starting point for programs involving development, technology comparison, or acquisition for RF and wireless semiconductor devices. The articles that comprise the handbook provide important information for practicing engineers in industry, government, and academia. The intended audience also includes microwave and other electrical engineers requiring information outside of their area of expertise as well as managers, marketers, and technical support workers who need better understanding of the fields driving their decisions.

The book covers every aspect of semiconductor device technology including basic material characteristics, system level concerns and constraints, simulation and modeling of devices, packaging, as well as detailed discussions of device operation. Individual chapters discuss details of the properties and characteristics of each semiconductor device type. Devices included in the text include: Varactors, Schottky diodes, transit-time devices, BJTs, HBTs, MOSFETs, MESFETs, and HEMTs. Additional chapters also examine wide bandgap devices as well as monolithic microwave integration. Finally, all of the articles provide the reader with additional references to related expert literature.

Acknowledgments

This handbook would simply never have been completed if it were not for the efforts of the managing editor, Janet Golio. I am also significantly indebted to the Handbook Editorial Board. This Board contributed to every phase of handbook development. Their efforts are reflected in the organization and outline of the material, selection and recruitment of authors, article contributions, and review of the articles. I am happy to acknowledge their help. I also thank the handbook professionals at CRC Press.

Editor-in-Chief

Dr. J. Michael Golio is the director, RF Technology Applications of Thoughtbeam, a Motorola company. His work focuses on the evaluation and commercialization of emerging compound semiconductor material technologies — especially for RF and microwave applications.

Dr. Golio received his BSEE degree from the University of Illinois in 1976. He worked for 2 years in the Microwave Tunable Devices Organization at Watkins-Johnson before returning to school to complete his MSEE and PhD degrees at North Carolina State University in 1980 and 1983, respectively. His graduate research focused on microwave devices, nonlinear models and carrier transport in compound semiconductors. Upon completion of his graduate work, he served as an assistant professor of electrical engineering at Arizona State University, later joining the Motorola Government Electronics Group in 1986. There he directed research on characterization, parameter extraction and modeling of nonlinear microwave devices. In 1991, he moved to Motorola's Semiconductor Products Sector to develop a GaAs fabrication facility to address commercial products including chips for cellular phones, digital pagers and wireless LANs. From 1996 to 2001 Dr. Golio was Director of the RF/Power Design Center at Rockwell Collins in Cedar Rapids, IA. The center conducted research and development efforts into RF, microwave and antenna technologies for commercial and military avionics applications.

Dr. Golio is the author of over 100 publications. He is editor of two successful books: *Microwave MESFETs and HEMTs*, Artech House, 1991, and *RF and Microwave Handbook*, CRC Press, Boca Raton, FL, 2000. He has served as organizer for several microwave conferences, workshops and panel sessions. In 1996 he was elected Fellow of the IEEE. He has served as the Distinguished Microwave Lecturer for the IEEE MTT Society and is currently co-editor of the *IEEE Microwave Magazine*.

Editorial Board

Managing Editor

Contributors

Avram Bar-Cohen
University of Minnesota
Minneapolis, Minnesota

Peter A. Blakey
Northern Arizona University
Flagstaff, Arizona

Prashant Chavarkar
CREE Lighting Company
Goleta, California

John C. Cowles
Analog Devices, Inc.
Beaverton, Oregon

Walter R. Curtice
W.R. Curtice Consulting
Washington Crossing, Pennsylvania

Lawrence P. Dunleavy
University of South Florida
Tampa, Florida

Jack East
University of Michigan
Ann Arbor, Michigan

Karl J. Geisler
Department of Mechanical Engineering
University of Minnesota
Minneapolis, Minnesota

Mike Golio
Motorola, Inc.
Tempe, Arizona

Mike Harris
Georgia Tech Research Institute
Atlanta, Georgia

Allan D. Kraus
Allan D. Kraus Associates
Beachwood, Ohio

William Liu
Texas Instruments
Dallas, Texas

Leonard MacEachern
Department of Electrical Engineering
Carleton University
Ottawa, Ontario, Canada

Tajinder Manku
Department of Electrical & Computer Engineering
University of Waterloo
Waterloo, Ontario, Canada

Umesh K. Mishra
University of California
Santa Barbara, California

Karen E. Moore
Department of Animal Sciences
University of Florida
Gainesville, Florida

Jeanne S. Pavio
Motorola SPS
Phoenix, Arizona

Michael S. Shur
Rensselaer Polytechnic Institute
Troy, New York

Jan Stake
Department of Microelectronics
Chalmers University of Technology
Goteborg, Sweden

Robert J. Trew
Virginia Tech University
Blacksburg, Virginia

Contents

1

Varactors

Jan Stake
Chalmers University of Technology

1.1 Introduction

A varactor is a nonlinear reactive device used for harmonic generation, parametric amplification, mixing, detection, and voltage-variable tuning.[1] However, present applications of varactors are mostly for harmonic generation at millimeter and submillimeter wave frequencies, and as tuning elements in various microwave applications. Varactors normally exhibit a voltage-dependent capacitance and can be fabricated from a variety of semiconductor materials.[2] A common varactor is the reverse biased Schottky diode. Advantages of varactors are low loss and low noise. The maximum frequency of operation is mainly limited by a parasitic series resistance (see Fig. 1.1).

1.2 Basic Concepts

Many frequencies may interact in a varactor, and of those, some may be useful inputs or outputs, while the others are *idlers* that, although they are necessary for the operation of the device, are not part of any input or output. For instance, to generate high harmonics in a frequency multiplier it is more or less necessary to allow current at intermediate harmonics (idlers) to flow. Such idler circuits are usually realized as short-circuit resonators, which maximize the current at idler frequencies.

1.2.1 Manley-Rowe Formulas

The Manley-Rowe formulas[3] for *lossless* nonlinear reactances are useful for intuitive understanding of multipliers, frequency converters, and dividers. Consider a varactor excited at two frequencies f_p and f_s; the corresponding general Manley-Rowe formulas are

0-8493-1562-X/03/$0.00+$1.50

$$\sum_{m=1}^{\infty}\sum_{n=-\infty}^{\infty}\frac{mP_{m,n}}{nf_p+mf_s}=0$$

$$\sum_{m=-\infty}^{\infty}\sum_{n=1}^{\infty}\frac{nP_{m,n}}{nf_p+mf_s}=0$$

where m and n are integers representing different harmonics and $P_{m,n}$ is the average power flowing into the nonlinear reactance at the frequencies nf_p and mf_s.

- Frequency multiplier (m = 0): if the circuit is designed so that only real power can flow at the input frequency, f_p, and at the output frequency, nf_p, the above equations predict a theoretical efficiency of 100%. The Manley-Rowe equation is $P_1 + P_n = 0$.
- Parametric amplifier and frequency converter: assume that the RF-signal at the frequency f_s is small compared to the pump signal at the frequency f_p. Then, the powers exchanged at sidebands of the frequencies nf_p and mf_s for m different from 1 and 0 are negligible. Furthermore, one of the Manley-Rowe formulas only involves the small-signal power as

$$\sum_{n=-\infty}^{\infty}\frac{P_{1,n}}{nf_p+f_s}=0$$

Hence, the nonlinear reactance can act as an amplifying upconverter for the input signal at frequency f_s and output signal extracted at $f_u = f_s + f_p$ with a gain of

$$\frac{P_u}{P_s}=\frac{P_{1,1}}{P_{1,0}}=-\left(1+\frac{f_p}{f_s}\right)=-\frac{f_u}{f_s}$$

1.2.2 Varactor Model

The intrinsic varactor model in Fig. 1.1 has a constant series resistance, R_s, and a nonlinear differential elastance, $S(V) = dV/dQ = 1/C(V)$, where V is the voltage across the diode junction. This simple model is used to describe the basic properties of a varactor and is adequate as long as the displacement current is much larger than any conduction current across the junction. A rigorous analysis should also include the effect of a frequency- and voltage-dependent series resistance, and the equivalent circuit of parasitic elements due to packaging and contacting. The differential elastance is the slope of the voltage-charge relation of the diode and the reciprocal of the differential capacitance, $C(V)$. Since the standard varactor model consists of a resistance in series with a nonlinear capacitance, the elastance is used rather than the capacitance. This simplifies the analysis and gives, generally, a better understanding of the varactor. The differential elastance can be measured directly and is used rather than the ratio of voltage to charge.

The elastance versus voltage for a conventional varactor and a symmetric varactor are shown in Fig. 1.2. In both cases, the maximum available elastance is achieved at the breakdown voltage. The conventional varactor is reverse biased in order to allow maximum elastance swing and avoid any forward conduction current. A symmetric varactor will only produce odd harmonics when a sinusoidal signal is applied. This means that a varactor frequency tripler can be realized without any second harmonic idler circuit or DC

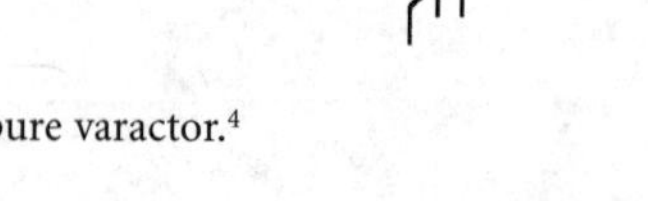

FIGURE 1.1 Equivalent circuit of a pure varactor.[4]

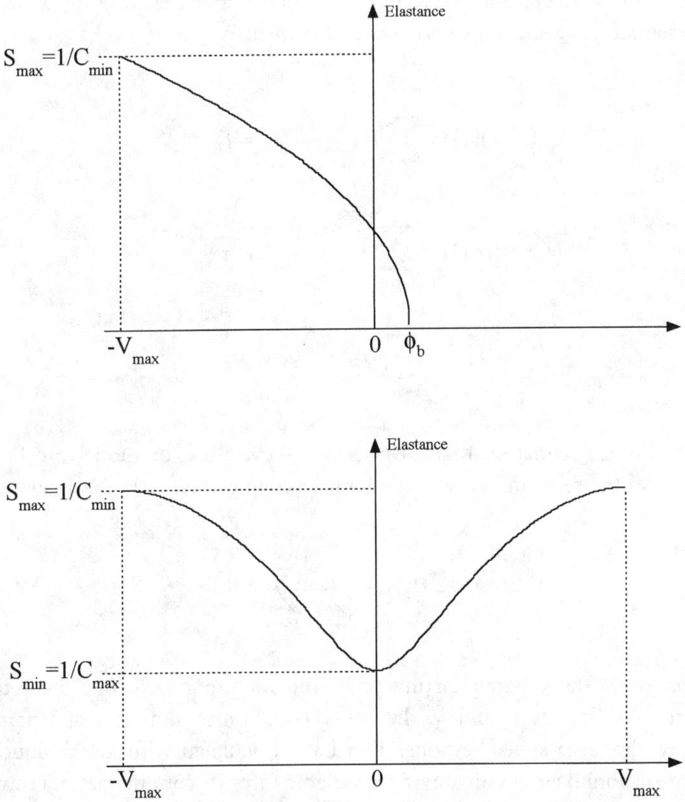

FIGURE 1.2 Elastance as a function of voltage for (top) one side junction diodes and (bottom) symmetric diodes.

bias. This simplifies the design of such circuits and, hence, many novel symmetric varactors have been proposed. Among these, the Heterostructure Barrier Varactor (HBV)[5] has so far shown the best performance.

1.2.3 Pumping

The pumping of a varactor is the process of passing a large current at frequency f_p through the varactor. The nonlinear varactor then behaves as a time-varying elastance, $S(t)$, and the series resistance dissipates power due to the large-signal current. The allowable swing in elastance is limited by the maximum elastance of the device used. Hence, the time domain equation describing the varactor model in Fig. 1.1 is given by:

$$V(t) = R_s i(t) + \int S(t) i(t) dt$$

The above equation, which describes the varactor in the time domain, must be solved together with equations describing the termination of the varactor. How the varactor is terminated at the input, output, and idler frequencies has a strong effect on the performance. The network has to terminate the varactor at some frequencies and couple the varactor to sources and loads at other frequencies. Since this embedding circuit is best described in the frequency domain, the above time domain equation is converted to the frequency domain. Moreover, the varactor is usually pumped strongly at one frequency, f_p, by a local oscillator. If there are, in addition, other small signals present, the varactor can be analyzed in two steps: (1) a large signal simulation of the pumped varactor at the frequency f_p, and (2) the varactor behaves like a time-varying linear elastance at the signal frequency, f_s. For the large signal analysis, the voltage, current, and differential elastance can be written in the forms

$$i(t)=\sum_{k=-\infty}^{\infty} I_k e^{jk\omega_p t},\ I_{-k}=I_k^*$$

$$v(t)=\sum_{k=-\infty}^{\infty} V_k e^{jk\omega_p t},\ V_{-k}=V_k^*$$

$$S(t)=\sum_{k=-\infty}^{\infty} S_k e^{jk\omega_p t},\ S_{-k}=S_k^*$$

Hence, the time domain equation that governs the above varactor model can be converted to the frequency domain and the relation between the Fourier coefficients, I_k, V_k, S_k, reads

$$V_k = R_s I_k + \frac{1}{jk\omega_p}\sum_{l=-\infty}^{\infty} I_l S_{k-l}$$

The above equation is the general starting point for analyzing varactors. Since there is a relation between the Fourier coefficients S_k and I_k, the above equation is nonlinear and hard to solve for the general case. Today, the large signal response is usually calculated with a technique called harmonic balance.[6] This type of nonlinear circuit solver is available in most commercial microwave CAD tools.

Assume that the varactor is fully pumped and terminated so that the voltage across the diode junction becomes sinusoidal. The corresponding elastance waveforms for the conventional and symmetrical varactor in Fig. 1.2 are shown in Fig. 1.3. It is important to note that the fundamental frequency of the symmetric elastance device is twice the pump frequency.

The nonlinear part of the elastance, $S(t) - S_{min}$, creates harmonics and its impedance should be large compared to the series resistance, R_s, for a good varactor. The impedance ratio of the series resistance and the nonlinear part of the elastance at the fundamental frequency can be written as

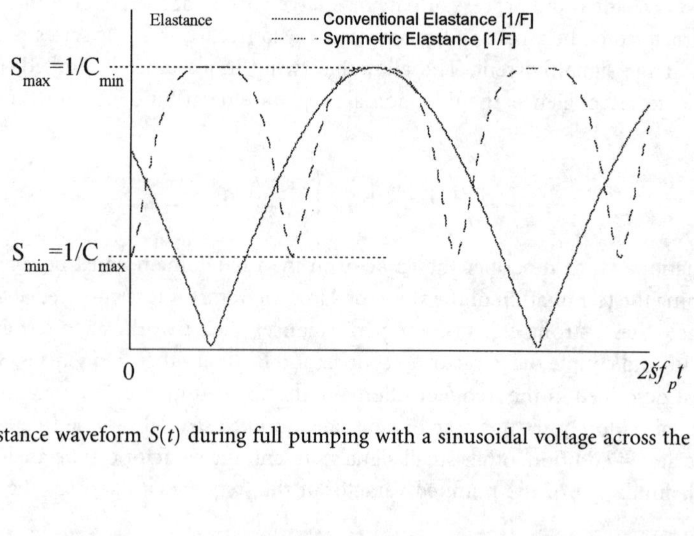

FIGURE 1.3 Elastance waveform $S(t)$ during full pumping with a sinusoidal voltage across the diode junction.

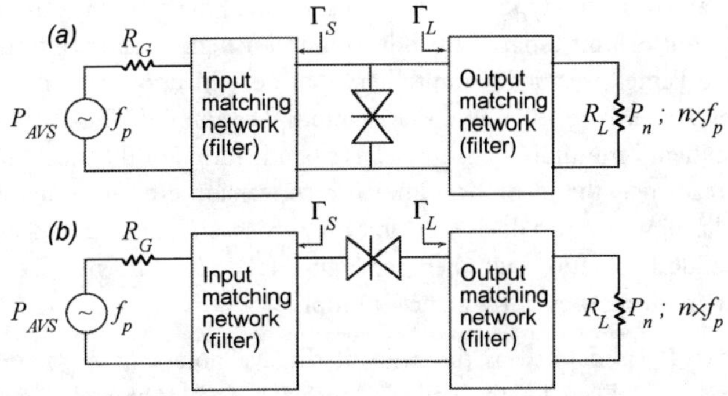

FIGURE 1.4 Block scheme of nth-order frequency multiplier circuit with (a) shunt-mounted and (b) series-mounted diodes.

$$\frac{R_s}{\dfrac{S_{max}-S_{min}}{2\pi f_p}}=\frac{f_p}{f_c}$$

and the dynamic cutoff frequency is introduced as

$$f_c=\frac{S_{max}-S_{min}}{2\pi R_s}$$

The dynamic cutoff frequency is an important figure of merit for many varactor applications and a typical value for a state-of-the-art varactor is more than 1 THz. The starting point for a varactor design is, hence, to maximize the elastance swing, $S_{max} - S_{min}$, and minimize any losses, R_s. For semiconductor varactors, the maximum elastance swing is limited by at least one of the following conditions:

- depletion layer punch-through,
- large electron conduction from impact ionization or forward conduction current,
- current saturation. The saturated electron velocity in the material determines the maximum length an electron can travel during a quarter of a pump cycle.[7]

Increasing the pump power beyond any of the above conditions will result in reduced performance and probably introduce extra noise.

1.3 Varactor Applications

1.3.1 Frequency Multipliers

Varactor frequency multipliers are extensively used to provide LO power to sensitive millimeter- and submillimeter-wavelength receivers. Today, frequency multipliers are the main application of varactors. Solid-state multipliers are relatively inexpensive, compact, lightweight, and reliable compared to vacuum tube technology, which makes them suitable for space applications at these frequencies. State-of-the-art balanced Schottky doublers can deliver 55 mW at 174 GHz[8] and state-of-the-art four-barrier HBV triplers deliver about 9 mW at 248 GHz.[9]

Frequency multiplication or harmonic generation in devices occur due to their nonlinearity. Based on whether the multiplication is due to a nonlinear resistance or a nonlinear reactance, one can differentiate between the varistor and varactor type of multipliers. Varactor type multipliers have a high potential conversion efficiency, but exhibit a narrow bandwidth and a high sensitivity to operating conditions. According to the Page-Pantell inequality, multipliers that depend upon a nonlinear resistance have at most an efficiency of $1/n^2$, where n is the order of multiplication.[10,11] The absence of reactive energy storage in varistor frequency multipliers ensures a large bandwidth. For the ideal varactor multiplier, i.e., a lossless nonlinear reactance, the theoretical limit is a conversion efficiency of 100% according to the Manley-Rowe formula. However, real devices exhibit properties and parameters that are a mixture of the ideal varistor and the ideal varactor multiplier (see Figure 1.4). The following set of parameters is used to describe and compare properties of frequency multipliers:

- Conversion loss, L_n, is defined as the ratio of the available source power, P_{AVS}, to the output harmonic power, P_n, delivered to the load resistance. It is usually expressed in decibels. The inverted value of L_n, i.e., the conversion efficiency, η_n, is often expressed as a percent.
- In order to minimize the conversion loss, the optimum source and load embedding impedances, Z_S and Z_L, should be provided to the diode. Optimum source and load impedances are found from maximizing, e.g., the conversion efficiency, and they depend on each other and on the input signal level. In a nonlinear circuit, such as a multiplier, it is not possible to define a true impedance. However, a "quasi-impedance", Z_n, can be defined for periodic signals as

$$Z_n = \frac{V_n}{I_n}$$

where V_n and I_n are the voltage and the current, respectively, at the nth harmonic.

Basic Principles of Single Diode Frequency Multipliers — Single diode frequency multipliers can either be shunt or series mounted. In both cases the input and the output filter should provide optimum embedding impedances at the input and output frequencies, respectively. The output filter should also provide an open circuit for the shunt-mounted varactor and a short circuit for the series-mounted varactor at the pump frequency. The same arguments apply to the input filter at the output frequency. Analysis and design of conventional doublers and high order varactor multipliers are described well in the book by Penfield et al.[1] and in Reference 12.

In addition to the above conditions, the correct impedances must be provided at the idler frequencies for a high order multiplier (e.g., a quintupler). In general, it is hard to achieve optimum impedances at the different harmonics simultaneously. Therefore, a compromise has to be found.

Performance of Symmetric Varactor Frequency Multipliers — In Fig. 1.5 a calculation of the minimum conversion loss for a tripler and a quintupler is shown. To systematically investigate how the tripler and quintupler performance depends on the shape of the *S-V* characteristic, a fifth degree polynomial model was employed by Dillner et al.[13] The best efficiency is obtained for a *S-V* characteristic with a large nonlinearity at zero volts or a large average elastance during a pump cycle. The optimum idler circuit for the quintupler is an inductance in resonance with the diode capacitance (i.e., maximized third harmonic current).

Practical Multipliers — Since frequency multipliers find applications mostly as sources at higher millimeter and submillimeter wave frequencies, they are often realized in waveguide mounts[14,15] (see Fig. 1.6). A classic design is the arrangement of crossed rectangular waveguides of widths specific for the input and output frequency bands. The advantages are:

- The input signal does not excite the output waveguide, which is cut off at the input frequency.
- Low losses.
- The height of the waveguide in the diode mounting plane may be chosen to provide the electrical matching conditions. Assuming a thin planar probe, the output embedding impedance is given by analytical expressions.[16]
- Movable short circuits provide input/output tunability.

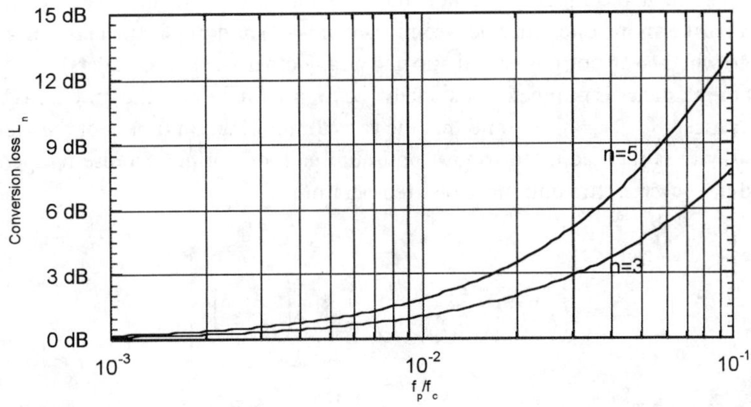

FIGURE 1.5 The minimum conversion loss for a tripler and a quintupler for the symmetric *S-V* characteristic shown in Fig. 1.2. The pump frequency is normalized to the dynamic cutoff frequency. For the quintupler case, the idler circuit is an inductance in resonance with the diode capacitance.

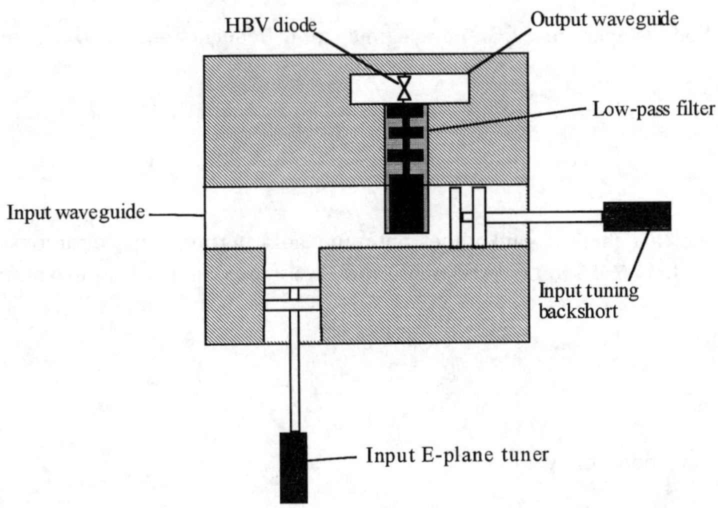

FIGURE 1.6 Schematic diagram of a crossed waveguide frequency multiplier. The output signal is isolated from the input waveguide with a low pass filter.

Today, whole waveguide mounts can be analyzed and designed using commercially available high frequency electromagnetic CAD tools. They either solve Maxwell's equations in the frequency domain or in the time domain using the FDTD method.

The inherently limited bandwidth of varactors can be improved by employing a transmission line periodically loaded with varactors.[17,18] These NonLinear Transmission Lines (NLTLs) can currently provide the largest bandwidth and still achieve a reasonable conversion efficiency as a frequency multiplier. Simultaneous effects of nonlinearity and dispersion may also be used for pulse compression (soliton propagation).

1.3.2 Frequency Converters

The varactor is useful as a frequency converter because of its good noise properties, and because gain can be achieved. The nonlinear reactance is pumped at a frequency f_p, and a small signal is introduced at a frequency f_s. Power will exchange at frequencies of the form $nf_p + f_s$ (n can be negative). If the output frequency is higher than the input frequency, the varactor acts as an upconverter, otherwise it is a downconverter. Furthermore, one differ between lower (n = –1) and upper sideband (n = 1) converters, and according to whether or not power is dissipated at any other sidebands (idlers).

Assume that the elastance is pumped sinusoidally (i.e., $S_k = 0$ for $|k| > 1$), the varactor is open-circuited at all frequencies except $f_s, f_p, f_u = f_p + f_s$, and that the varactor termination tunes out the average elastance. The source resistance is then adjusted to give optimum gain or optimum noise temperature. For the upper sideband upconverter, the minimal noise temperature is

$$T_{min} = T_d \frac{2f_s}{m_1 f_c}\left[\frac{f_s}{m_1 f_c} + \sqrt{1+\left(\frac{f_s}{m_1 f_c}\right)^2}\right]$$

when the source resistance is

$$R_o^T = R_s \sqrt{1+\left(\frac{m_1 f_c}{f_s}\right)^2}$$

where T_d is the diode temperature, f_c is the dynamic cutoff frequency, and m_1 is the modulation ratio defined as

$$m_1 = \frac{|S_1|}{S_{max} - S_{min}}$$

It can be shown that there is gain under optimum noise termination conditions only for signal frequencies smaller than $0.455 m_1 f_c$.[1] A different source resistance will result in maximum gain

$$R_o^G = R_s \sqrt{1+\frac{m_1^2 f_c^2}{f_s f_u}}$$

The corresponding optimum gain is

$$G_{MAX} = \left(\frac{\dfrac{m_1 f_c}{f_s}}{1+\sqrt{1+\dfrac{m_1^2 f_c^2}{f_s f_u}}}\right)^2$$

As predicted by the Manley-Rowe formula for a lossless varactor, the gain increases as the output frequency, f_u, increases. The effect of an idler termination at $f_l = f_p - f_s$ can further increase the gain and reduce the noise temperature.

The above expressions for optimum noise and the corresponding source impedance are valid for the lower sideband upconverter as well. However, the lower sideband upconverter may have negative input and output resistances and an infinite gain causing stability problems and spurious oscillations. All pumped varactors may have such problems. With a proper choice of source impedance and pump frequency, it is possible to simultaneously minimize the noise and make the exchangeable gain infinite. This occurs for an "optimum" pump frequency of $f_p = \sqrt{m_1^2 f_c^2 + f_s^2}$ or approximately $m_1 f_c$ if the signal frequency is small. Further information on how to analyze, design, and optimize frequency converters can be found in the book by Penfield et al.[1]

1.3.3 Parametric Amplifiers

The parametric amplifier is a varactor pumped strongly at frequency f_p, with a signal introduced at frequency f_s. If the generated sidebands are terminated properly, the varactor can behave as a negative resistance at f_s. Especially the termination of the idler frequency, $f_p - f_s$, determines the real part of the impedance at the signal frequency. Hence, the varactor can operate as a negative resistance amplifier at the signal frequency, f_s. The series resistance limits the frequencies f_p and f_c for which amplification can be achieved and it also introduces noise.

The explanation of the effective negative resistance can be described as follows: The application of signal plus pump power to the nonlinear capacitance causes frequency mixing to occur. When current is allowed to flow at the idler frequency $f_p - f_s$, further frequency mixing occurs at the pump and idler frequencies. This latter mixing creates harmonics of f_p and $f_p - f_s$, and power at f_s is generated. When the power generated through mixing exceeds that being supplied at the signal frequency f_s, the varactor appears to have a negative resistance. If idler current is not allowed to flow, the negative resistance vanishes. Assuming that the elastance is pumped sinusoidally (i.e., $S_k = 0$ for $|k| > 1$), and the varactor is open circuited at all frequencies except f_s, f_p, $f_i = f_p - f_s$, and that the varactor termination tunes out the average elastance, gain can only be achieved if

$$f_s f_i \left(R_s + R_i\right) < R_s m_1^2 f_c^2$$

where R_i is the idler resistance. By terminating the varactor reactively at the idler frequency, it can be shown that a parametric amplifier attains a minimum noise temperature when pumped at the optimum pump frequency, which is exactly the same as for the simple frequency converter. This is true for nondegenerated amplifiers where the frequencies are well separated. The degenerate parametric amplifier operates with f_i close to f_s, and can use the same physical circuit for idler and signal frequencies. The degenerate amplifier is easier to build, but ordinary concepts of noise figure, noise temperature, and noise measure do not apply.

1.3.4 Voltage Tuning

One important application of varactors is voltage tuning. The variable capacitance is used to tune a resonant circuit with an externally applied voltage. This can be used to implement a Voltage Controlled Oscillator (VCO), since changing the varactor capacitance changes the frequency of oscillation within a certain range. As the bias is increased, the resonant frequency f_o increases from $f_{o,\min}$ to $f_{o,\max}$ as the elastance changes from $S_{\min}$ to $S_{\max}$. If the present RF power is low, the main limitations are the finite tuning range implied by the minimum and maximum elastance and the fact that the series resistance degrades the quality factor, Q, of the tuned circuit. The ratio of the maximum and minimum resonant frequency gives a good indication of the tunability

$$\frac{f_{o,\max}}{f_{o,\min}} \le \sqrt{\frac{S_{\max}}{S_{\min}}}$$

However, if the present RF power level is large, the average elastance, which determines the resonant frequency, depends upon drive level as well as bias. Second, the allowed variation of voltage is reduced for large RF power levels.

Since the varactor elastance is nonlinear, quite steep at low voltages, and almost flat at high voltages, the VCO tuning range is not naturally linear. However, an external bias circuit can improve the linearity of the VCO tuning range. It is also possible to optimize the doping profile of the varactor in terms of linearity, *Q*-value, or elastance ratio.

1.4 Varactor Devices

1.4.1 Conventional Diodes

Common conventional varactors at lower frequencies are reverse biased semiconductor abrupt p^+-n junction diodes made from GaAs or silicon.[2] However, metal-semiconductor junction diodes (Schottky diodes) are superior at high frequencies since the carrier transport only relies on electrons (unipolar device). The effective mass is lower and the mobility is higher for electrons compared to holes. Furthermore, the metal-semiconductor junction can be made very precisely even at a submicron level. A reverse biased Schottky diode exhibits a nonlinear capacitance with a very low leakage current. High frequency diodes are made from GaAs since the electron mobility is much higher than for silicon.

The hyperabrupt p^+-n junction varactor diode has a nonuniform n-doping profile and is often used for voltage tuning. The n-doping concentration is very high close to the junction and the doping profile is tailored to improve elastance ratio and sensitivity. Such doping profiles can be achieved with epitaxial growth or by ion implantation.

1.4.2 The Heterostructure Barrier Varactor Diode

The Heterostructure Barrier Varactor (HBV), first introduced in 1989 by Kollberg et al.,[5] is a symmetric varactor. The main advantage compared to the Schottky diode is that several barriers can be stacked epitaxially. Hence, an HBV diode can be tailored for a certain application in terms of both frequency and power handling capability. Moreover, the HBV operates unbiased and is a symmetric device, thus generating only odd harmonics. This greatly simplifies the design of high order and broadband multipliers.

The HBV diode is an unipolar device and consists of a symmetric layer structure. An undoped high band gap material (barrier) is sandwiched between two moderately n-doped, low band gap materials. The barrier prevents electron transport through the structure. Hence, the barrier should be undoped (no carriers), high and thick enough to minimize thermionic emission and tunnelling of carriers. When the diode is biased a depleted region builds up (Fig. 1.7), causing a nonlinear CV curve.

Contrary to the Schottky diode, where the barrier is formed at the interface between a metallic contact and a semiconductor, the HBV uses a heterojunction as the blocking element. A heterojunction, i.e., two

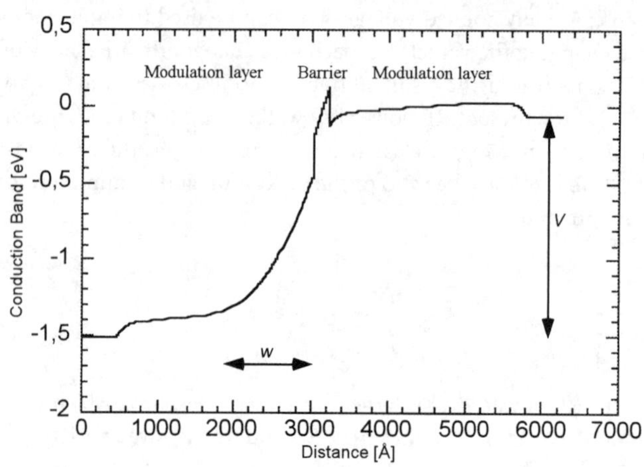

FIGURE 1.7 Conduction band of a biased GaAs/$Al_{0.7}$GaAs HBV.

TABLE 1.1 Generic Layer Structure of an HBV

Layer		Thickness [Å]	Doping Level [cm^{-3}]	
7	Contact	~3000	n^{++}	
6	Modulation	$l = 3000$	$N_d \sim 10^{17}$	
5	Spacer	$s = 50$	undoped	
4	Barrier	$b \geq 100$	undoped	xN
3	Spacer	$s = 50$	undoped	
2	Modulation	$l = 3000$	$N_d \sim 10^{17}$	
1	Buffer	—	n^{++}	
0	Substrate		n^{++} or SI	

Note: For N epitaxially stacked barriers, the layer sequence 2–5 is repeated N times.

adjacent epitaxial semiconductor layers with different band gaps, exhibits band discontinuities both in the valence and in the conduction band. Since the distance between the barriers (>1000 Å) is large compared to the de Broglie wavelength of the electron, it is possible to understand the stacked barrier structure as a series connection of N individual barriers. A generic layer structure of an HBV is shown in Table 1.1.

The HBV Capacitance — The parallel plate capacitor model, where the plate separation should be replaced with the sum of the barrier thickness, b, the spacer layer thickness, s, and the length of the depleted region, w, is normally an adequate description of the (differential) capacitance. The depletion length is bias dependent and the layer structure is symmetric, therefore the elastance is an even function of applied voltage and is given by

$$S = \frac{1}{C} = \frac{N}{A}\left(\frac{b}{\varepsilon_b} + \frac{s}{\varepsilon_d} + \frac{w}{\varepsilon_d}\right)$$

$$w = \sqrt{\frac{2\varepsilon_d |V_d|}{qN_d}}$$

where V_d is the voltage across the depleted region, N_d is the doping concentration in the modulation layers, b is the barrier thickness, s is the undoped spacer layer thickness, A is the device area, and ε_b and ε_d are the dielectric constants in the barrier material and modulation layers, respectively. The maximum capacitance or the minimum elastance, S_{min}, occurs at zero bias. However, due to screening effects, the minimum elastance, S_{min}, must include the extrinsic Debye length, L_D, as:

$$S_{min} = \frac{1}{C_{max}} = \frac{N}{A}\left(\frac{b}{\varepsilon_b} + \frac{2s}{\varepsilon_d} + \frac{2L_D}{\varepsilon_d}\right)$$

$$L_D \equiv \sqrt{\frac{\varepsilon_d kT}{q^2 N_d}}$$

To achieve a high C_{max}/C_{min} ratio, the screening length can be minimized with a sheet doping, N_s, at the spacer/depletion layer interface. The minimum capacitance, C_{min}, is normally obtained for punch through condition, i.e., $w = l$, or when the breakdown voltage, V_{max}, is reached.

FIGURE 1.8 Planar four-barrier HBV (37 μm²/CTH-NU2003J).

An accurate quasi-empirical expression for the *C-V* characteristic of homogeneously doped HBVs has been derived by Dillner et al.[19] The voltage across the nonlinear capacitor is expressed as a function of its charge as

$$V(Q) = N\left(\frac{bQ}{\varepsilon_b A} + 2\frac{sQ}{\varepsilon_d A} + Sign(Q)\left(\frac{Q^2}{2qN_d\varepsilon_d A^2} + \frac{4kT}{q}\left(1 - e^{\frac{|Q|}{2L_D A q N_d}}\right)\right)\right)$$

where T is the device temperature, q is the elementary charge, and Q is the charge stored in the HBV.

The substrate is either highly doped or semi-insulating (SI), depending on how the device is intended to be mounted. The contact layers (Nos. 1 and 7) should be optimized for low losses. Therefore, the buffer layer (No. 1) must be relatively thick ($\delta \sim 3$ μm) and highly doped for planar HBVs (see Fig. 1.8). The barrier itself can consist of different layers to further improve the blocking characteristic. The spacer prevents diffusion of dopants into the barrier layer and increases the effective barrier height. The thickness of the barrier layer will not influence the cutoff frequency, but it has some influence on the optimum embedding impedances. Hence, the thickness is chosen to be thick enough to avoid tunneling of carriers. Several III-V semiconductor material systems have been employed for HBVs. The best choices to date for HBVs are the lattice matched $In_{0,53}Ga_{0,47}As/In_{0,52}Al_{0,48}As$ system grown on InP substrate and the lattice matched GaAs/AlGaAs system grown on GaAs substrate. High dynamic cutoff frequencies are achieved in both systems. However, the GaAs/AlGaAs system is well characterized and relatively easy to process, which increases the probability of reproducible results. The $In_{0,53}GaAs/In_{0,52}AlAs$ system exhibits a higher electron barrier and is therefore advantageous from a leakage current point of view. The thickness and doping concentration of the modulation layers should be optimized for maximal dynamic cutoff frequency.[20]

In the future, research on wide bandgap semiconductors (e.g., InGaN) could provide solutions for very high power HBVs, a combination of a II-VI barrier for low leakage current and a III-V modulation layer for high mobility and peak velocity. Today, narrow bandgap semiconductors from the III-V groups (e.g., $In_xGa_{1-x}As$) seem to be the most suitable for submillimeter wave applications.

1.4.3 The Si/SiO$_2$/Si Varactor

By bonding two thin silicon wafers, each with a thin layer of silicon dioxide, it is possible to form a structure similar to HBV diodes from III-V compounds. The SiO_2 layer blocks the conduction current very efficiently, but the drawback is the relatively low mobility of silicon. If a method to stack several barriers can be developed, this material system may be interesting for lower frequencies where the series resistance is less critical.

1.4.4 The Ferroelectric Varactor

Ferroelectrics are dielectric materials characterized by an electric field and temperature-dependent dielectric constant. Thin films of $Ba_xSr_{1-x}TiO_3$ have been proposed to be used for various microwave applications. Parallel plate capacitors made from such films can be used in varactor applications. However, the loss mechanisms at strong pump levels and high frequencies have not yet been fully investigated.

References

1. P. Penfield and R. P. Rafuse, *Varactor Applications.* Cambridge: M.I.T. Press, 1962.
2. S. M. Sze, *Physics of Semiconductor Devices,* 2nd ed. Singapore: John Wiley & Sons, 1981.
3. J. M. Manley and H. E. Rowe, Some General Properties of Nonlinear Elements, *IRE Proc.,* 44, 78, 904–913, 1956.
4. A. Uhlir, The potential of semiconductor diodes in high frequency communications, *Proc. IRE,* 46, 1099–1115, 1958.
5. E. L. Kollberg and A. Rydberg, Quantum-barrier-varactor diode for high efficiency millimeter-wave multipliers, *Electron. Lett.,* 25, 1696–1697, 1989.
6. S. A. Maas, Harmonic Balance and Large-Signal-Small-Signal Analysis, in *Nonlinear Microwave Circuits.* Artech House, Norwood, MA, 1988.
7. E. L. Kollberg, T. J. Tolmunen, M. A. Frerking, and J. R. East, Current saturation in submillimeter wave varactors, *IEEE Trans. Microwave Theory and Techniques,* 40, 5, 831–838, 1992.
8. B. J. Rizzi, T. W. Crowe, and N. R. Erickson, A high-power millimeter-wave frequency doubler using a planar diode array, *IEEE Microwave and Guided Wave Letters,* 3, 6, 188–190, 1993.
9. X. Mélique, A. Maestrini, E. Lheurette, P. Mounaix, M. Favreau, O. Vanbésien, J. M. Goutoule, G. Beaudin, T. Nähri, and D. Lippens, 12% Efficiency and 9.5 dBm Output Power from InP-based Heterostructure Barrier Varactor Triplers at 250 GHz, presented at IEEE-MTT Int. Microwave Symposium, Anaheim, CA, 1999.
10. R. H. Pantell, General power relationship for positive and negative nonlinear resistive elements, *Proceedings IRE,* 46, 1910–1913, December 1958.
11. C. H. Page, Harmonic generation with ideal rectifiers, *Proceedings IRE,* 46, 1738–1740, October 1958.
12. C. B. Burckhardt, Analysis of varactor frequency multipliers for arbitrary capacitance variation and drive level, *The Bell System Technical Journal,* 675–692, April 1965.
13. L. Dillner, J. Stake, and E. L. Kollberg, Analysis of symmetric varactor frequency multipliers, *Microwave Opt. Technol. Lett.,* 15, 1, 26–29, 1997.
14. J. W. Archer, Millimeter wavelength frequency multipliers, *IEEE Trans. on Microwave Theory and Techniques,* 29, 6, 552–557, 1981.
15. J. Thornton, C. M. Mann, and P. D. Maagt, Optimization of a 250-GHz Schottky tripler using novel fabrication and design techniques, *IEEE Trans. on Microwave Theory and Techniques,* 46, 8, 1055–1061, 1998.
16. E. L. Eisenhart and P. J. Khan, Theoretical and experimental analysis for a waveguide mounting structure, *IEEE Trans. on Microwave Theory and Techniques,* 19, 8, 1971.
17. S. Hollung, J. Stake, L. Dillner, M. Ingvarson, and E. L. Kollberg, A distributed heterostructure barrier varactor frequency tripler, *IEEE Microwave and Guided Wave Letters,* 10, 1, 24–26, 2000.
18. E. Carman, M. Case, M. Kamegawa, R. Yu, K. Giboney, and M. J. Rodwell, V-band and W-band broadband, monolithic distributed frequency multipliers, *IEEE Microwave and Guided Wave Letters,* 2, 6, 253–254, 1992.
19. L. Dillner, J. Stake, and E. L. Kollberg, Modeling of the Heterostructure Barrier Varactor Diode, presented at 1997 International Semiconductor Device Research Symposium, Charlottesville, SC, 1997.
20. J. Stake, S. H. Jones, L. Dillner, S. Hollung, and E. Kollberg, Heterostructure barrier varactor design, *IEEE Trans. on Microwave Theory and Techniques,* 48, 4, Part 2, 2000.

Further Information

M. T. Faber, J. Chramiec, and M. E. Adamski, *Microwave and Millimeter-Wave Diode Frequency Multipliers.* Artech House Publishers, Norwood, MA, 1995.

S. A. Maas, *Nonlinear Microwave Circuits.* Artech House, Norwood, MA, 1988.

P. Penfield and R. P. Rafuse, *Varactor Applications.* M.I.T. Press, Cambridge, 1962.

S. Yngvesson, *Microwave Semiconductor Devices.* Kluwer Academic Publishers, Boston, 1991.

2

Schottky Diode Frequency Multipliers

Jack East
University of Michigan

2.1 Introduction

Heterodyne receivers are an important component of most high frequency communications systems and other receivers. In its simplest form a receiver consists of a mixer being pumped by a local oscillator. At lower frequencies a variety of local oscillator sources are available, but as the desired frequency of operation increases, the local oscillator source options become more limited. The "lower frequencies" limit has increased with time. Early transistor oscillators were available in the MHz and low GHz range. Two terminal transit time devices such as IMPATT and Gunn diodes were developed for operation in X and Ka band in the early 1970s. However, higher frequency heterodyne receivers were needed for a variety of communications and science applications, so alternative local oscillator sources were needed. One option was vacuum tubes. A variety of vacuum tubes such as klystrons and backward wave oscillators grew out of the radar effort during the Second World War. These devices were able to produce large amounts of power over most of the desired frequency range. However, they were large, bulky, expensive, and suffered from modest lifetimes. They were also difficult to use in small science packages.

An alternative solid state source was needed and the technology of the diode frequency multiplier was developed beginning in the 1950s. These devices use the nonlinear reactance or resistance of a simple semiconductor or diode to produce high frequency signals by frequency multiplication. These multipliers have been a part of many high frequency communications and science applications since that time. As time passed the operating frequencies of both transistors and two-terminal devices increased. Silicon and GaAs transistors have been replaced by much higher frequency HFETs and HBTs with f_{max} values of hundreds of GHz. Two-terminal IMPATT and Gunn diodes can produce more than 100 milliwatts at frequencies above 100 GHz. However, the desired operating frequencies of communications and scientific applications have also increased. The most pressing needs are for a variety of science applications in the frequency range between several hundred GHz and several THz. Applications include space-based remote sensing of the earth's upper atmosphere to better understand the chemistry of ozone depletion and basic astrophysics to investigate the early history of the universe. Both missions will require space-based heterodyne receivers with near THz local oscillators. Size, weight, and prime power will be important parameters. Alternative approaches include mixing of infrared lasers to produce the desired local oscillator frequency from higher frequencies, and a multiplier chain to produce the desired frequency from lower

0-8493-1562-X/03/$0.00+$1.50

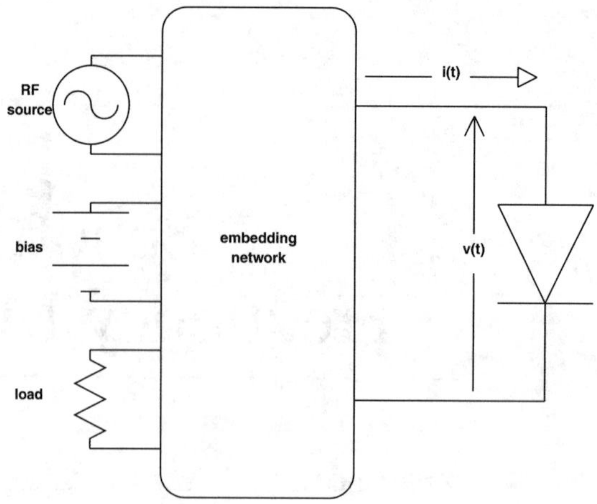

FIGURE 2.1 Schottky diodes (a) structure, (b) electric field, (c) capacitance vs. voltage, and (d) equivalent circuit.

frequencies. Laser-based systems with the desired output frequencies and powers are available, but not with the desired size and weight. Semiconductor diode-based frequency multipliers have the modest size and weight needed, but as of now cannot supply the required powers, on the order of hundreds of microwatts, needed for the missions. This is the subject of ongoing research.

The goal of this chapter is to briefly described the performance of diode frequency multipliers in order to better understand their performance and limitations. The chapter is organized as follows. The next section will describe the properties of Schottky barrier diodes, the most useful form of a varactor multiplier. The following section will describe the analytic tools developed to predict multiplier operation. Two limitations, the reactive multiplier described by Manley and Rowe and the resistive multiplier discussed by Page will be discussed. The results of these two descriptions can be used to understand the basic limits of multiplier operation. However, these analytic results do not provide enough information to design operating circuits. A more realistic computer-based design approach is needed. This will be discussed in the next section. Limitations on realistic devices and some alternative structures will then be described followed by a look at one future application and a brief summary.

2.2 Schottky Diode Characteristics

Multiplier operation depends on the nonlinear properties of Schottky diodes. The diode has both a capacitive and a resistive nonlinearity. Consider the simple representation of a uniformly Schottky diode shown in Fig. 2.1. Figure 2.3(a) shows a semiconductor with a Schottky barrier contact on the right and an ohmic contact on the left. The semiconductor has a depletion layer width w, with the remaining portion of the structure undepleted. The depletion layer can be represented as a capacitor and the undepleted portion can be represented as a resistor. The depletion layer will act as a parallel plate capacitor with a capacitance of

$$C = \frac{\epsilon A}{w}, \tag{2.1}$$

where C is the capacitance, ϵ is the dielectric constant, A is the area, and W is the depletion width w from Fig. 2.1(a). The electric field vs. distance for this structure is shown in Fig. 2.1(b). For reasonable conditions the electric field in the undepleted region is small. The field in the depletion regions extends over the width of the depletion region and depends linearly on x. The area under the electric field curve in Fig. 2.1(b) is the depletion layer voltage, the sum of the applied voltage V_b and the built-in potential ϕ_{bi}. The resulting depletion width vs. applied reverse voltage is

$$w = \sqrt{\frac{2\epsilon\left(\phi_{bi} + V_{bios}\right)}{qN_d}}, \tag{2.2}$$

where ϕ_{bi} is the built-in potential of the metal semiconductor or junction, V_{bios} is the applied reverse bias, q is the electronic charge, and N_d is the uniform semiconductor doping. This width vs. applied bias will result in the capacitance vs. applied voltage of the form

$$C(V) = \frac{C_{i0}}{\sqrt{\phi_{bi} + V_{bios}}} \tag{2.3}$$

where C_{i0} is the capacitance at zero applied bias. The resulting capacitance vs. bias voltage is shown in Fig. 2.1(c). This capacitance characteristic is the starting point for the analytic models in the next section. However, other effects are also present. Under realistic pumping conditions, the diode can also be forward biased allowing forward conduction current flow. This can be approximated with a nonlinear voltage-dependent resistance. The resulting equivalent circuit then becomes the right 2 nonlinear elements in Fig. 2.1(c). There are also parasitic elements. The undepleted region of the device and various contact and package resistance will appear in series with the nonlinear junction. Although the undepleted region width is voltage dependent, this series resistance is usually modeled with a constant value. However, at very high frequencies the current through the undepleted region can crowd to the outside edge of the material due to the skin effect, increasing the resistance. This frequency-dependent resistance is sometimes included in multiplier simulations [9]. The varactor diode must be connected to the external circuit. This resulting physical connection usually results in a parasitic shunt capacitance associated with the connection and a parasitic series inductance associated with the current flow through the wire connection. A major part of multiplier research over the past decade has involved attempts to reduce these parasitic effects.

This nonlinear capacitance vs. voltage characteristic can be used as a frequency multiplier. Consider the charge Q in the nonlinear capacitance as a function of voltage

$$Q(v) = \sqrt{a\left(\phi_{bi} - V\right)}, \tag{2.4}$$

where a is a constant. This function can be expanded in a series

$$Q(V) = a_0 + a_1 V + a_2 V^2 + a_3 V^3 + \ldots, \tag{2.5}$$

The current $I(t)$ is the time derivative of the charge,

$$I(t) = \frac{dQ}{dt} = \left[a_1 + 2a_2 V + 3a_3 V^2 + \ldots\right]\frac{dV}{dt} \tag{2.6}$$

If $V(t)$ is of the form $V_{rf}\, sin\, \omega t$, then the higher order V terms will produce harmonics of the input pump frequency.

2.3 Analytic Descriptions of Diode Multipliers

Equation (2.6) shows that we can get higher order frequencies out of a nonlinear element. However, it is not clear what the output power or efficiency will be. The earliest research on frequency multipliers were based on closed form descriptions of multiplier operation to investigate this question. This section will discuss the ideal performance of reactive and resistive frequency multipliers.

A nonlinear resistor or capacitor, when driven by a pump source, can generate a series of harmonic frequencies. This is the basic form of a harmonic multiplier. The Manley-Rowe relations are a general description of power and frequency conversion relations in nonlinear reactive elements [1, 2]. They describe the properties of frequency conversion and general in nonlinear reactances. The earliest work on these devices sometimes used nonlinear inductances, but all present work involves the nonlinear capacitance vs. voltage characteristic of a reverse-biased Schottky barrier diode. Although the Manley Rowe equations describe frequency multiplication, mixer operation, and parametric amplification, they are also useful as an upper limit on multiplier operation. If an ideal nonlinear capacitance is a pump with a local oscillator at frequency f_0, and an embedding circuit allows power flow at harmonic frequencies, then the sum of the powers into and out of the capacitor is zero,

$$\sum_{m=0}^{\infty} P_m = 0. \tag{2.7}$$

This expression shows that we can have an ideal frequency multiplier at 100% efficiency converting input power to higher frequency output power if we properly terminate all the other frequencies. Nonlinear resistors can also be used as frequency multipliers [3, 4]. For a nonlinear resistor pumped with a local oscillator at frequency f_0, the sum of the power is

$$\sum_{m=0}^{\infty} m^2 P_m \geq 0. \tag{2.8}$$

For an mth order resistive harmonic generator with only an input and output signal, the efficiency is at best $\frac{1}{m^2}$, 25% for a doubler and 11% for a tripler.

Although Eqs. (2.7) and (2.8) give upper limits on the efficiency to be expected from a multiplier, they provide little design information for real multipliers. The next step in multiplier development was the development of closed form expressions for design based on varactor characteristics [5, 6]. Burckardt [6] gives design tables for linear and abrupt junction multipliers based on closed form expressions for the charge in the diode. These expressions form the starting point for waveform calculations. Computer simulations based on the Fourier components of these waveforms give efficiency and impedance information from 2nd to 8th order. However, these approximations limit the amount of design information available. A detailed set of information on multiplier design and optimization requires a computer-based analysis.

2.4 Computer-Based Design Approaches

The analytic tools discussed in the last section are useful to predict the ideal performance of various frequency multipliers. However, more exact techniques are needed for useful designs. Important information such as input and output impedances, the effects of series resistance, and the effect of harmonic terminations at other harmonic frequencies are all important in multiplier design. This information requires detailed knowledge of the current and voltage information at the nonlinear device. Computer-based simulations are needed to provide this information. The general problem can be described with the help of Fig. 2.2. The multiplier consists of a nonlinear microwave diode, an embedding network that

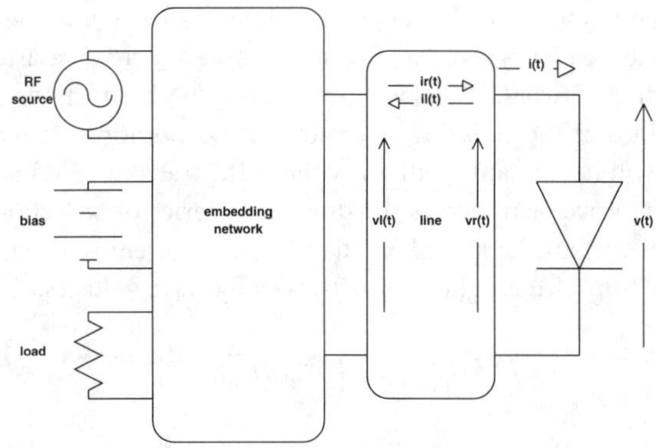

FIGURE 2.2 Generalized frequency multiplier.

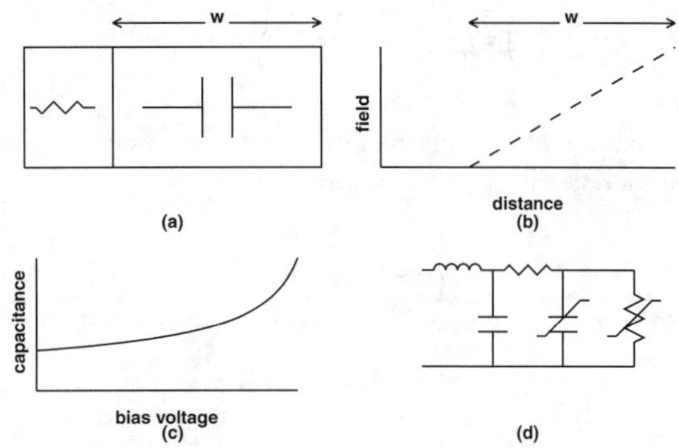

FIGURE 2.3 Multiple reflection circuit.

provides coupling between the local oscillator source and the output load, provisions for DC bias and terminations for all other non-input or output frequencies. Looking toward the embedding network from the diode, the device sees embedding impedances at the fundamental frequency and each of the harmonic frequencies. The local oscillator power available is usually specified along with the load impedances and the other harmonic frequencies under consideration. The goal is to obtain the operating conditions, the output power and efficiency, and the input and output impedances of the overall circuit. The nonlinear nature of the problem makes the solution more difficult. Within the context of Fig. 2.2, the current as a function of time is a nonlinear function of the voltage. Device impedances can be obtained from ratios of the Fourier components of the voltage and current. The impedances of the diode must match the embedding impedances at the corresponding frequency. A "harmonic balance" is required for the correct situation. Many commercial software tools use nonlinear optimization techniques to solve this "harmonic balance" optimization problem. An alternative approach is the multiple reflection algorithm. This solution has the advantage of a very physical representation of the actual transient response of the circuit. This technique is discussed in References 7, 8, and 9 and will be described here.

The basic problem is to calculate the diode current and voltage waveforms when the device is embedded in an external linear circuit. The diode waveforms are best represented in the time domain. However, the embedding circuit consists of linear elements and is best represented in the frequency or impedance domain. One approach is the multiple reflection technique [7, 8, 9]. This technique splits the simulation into two parts, a nonlinear time domain description of the diode multiplier and a linear frequency domain description of the embedding network. The solution goal is to match the frequency domain impedances of the embedding circuit with the time domain impedances of the nonlinear device. The circuit is shown in Fig. 2.3. The initial circuit is modified by including a long transmission line with a length l equal to an integral number of pump frequency wavelengths and an arbitrary characteristic impedance Z_0 between the diode and the embedding network. Waves will propagate in both directions on this transmission line depending on the conditions at the ends. When steady-state conditions are reached, the waveforms at the circuit and the diode will be the same, with or without the transmission line. This transmission line allows us to determine the waveforms across the diode as a series of reflections from the circuit. The signals on the transmission line are composed of left- and right-traveling current and voltage waves. The voltage at the diode is the sum of the right- and left-traveling wave voltages,

$$v(x) = v_r(x) + v_l(x) \tag{2.9}$$

and the current at the diode is the difference

$$i(x) = i_r(x) - i_l(x) = \frac{v_r(x) - v_l(x)}{Z_0} \tag{2.10}$$

Since the transmission line is an integral number of pump frequency wavelengths long, the conditions are the same at each end under steady-state conditions,

$$v(x=0) = v(x=l) \tag{2.11}$$

$$i(x=0) = i(x=l) \tag{2.12}$$

At the start of the simulation, we assume that there is a right-traveling wave associated with the DC bias at frequency 0 and the pump at frequency 1. This wave with a DC component and a component at the local oscillator frequency will arrive at the diode at $x = 0$ as a $V_r(t)$. The resulting voltage across the diode will produce a diode current. This current driving the transmission line will produce a first reflected voltage,

$$v^1_{reflected} = v^1_l = \left(v^1_{diode} - i^1_{diode}(t) Z_0\right)/2 \tag{2.13}$$

This time domain reflected or left-traveling wave will then propagate to the embedding network. Here is can be converted into the frequency domain with a Fourier transform. The resulting signal will contain harmonics of the local oscillator signal due to the nonlinear nature of the diode current vs. voltage and charge vs. voltage characteristic. The resulting frequency domain information can then be used to construct a new reflected voltage wave from the embedding network. This process of reflections from the diode and the circuit or "multiple reflections" continues until a steady-state solution is reached.

This computer-based solution has several advantages over the simpler analysis-based solutions. It can handle combinations of resistive and reactive nonlinearities. Most high-efficiency multipliers are pumped into forward conduction during a portion of the RF cycle so this is an important advantage for accurate performance predictions. In practice the varactor diode will also have series resistance, parasitic capacitances, and series inductances. These additional elements are easily included in the multiplier simulation. At very high frequencies the series resistance can be frequency dependent, due to current crowding in the semiconductor associated with the skin effect. This frequency dependence loss can also be included in simulations. Computer programs with these features are widely used to design high performance frequency multipliers.

An alternative solution technique is the fixed point method [10]. This technique uses a circuit similar to Fig. 2.3, with a nonlinear device and an embedding network connected with a transmission line. However, this approach uses a fixed point iteration to arrive at a converged solution. The solution starts with arbitrary voltages at the local oscillator frequency at the harmonics. With these starting conditions, a new voltage is obtained from the existing conditions using

$$V_{n,k+1} = \frac{Z_0 V_{rf}}{Z_n^L + Z_0} + \frac{Z_n^L\left(V_{n,k} - I_{n,k}Z_0\right)}{Z_n^L + Z_0} \tag{2.14}$$

for the driven local oscillator frequency and

$$V_{n,k+1} = \frac{Z_0}{Z_n^L + Z_0}\left(V_{n,k} - I_{n,k}Z_0\right) \tag{2.15}$$

for the remaining frequencies, where Z_n^L are the embedding impedances at frequency n, Z_0 is the same line characteristic impedance used in the multiple reflection simulation, $Z_{n,k}^{NL}$ is the nonlinear device impedance at frequency n and iteration number k, $V_{n,k}$, and $I_{n,k}$ are the frequency domain voltage and current at iteration k, and V_n^S is the RF voltage at the pump source. These two equations provide an iterative solution of the nonlinear problem. They are particularly useful when a simple equivalent circuit for the nonlinear device is not available. We now have the numerical tools to investigate the nonlinear operation of multipliers. However, there are some operating conditions where this equivalent circuit approach breaks down. Some of these limitations will be discussed in the next section.

2.5 Device Limitations and Alternative Device Structures

The simulation tools and simple device model do a good job of predicting the performance of low frequency diode multipliers. However, many multiplier applications require very high frequency output signals with reasonable output power levels. Under these conditions, the output powers and efficiencies predicted are always higher than the experimental results. There are several possible reasons. Circuit loss increases with frequency, so the loss between the diode and the external connection should be higher. Measurements are less accurate at these frequencies, so the differences between the desired designed circuit embedding impedances and the actual values may be different. Parasitic effects are also more important, degrading the performance. However, even when all these effects are taken into account, the experimental powers and efficiencies are still low. The problem is with the equivalent circuit of the diode in Fig. 2.1(d). It does not correctly represent the high frequency device physics [11]. The difficulty can be explained by referring back to Fig. 2.1(a). The device is a series connection of the depletion layer capacitance and a bulk series resistance. The displacement current flowing through the capacitor must be equal to the conduction current flowing through the undepleted resistive region. The capacitor displacement current is

$$I(t) = \frac{dC(V)V(t)}{dt} \tag{2.16}$$

If we approximate $V(t)$ with $V_{rf}\cos(\omega t)$, then the current becomes

$$I(t) = V_{rf}\omega C(V)\sin\omega t + V(t)\frac{dC(V)}{dt} \tag{2.17}$$

For a given device, the displacement current and the resulting current through the resistor increase with the frequency and the RF voltage. At modest drive levels and frequencies the undepleted region can support this current and the equivalent resistance remains constant. However, the current density through the undepleted region is

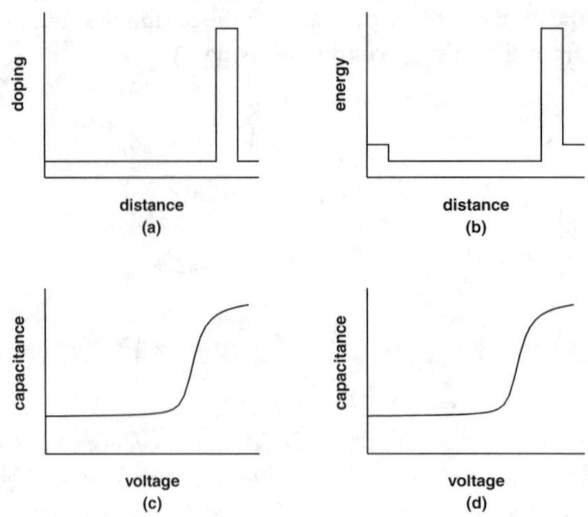

FIGURE 2.4 Alternative structures (a) hyperabrupt doping, (b) BNN energy, (c) hyperabrupt capacitance, and (d) BNN capacitance.

$$J_{dep} = qN_d v(E), \tag{2.18}$$

where J_{dep} is the conduction current density in the undepleted region and $v(E)$ is the carrier velocity vs. electric field. At low electric fields the slope of the velocity field curve is constant with a value equal to the carrier mobility μ. However at higher electric fields, the velocity begins to saturate at a constant value. Additional increases in the electric field do not increase the conduction current through the varactor. This current saturation is a fundamental limit on the multiplier performance. A more physical explanation also uses Fig. 2.1(a). A nonlinear capacitor implies a change in the depletion layer width with voltage. However, changing the depletion layer width involves moving electrons from the depletion layer edge. These electrons are limited by their saturated velocity, so the time rate of change of the capacitance is also unlimited.

This simple theory allows a modified device design. Equations (2.18) and (2.17) are the starting point. Our goal is usually to optimize the power or efficiency available from a multiplier circuit. Clearly one option is to increase the doping N_d in Eq. (2.18) to increase the available J_{dep}. However, increasing the doping decreases the breakdown voltage and thus the maximum RF voltage that can be present across the reverse biased depletion layer. The device doping design becomes a parameter in the overall multiplier design. The optimum efficiency and optimum power operating points for the same input and output frequency are usually different. Although this simple description provides useful information, a detailed physical model is usually needed for best results [12].

The discussion so far has been based on a uniformly doped abrupt junction varactor. However, other doping or material layer combinations are possible [13]. One option is to tailor the doping or material profile to obtain a capacitance vs. voltage that is more nonlinear than the $1/\sqrt{V_{bios}}$ dependence in Eq. (2.2). Two options are the hyperabrupt varactor and the BNN structure. These devices are shown in Fig. 2.4. The doping profile of a hyperabrupt varactor is shown in Fig. 2.4(a). Instead of a uniform doping, this structure has a much smaller doping over most of the structure with a high doping or doping spike near the metal semiconductor junction. The corresponding capacitance vs. voltage characteristic is shown in Fig. 2.4(c). At modest reverse biases the depletion layer extends from the metal contact to the doping spike. The resulting narrow depletion layer produces a high capacitance. Higher applied voltages begin to deplete the charge in the doping spike. When the spike charge is depleted, there is a rapid increase in the depletion layer width through the second lightly doped region and a corresponding decrease in the capacitance. This structure can produce a more nonlinear capacitance variation than a uniformly doped device. However, the structure combines both lightly doped and heavily doped regions, so saturation

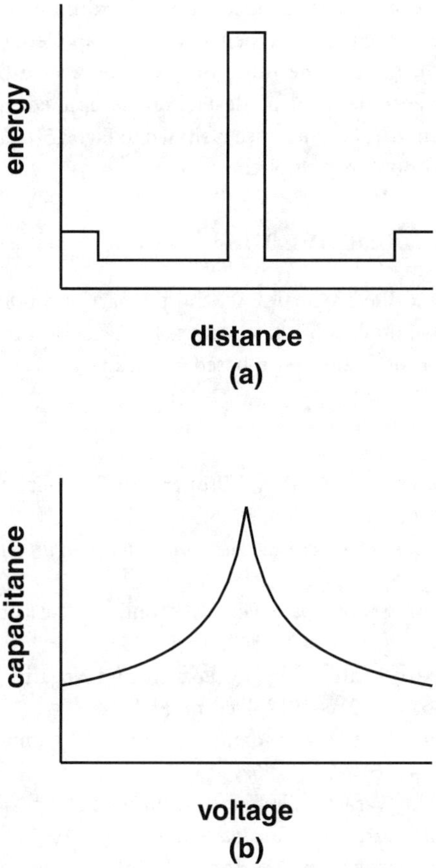

FIGURE 2.5 Structure and associated capacitance voltage characteristic.

effects can occur in the lightly doped portion. An alternative structure is the BNN or BIN device. This structure uses combinations of **B**arriers **I**ntrinsic and **N** doped regions formed with combinations of different epitaxial materials and doping to produce optimized capacitance structures. This structure can have either ohmic or Schottky contacts. A typical structure is shown in Fig. 2.4(b). Notice that conduction band energy rather than doping is being plotted. The structure consists of an *n*+ ohmic contact on the right followed by an *n* region that can be depleted, a wide bandgap barrier region, and a second *n*+ ohmic contact on the left. This structure can have a highly nonlinear capacitance characteristics as shown in Fig. 2.4(d), depending on the choice of layer doping and energy band offsets. These BNN structures have potential advantages in monolithic integration and can be fabricated in a stacked or series configuration for higher output powers.

Since the major application of frequency multipliers is for high frequency local oscillator sources, it would be reasonable to try to fabricate higher order multipliers, triplers for example, instead of doublers. Although based on Eq. (2.7), this is possible, there are some problems. Efficient higher order multiplication requires currents and voltages at intermediate frequencies, at the second harmonic in a tripler for example. This is an "idler" frequency or circuit. However, in order to avoid loss, this frequency must have the correct reactive termination. This adds to the complexity of the circuit design. Some details of doublers and triplers are given in Reference 14. An alternative that avoids the idlers is an even or symmetrical capacitance voltage characteristic. One possibility is the single barrier or quantum barrier varactor [15]. The structure and associated capacitance voltage characteristic are shown in Fig. 2.5. This structure, shown in Fig. 2.5(a), is similar to the BNN expect the barrier is the middle with lightly doped regions on either side. This will be a series connection of a depletion layer, a barrier and a depletion region, with ohmic contacts on each end. The capacitance is maximum at zero applied bias, with a built-in depletion layer on each side of the barrier. When a voltage is applied, one to the depletion layers will become forward biased and shrink and the other one will be reverse biased and expand. The series combination capacitance will become smaller. Reversing the applied voltage will produce the same capacitance. The resulting symmetrical capacitance is shown in Fig. 2.5(b). This capacitance characteristic is a useful starting point for odd order multipliers.

2.6 Summary and Conclusions

This chapter has briefly discussed the properties of Schottky barrier diodes that are useful for frequency multipliers. Although a variety of other solid state sources are available for lower frequency sources, these devices are a critical component of future space-based applications.

References

1. J.M. Manley and H.E. Rowe, Some General Properties of Nonlinear Elements, I: General Energy Relations, *Proceedings of the IRE,* 44, 904, 1956.
2. H.A. Watson, *Microwave Semiconductor Devices and Their Circuit Applications,* McGraw-Hill, New York, 1969, chap. 8.
3. C.H. Page, Frequency Conversion With Positive Nonlinear Resistors, *Journal of Research of the National Bureau of Standards,* 56, 4, 179–182, April 1956.
4. R.H. Pantell, General Power Relationships for Positive and Negative Nonlinear Resistive Elements, *Proceedings of the IRE,* 46, 12, 1910–1913, December 1958.
5. J.A. Morrison, Maximization of the Fundamental Power in Nonlinear Capacitance Diodes, *Bell System Technical Journal,* 41, 677–721, 1962.
6. C.B. Burckardt, Analysis of Varactor Frequency Multipliers for Arbitrary Capacitance Variation and Drive Level, *Bell System Technical Journal,* 44, 675–692, 1965.
7. D.N. Held and A.R. Kerr, Conversion Loss and Noise of Microwave and Millimeter Wave Mixers: Part 1-Theory; Part 2 Experiment, *IEEE Transactions on Microwave Theory and Techniques,* MTT-26, 49, 1978.
8. S.A. Maas, *Microwave Mixers,* Artech House, Norwood, MA, 1986.
9. P. Siegel, A. Kerr and W. Hwang, Topics in the Optimization of Millimeter Wave Mixers, NASA Technical Paper 2287, March 1984.
10. G.B. Tait, Efficient Solution Method Unified Nonlinear Microwave Circuit and Numerical Solid-State Device Simulation, *IEEE Microwave and Guided Wave Letters,* 4, 12, 420–422, December 1994.
11. E.L. Kollberg, T.J. Tolmunen, M.A. Frerking, and J.R. East, Current Saturation in Submillimeter Wave Varactors, *IEEE Transactions on Microwave Theory and Techniques,* MTT-40, 5, 831–838, May 1992.
12. J.R. East, E.L. Kollberg and M.A. Frerking, Performance Limitations of Varactor Multipliers, Fourth International Conference on Space Terahertz Technology, Ann Arbor, MI, March 1993.
13. M. Frerking and J.R. East, Novel Heterojunction Varactors, *Proceedings of the IEEE Special Issue on Terahertz Technology,* 80, 11, 1853–1860, November 1992.
14. A. Raisanen, Frequency Multipliers for Millimeter and Submillimeter Wavelengths, *Proceedings of the IEEE Special Issue on Terahertz Technology,* 80, 11, 1842–1852, November 1992.
15. E. Kollberg and A. Rydberg, Quantum Barrier Varactor Diodes for High-Efficiency Millimeter-Wave Multipliers, *Electronics Letters,* 25, 1696–1697, December 1989.

3

Transit Time Microwave Devices

Robert J. Trew
Virginia Tech University

3.1 Introduction

There are several types of active two-terminal diodes that can oscillate or supply gain at microwave and millimeter-wave frequencies. These devices can be fabricated from a variety of semiconductor materials, but Si, GaAs, and InP are generally used. The most common types of active diodes are the IMPATT (an acronym for IMPact Avalanche Transit-Time) diode, and the Transferred Electron Device (generally called a Gunn diode). Tunnel diodes are also capable of producing active characteristics at microwave and millimeter-wave frequencies, but have been replaced in most applications by three-terminal transistors (such as GaAs MESFETs and AlGaAs/GaAs HEMTs), which have superior RF and noise performance, and are also much easier to use in systems applications. The IMPATT and Gunn diodes make use of a combination of internal feedback mechanisms and transit-time effects to create a phase delay between the RF current and voltage that is more than 90°, thereby generating active characteristics. These devices have high frequency capability since the saturated velocity of an electron in a semiconductor is high (generally on the order of $\sim 10^7$ cm/sec) and the transit time is short since the length of the region over which the electron transits can be made on the order of a micron (i.e., 10^{-4} cm) or less. The ability to fabricate devices with layer thicknesses on this scale permits these devices to operate at frequencies well into the millimeter-wave region. Oscillation frequency on the order of 400 GHz has been achieved with IMPATT diodes, and Gunn devices have produced oscillations up to about 150 GHz. These devices have been in practical use since the 1960s and their availability enabled a wide variety of solid-state system components to be designed and fabricated.

3.2 Semiconductor Material Properties

Active device operation is strongly dependent upon the charge transport, thermal, electronic breakdown, and mechanical characteristics of the semiconductor material from which the device is fabricated. The charge transport properties describe the ease with which free charge can flow through the material. This is described by the charge velocity-electric field characteristic, as shown in Fig. 3.1 for several commonly used semiconductors. At low values of electric field, the charge transport is ohmic and the charge velocity is directly proportional to the magnitude of the electric field. The proportionality constant is called the

0-8493-1562-X/03/$0.00+$1.50

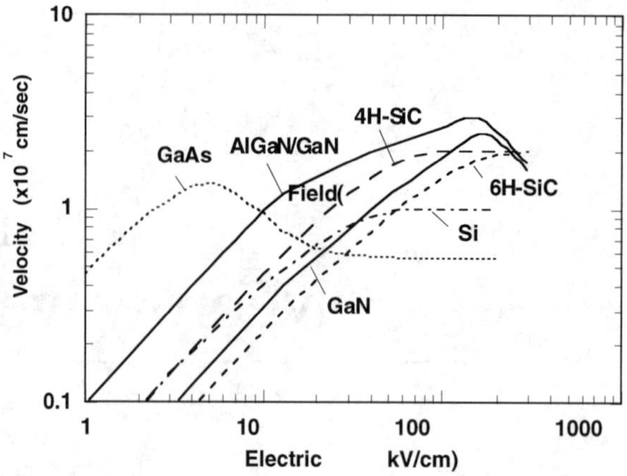

FIGURE 3.1 Electron velocity versus electric field characteristics for several semiconductors.

mobility and has units of cm^2/V-sec. Above a critical value for the electric field, the charge velocity (units of cm/sec) saturates and either becomes constant (e.g., Si) or decreases with increasing field (e.g., GaAs). Both of these behaviors have implications for device fabrication, especially for devices intended for high frequency operation. Generally, for transit time devices, a high velocity is desired since current is directly proportional to velocity. The greatest saturated velocity is demonstrated for electrons in the wide bandgap semiconductors, SiC and GaN. Both of these materials have saturated electron velocities on the order of $v_s \sim 2 \times 10^7$ cm/sec. This is one of the main reasons these materials are being developed for high frequency electronic devices. Also, a low value for the magnitude of the electric field at which velocity saturation occurs is desirable since this implies high charge mobility. High mobility produces low resistivity, and therefore low values for parasitic and access resistances for semiconductor devices.

The decreasing electron velocity with electric field characteristic for compound semiconductors such as GaAs and InP makes active two-terminal devices called Transferred Electron Devices (TED's) or Gunn diodes possible. The negative slope of the velocity versus electric field characteristic implies a decreasing current with increasing voltage. That is, the device has a negative resistance. When a properly sized piece of these materials is biased in the region of decreasing current with voltage, and placed in a resonant cavity, the device will be unstable up to very high frequencies. By proper selection of embedding impedances, oscillators or amplifiers can be constructed.

Other semiconductor material parameters of interest include thermal conductivity, dielectric constant, energy bandgap, electric breakdown critical field, and minority carrier lifetime. The thermal conductivity of the material is important because it describes how easily heat can be extracted from the device. The thermal conductivity has units of W/cm-°K, and in general, high thermal conductivity is desirable. Compound semiconductors, such as GaAs and InP, have relatively poor thermal conductivity compared to elemental semiconductors such as Si. Materials such as SiC have excellent thermal conductivity and are used in high power electronic devices. The dielectric constant is important since it represents capacitive loading and, therefore, affects the size of the semiconductor device. Low values of dielectric constant are desirable since this permits larger device area, which in turn results in increased RF current and increased RF power that can be developed. Electric breakdown characteristics are important since electronic breakdown limits the magnitudes of the DC and RF voltages that can be applied to the device. A low magnitude for electric field breakdown limits the DC bias that can be applied to a device, and thereby limits the RF power that can be handled or generated by the device. The electric breakdown for the material is generally described by the critical value of electric field that produces avalanche ionization. Minority carrier lifetime is important for bipolar devices, such as pn junction diodes, rectifiers, and bipolar junction transistors (BJTs). A low value for minority carrier lifetime is desirable for devices such

TABLE 3.1 Material Parameters for Several Semiconductors

Material	E_g(eV)	ε_r	κ(W/K-cm) @ 300°K	E_c(V/cm)	$\tau_{minority}$ (sec)
Si	1.12	11.9	1.5	3×10^5	2.5×10^{-3}
GaAs	1.42	12.5	0.54	4×10^5	$\sim 10^{-8}$
InP	1.34	12.4	0.67	4.5×10^5	$\sim 10^{-8}$
6H-SiC	2.86	10.0	4.9	3.8×10^6	$\sim(10\text{–}100) \times 10^{-9}$
4H-SiC	3.2	10.0	4.9	3.5×10^6	$\sim(10\text{–}100) \times 10^{-9}$
3C-SiC	2.2	9.7	3.3	$(1\text{–}5) \times 10^6$	$\sim(10\text{–}100) \times 10^{-9}$
GaN	3.4	9.5	1.3	2×10^6	$\sim(1\text{–}100) \times 10^{-9}$

as diode temperature sensors and switches where low reverse bias leakage current is desirable. A long minority carrier lifetime is desirable for devices such as bipolar transistors. For materials such as Si and SiC the minority carrier lifetime can be varied by controlled impurity doping. A comparison of some of the important material parameters for several common semiconductors is presented in Table 3.1. The large variation for minority lifetime shown in Table 3.1 for SiC and GaN is due to relatively immature materials growth technology for these wide bandgap semiconductors.

3.3 Two-Terminal Active Microwave Devices

The IMPATT diode, transferred electron device (often called a Gunn diode), and tunnel diode are the most commonly used two-terminal active devices. These devices can operate from the low microwave through high mm-wave frequencies, extending to several hundred GHz. They were the first semiconductor devices that could provide useful RF power levels at microwave and mm-wave frequencies and were extensively used in early systems as solid-state replacements for vacuum tubes. The three devices are similar in that they are fabricated from diode or diode-like semiconductor structures. DC bias is applied through two metal contacts that form the anode and cathode electrodes. The same electrodes are used for both the DC and RF ports and since only two electrodes are available, the devices must be operated as a one-port RF network, as shown in Fig. 3.2. This causes little difficulty for oscillator circuits, but is problematic for amplifiers since a means of separating the input RF signal from the output RF signal must be devised. The use of a nonreciprocal device, such as a circulator can be used to accomplish the task. Circulators, however, are large, bulky, and their performance is sensitive to thermal variations. In general, circulators are difficult to use for integrated circuit applications. The one-port character of diodes has limited their use in modern microwave systems, particularly for amplifiers, since transistors, which have three terminals and are two-port networks, can be designed to operate with comparable RF performance, and are much easier to integrate. Diodes, however, are often used in oscillator circuits since these components are by nature one-port networks.

IMPATT and Gunn diodes require a combination of charge injection and transit time effects to generate active characteristics and they operate as negative immittance components (the term "immittance" is a general reference that includes both "impedance" and "admittance"). When properly designed and biased,

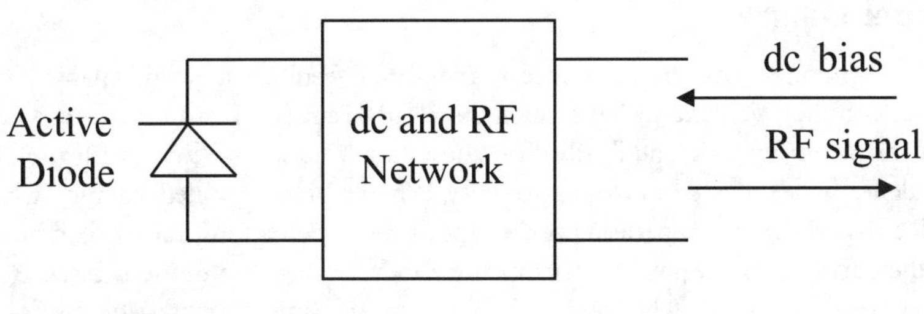

FIGURE 3.2 One-port network.

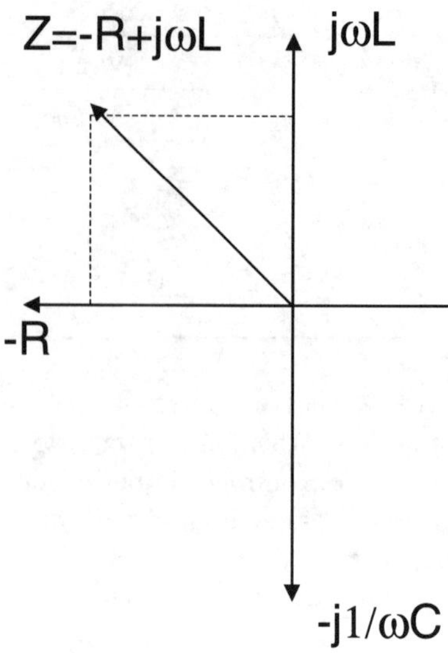

FIGURE 3.3 Complex impedance plane showing active characteristic.

the active characteristics of the diodes can be described as either a negative resistance or a negative conductance. Which description to use is determined by the physical operating principles of the particular device, and the two descriptions are, in general, not interchangeable. Bias and RF circuits for the two active characteristics must satisfy different stability and impedance matching criteria.

Transit time effects alone cannot generate active characteristics. This is illustrated in Fig. 3.3, which shows a general impedance plane. All passive circuits, no matter how complex or how many circuit elements are included, when arranged into a one-port network as shown in Fig. 3.2, and viewed from an external vantage point, will have an input impedance that lies in the right-hand plane of Fig. 3.3. The network resistance will be positive and real, and the reactance will be inductive or capacitive. This type of network is not capable of active performance and cannot add energy to a signal. Transit time effects can only produce terminal impedances with inductive or capacitive reactive effects, depending upon the magnitude of the delay relative to the RF period of the signal. In order to generate active characteristics it is necessary to develop an additional delay that will result in a phase delay between the terminal RF voltage and current that is greater than 90° and less than 270°. The additional delay can be generated by feedback that can be developed by physical phenomena internal to the device structure, or created by circuit design external to the device. The IMPATT and Gunn diodes make use of internal feedback resulting from electronic charge transfer within the semiconductor structure. The internal feedback generally produces a phase delay of ~90°, which when added to the transit time delay will produce a negative real component to the terminal immittance.

3.3.1 Tunnel Diodes

Tunnel diodes [1] generate active characteristics by an internal feedback mechanism involving the physical tunneling of electrons between energy bands in highly doped semiconductors, as illustrated in the energy band diagram shown in Fig. 3.4. The illustration shows a p^+n junction diode with heavily doped conduction and valence bands located in close proximity. When a bias is applied, charge carriers can tunnel through the electrostatic barrier separating the p-type and n-type regions, rather than be thermionically emitted over the barrier, as generally occurs in most diodes. When the diode is biased (either forward or reverse bias) current immediately flows and ohmic conduction characteristics are obtained. In the forward bias direction conduction occurs until the applied bias forces the conduction and valence bands

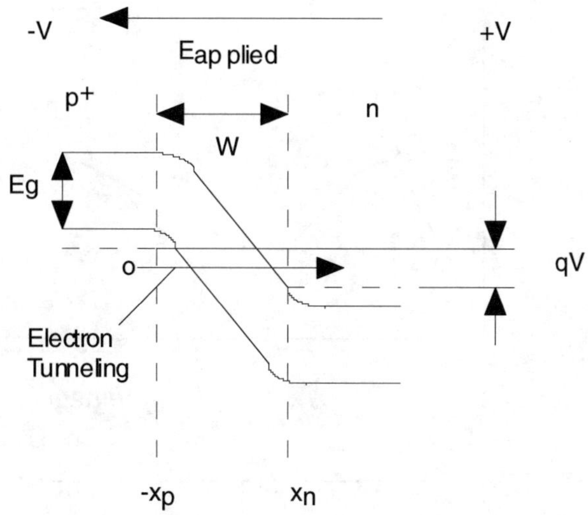

FIGURE 3.4 Energy-band diagram for a p$^+$n semiconductor junction showing electron tunneling behavior.

to separate. The tunnel current then decreases and normal, thermionic junction conduction occurs. In the forward bias region where the tunnel current is decreasing with increasing bias voltage an N-type negative immittance characteristic is generated, as shown in Fig. 3.5a. The immittance is called "N-type" because the I-V characteristic looks like the letter N. This type of active element is current driven and is short-circuit stable. It is described by a negative conductance in shunt with a capacitance, as shown in Fig. 3.5b. Tunnel diodes are limited in operation frequency by the time it takes for charge carriers to tunnel through the junction. Since this time is very short (on the order of 10^{-12} s) operation frequency can be very high, approaching 1000 GHz.

Tunnel diodes have been operated at 100s of GHz, and are primarily limited in frequency response by practical packaging and parasitic impedance considerations. The RF power available from a tunnel diode is limited (~100s of mW level) since the maximum RF voltage swing that can be developed across the junction is limited by the forward turn-on characteristics of the device (typically 0.6 to 0.9 v). Increased RF power can only be obtained by increasing device area to increase RF current. However, increases in diode area will limit operation frequency due to increased diode capacitance. Tunnel diodes have moderate DC-to-RF conversion efficiency (<10%) and very low noise figures and have been used in low noise systems applications, such as microwave and mm-wave receivers used for radioastronomy.

3.3.2 Transferred Electron Devices

Transferred electron devices (i.e., Gunn diodes) [2] also have N-type active characteristics and can be modeled as a negative conductance in parallel with a capacitance, as shown in Fig. 3.5b. Device operation, however, is based upon a fundamentally different principle. The negative conductance derives from the complex conduction band structure of certain compound semiconductors, such as GaAs and InP. In these direct bandgap materials the lower valley central (or Γ) conduction band is in close energy-momentum proximity to secondary, higher order conduction bands (i.e., the X and L) valleys (illustrated schematically as the L upper valley in Fig. 3.6). The electron effective mass is determined by the shape of the conduction bands and the effective mass is "light" in the Γ valley, but "heavy" in the higher order X and L upper valleys. When the crystal is biased, current flow is initially due to electrons in the light effective mass Γ valley and conduction is ohmic. However, as the bias field is increased, an increasing proportion of the free electrons are transferred into the X and L valleys where the electrons have heavier effective mass. The increased effective mass slows down the electrons, with a corresponding decrease in conduction current through the crystal. The net result is that the crystal displays a region of applied bias

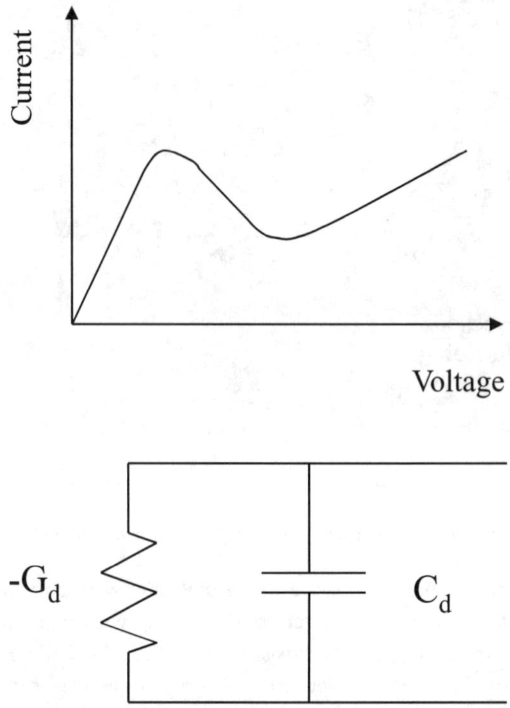

FIGURE 3.5 (a) Current-voltage characteristic for an N-type active device. (b) Small-signal RF equivalent circuit for an N-type active device.

voltages where current decreases with increasing voltage. That is, a negative resistance is generated. A charge dipole domain is formed in the device, and this domain will travel through the device generating a transit time effect. The combination of the transferred electron effect and the transit time delay will produce a phase shift between the terminal RF current and voltage that is greater than 90°. The device is unstable and when placed in an RF circuit or resonant cavity, oscillators or amplifiers can be fabricated.

The Gunn device is not actually a diode since no pn or Schottky junction is used. The transferred electron phenomenon is a characteristic of the bulk material and the special structure of the conduction bands in certain compound semiconductors. In order to generate a transferred electron effect, a semiconductor must have Γ, X, and L valleys in the conduction bands in close proximity so that charge can be transferred from the lower Γ valley to the upper valleys at reasonable magnitude of applied electric field. It is desirable that the charge transfer occur at low values of applied bias voltage in order for the device to operate with good DC-to-RF conversion efficiency. Most semiconductors do not have the conduction band structure necessary for the transferred electron effect, and in practical use, Gunn diodes have only been fabricated from GaAs and InP. It should be noted that the name "Gunn diode" is actually a misnomer since the device is not actually a diode, but rather a piece of bulk semiconductor.

TEDs are widely used in oscillators from the microwave through high mm-wave frequency bands. They can be fabricated at low cost and provide an excellent price-to-performance ratio. They are, for example, the most common oscillator device used in police automotive radars. They have good RF output power capability (mW to W level), moderate efficiency (< 20%), and excellent noise and bandwidth capability. Octave or multi-octave band tunable oscillators are easily fabricated using devices such as YIG (yttrium iron garnet) resonators. High tuning speed can be achieved by using varactors as the tuning element. Many commercially available solid-state sources for 60 to 100 GHz operation (for example, automotive collision-avoidance radars) often use InP TEDs.

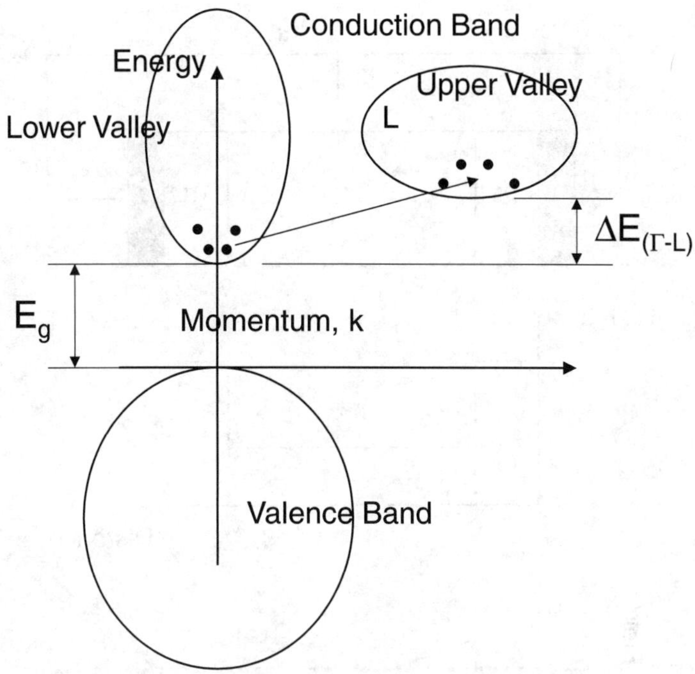

FIGURE 3.6 Energy band diagram for a semiconductor that demonstrates the transferred electron effect.

3.3.3 IMPATT Diodes

IMPATT (IMPact Avalanche Transit Time) diodes [3] are fabricated from pn or Schottky junctions. The doping profile in the device is generally tailored for optimum performance and a typical p^+nn^+ junction device structure as shown in Fig. 3.7. The diode is designed so that when it is reverse biased, the n-region is depleted of free electrons. The electric field at the p^+n junction exceeds the critical magnitude for avalanche breakdown, and the electric field exceeds the magnitude required to maintain electron velocity saturation throughout the n-region. Saturated charge carrier velocity must be maintained throughout the RF cycle in order for the device to operate with maximum efficiency. In operation, the high electric field region at the p^+n interface will generate charge when the sum of the RF and DC bias voltage produces an electric field that exceeds the critical value. A pulse of free charge (electrons and holes) will be generated. The holes will be swept into the p^+ region and the electrons will be injected into the depleted n-region, where they will drift through the diode, inducing a current in the external circuit as shown in Fig. 3.8. Due to the avalanche process, the RF current across the avalanche region lags the RF voltage by 90°. This inductive delay is not sufficient, by itself, to produce active characteristics. However, when the 90° phase shift is added to that arising from an additional inductive delay caused by the transit time of the carriers drifting through the remainder of the diode external to the avalanche region, a phase shift between the terminal RF voltage and current greater than 90° is obtained. A Fourier analysis of the resulting waveforms reveals a device impedance with a negative real part. That is, the device is active and can be used to generate or amplify RF signals. The device impedance has an "S-type" active i-v characteristic, as shown in Fig. 3.9a, and the device equivalent circuit consists of a negative resistance in series with an inductor, as shown in Fig. 3.9b. An S-Type active device is voltage driven and is open-circuit stable. For IMPATT diodes, the active characteristics only exist under RF conditions. That is, there is a lower frequency below which the diode does not generate a negative resistance. Also, the negative resistance is generally small in magnitude, and on the order of $-1\ \Omega$ to $-10\ \Omega$. Therefore, it is necessary to reduce all parasitic resistances in external circuits to the maximum extent possible since parasitic series

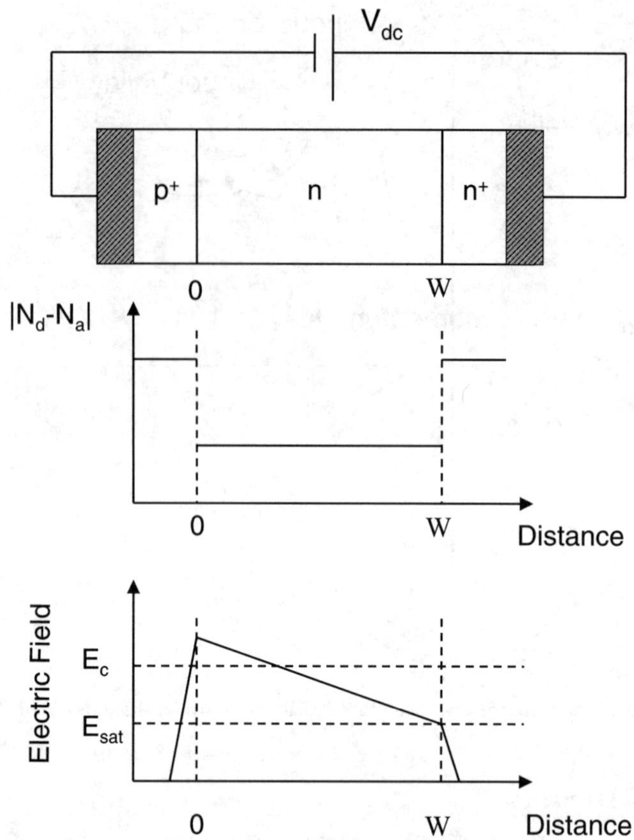

FIGURE 3.7 Diode structure, doping profile, and internal electric field for a p⁺nn⁺ IMPATT diode.

resistance will degrade the device's negative resistance, thereby limiting device performance. An IMPATT diode has significant pn junction capacitance that must be considered and a complete equivalent circuit includes the device capacitance in parallel with the series negative resistance-inductance elements, as shown in Fig. 3.9b.

For optimum performance the drift region is designed so that the electric field throughout the RF cycle is sufficiently high to produce velocity saturation for the charge carriers. In order to achieve this it is common to design complex structures consisting of alternating layers of highly doped and lightly doped semiconductor regions. These structures are called "high-low," "low-high-low," or "Read" diodes, after the man who first proposed their use [1]. They can also be fabricated in a back-to-back arrangement to form double-drift structures [4]. These devices are particularly attractive for mm-wave applications since the back-to-back arrangement permits the device to generate RF power from series-connected diodes, but each diode acts independently with regard to frequency response.

IMPATT diodes can be fabricated from most semiconductors, but are generally fabricated from Si or GaAs. The devices are capable of good RF output power (mW to tens of W) and good DC-to-RF conversion efficiency (~10 to 20%). They operate well into the mm-wave region and have been operated at frequencies in excess of 400 GHz. They have moderate bandwidth capability, but have relatively poor noise performance due to the impact ionization process associated with avalanche breakdown. The power-frequency performance of IMPATT diodes fabricated from Si, GaAs, and SiC is shown in Fig. 3.10. The Si and GaAs data are experimental, and the SiC data are predicted from simulation. The numbers associated with the data points are the conversion efficiencies for each point.

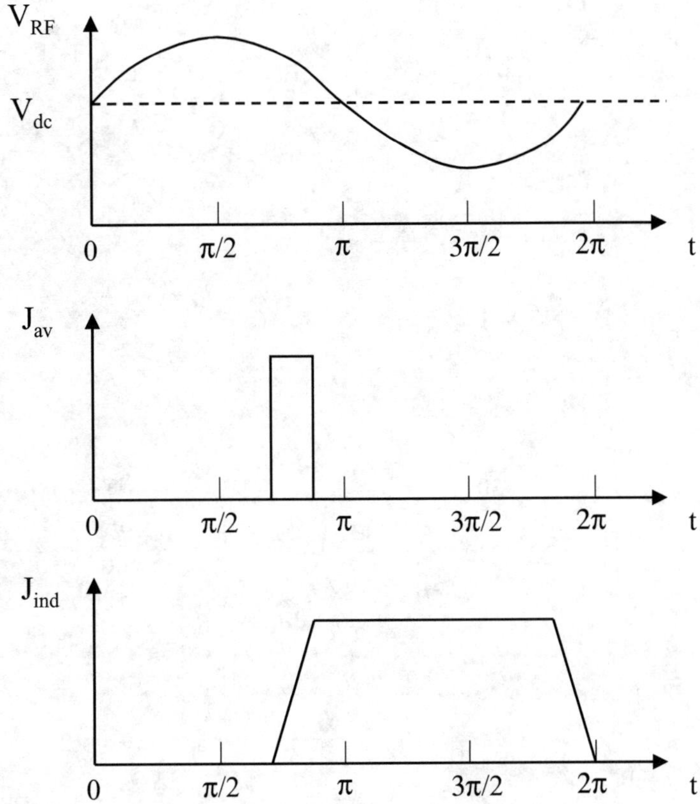

FIGURE 3.8 Terminal voltage, avalanche current, and induced current for an IMPATT oscillator.

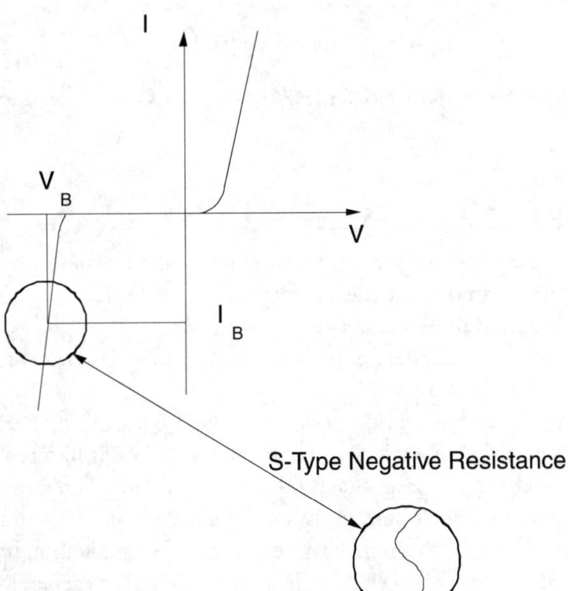

FIGURE 3.9 (a) Current-voltage characteristic for an S-type active device. The S-characteristic behavior does not exist under DC conditions for an IMPATT diode. The inset shows the dynamic i-v behavior that would exist about the DC operation point under RF conditions. (b) Small-signal RF equivalent circuit for an S-type active device.

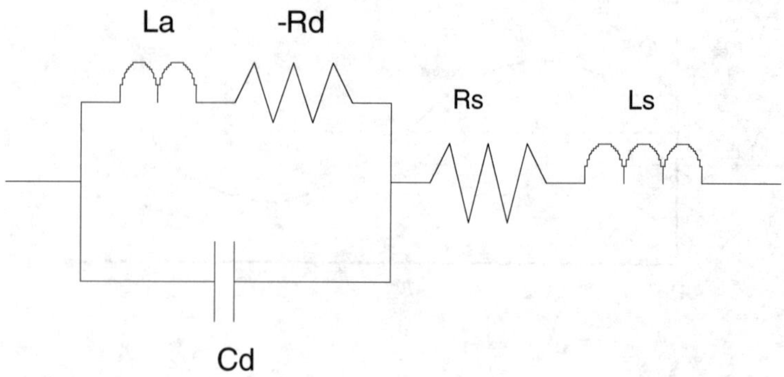

FIGURE 3.9 (continued)

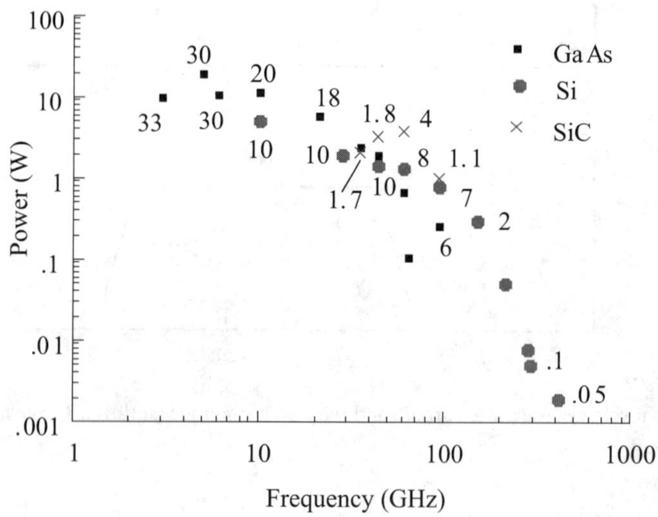

FIGURE 3.10 RF power-frequency performance for IMPATT oscillators (numbers associated with data points are DC-to-RF conversion efficiency).

Defining Terms

Immittance: A general term that refers to both impedance and admittance.

Active device: A device that can convert energy from a DC bias source to a signal at an RF frequency. Active devices are required in oscillators and amplifiers.

Two-terminal device: An electronic device, such as a diode, that has two contacts. The contacts are usually termed the cathode and anode.

One-port network: An electrical network that has only one RF port. This port must be used as both the input and output to the network. Two-terminal devices result in one-port networks.

Charge carriers: Units of electrical charge that produce current flow when moving. In a semiconductor two types of charge carriers exist: electrons and holes. Electrons carry unit negative charge and have an effective mass that is determined by the shape of the conduction band in energy-momentum space. The effective mass of an electron in a semiconductor is generally significantly less than an electron in free space. Holes have unit positive charge. Holes have an effective mass that is determined by the shape of the valence band in energy-momentum space. The effective mass of a hole is generally significantly larger than that for an electron. For this reason electrons generally move much faster than holes when an electric field is applied to the semiconductor.

References

1. Sze, S.M., 1981. *Physics of Semiconductor Devices*, 2nd Edition, Wiley-Interscience, New York.
2. Bosch, B.G., Engelmann, R.W. 1975. *Gunn-Effect Electronics*, Halsted Press, New York.
3. S.M. Sze, 1998. *Modern Semiconductor Device Physics*, Wiley-Interscience, New York.
4. P. Bhartia and I.J. Bahl, 1984. *Millimeter Wave Engineering and Applications*, Wiley-Interscience, New York.

4

Bipolar Junction Transistors

John C. Cowles
Analog Devices, Inc.

4.1 Introduction

The topic of bipolar junction transistors (BJTs) is obviously quite broad and a full treatment would consume volumes. This work focuses on basic principles to develop an intuitive feel for the transistor behavior and its application in contemporary high-speed integrated circuits (IC). The exponential growth in high bandwidth wired, wireless, and fiber communication systems coupled with advanced IC technologies has created an interesting convergence of two disparate worlds: the microwave and analog domains. The traditional microwave IC consists of a few transistors in discrete form or in low levels of integration surrounded by a sea of transmission lines and passive components. The modern high-speed analog IC usually involves 10s to 1000s of transistors with few passive components. The analog designer finds it a more cost-effective solution to use extra transistors rather than passive components to resolve performance issues. The microwave designer speaks in terms of noise figure, IP3, power gain, stability factor, VSWR, and s-parameters while the analog designer prefers noise voltages, harmonic distortion, voltage gain, phase margin, and impedance levels. Present BJT IC technologies fall exactly in this divide and thus force the two worlds together. Since parasitics within an IC are significantly lower than those associated with packages and external interconnects, analog techniques can be applied well into the microwave region and traditional microwave techniques such as impedance matching are only necessary when interfacing with the external world where reference impedances are the rule. This symbiosis has evolved into what is now termed radio frequency IC (RFIC) design. With this in mind, both analog and microwave aspects of bipolar transistors will be addressed. Throughout the discussion, the major differences between BJTs and field-effect transistors (FETs) will be mentioned.

4.1.1 A Brief History

The origin of the BJT was more a fortuitous discovery rather than an invention. The team of Shockley, Brattain, and Bardeen at Bell Labs had been pursuing a field effect device in the 1940s to replace vacuum tubes in telephony applications when they stumbled upon bipolar transistor action in their experimental point-contact structure. The point contact transistor consisted of a slab of n-type germanium sitting on a metal "base" with two metal contacts on either side. With proper applied bias, the "emitter" contact would inject current into the bulk and then the "collector" contact would sweep it up. They announced

0-8493-1562-X/03/$0.00+$1.50

their discovery in 1948 for which they later received the Nobel Prize in 1956. At that point, the device was described as a current-controlled voltage source. This misinterpretation of the device as a trans-resistor led to coining of the universal term transistor that today applies to both BJTs and their field-effect brethren.

Within a few years of the discovery of the transistor, its theory of operation was developed and refined. The decade of the 1950s was a race toward the development of a practical technique to fabricate them. Research into various aspects of the material sciences led to the development of the planar process with silicon as the cornerstone element. Substrate and epitaxial growth, dopant diffusion, ion-implantation, metalization, lithography, and oxidation had to be perfected and integrated into a manufacturable process flow. These efforts culminated with Robert Noyce's patent application for the planar silicon BJT IC concept that he invented while at Fairchild. In the 1960s, the first commercial ICs appeared on the market, launching the modern age of electronics. Since then, transistor performance and integration levels have skyrocketed driven primarily by advances in materials, metrology, and process technology as well as competition, cost, and opportunity.

4.2 Basic Operation

The basic structure of the bipolar transistor given in Fig. 4.1a consists of two back-to-back intimately coupled p-n junctions. While the npn transistor will be chosen as the example, note that the pnp structure operates identically, albeit in complementary fashion. Bipolar transistor action is based on the injection of minority carrier electrons from the emitter into the base as dictated by the base-emitter voltage. These carriers diffuse across the thin base region and are swept by the collector, which is normally reverse biased. In high speed ICs, BJTs are nearly always biased in this way, which is known as the forward active mode. The net effect is that an input signal voltage presented across the base-emitter terminals causes a current to flow into the collector terminal. The relationship between collector current, I_c, and base-emitter voltage, V_{be}, originates from fundamental thermodynamic arguments via the statistical Boltzmann distribution of electrons and holes in solids. It is given by,

$$I_c = I_s \exp\left(V_{be}/V_t\right) \tag{4.1}$$

where I_s is called the saturation current and V_t is Boltzmann's thermal voltage kT/q, which is 26 mV at room temperature (RT). This expression plotted in Fig. 4.1b is valid over many decades of current and its implications are far reaching in BJT IC design. Consider the small-signal transfer characteristics relating the output current to input voltage at a given bias current, I_c, i.e., the transconductance, g_m. It is given by the remarkably simple expression,

$$\delta I_c/\delta V_{be}\big|_{I_c} = g_m = I_c/V_t . \tag{4.2}$$

The notion that transconductance is precisely and linearly proportional to the bias current, I_c and inversely proportional to temperature through V_t is known as translinearity. Notice that I_s, which has a strong dependence on temperature, base-width, and other physical device parameters, has dropped out of the picture. This view of bipolar transistors as a voltage-controlled current source suggests that it should have been named the transductor rather than the transistor. Although bipolar transistors are usually considered hopelessly nonlinear because of the exponential law in Eq. (4.1), extremely precise and robust linear and nonlinear functions can be realized at very high frequencies by proper circuit techniques that exploit the translinear property embodied in Eq. (4.2).

The intrinsic bandwidth of high frequency BJTs is ultimately limited by minority carrier storage primarily in the base region. Current is conveyed from the emitter to the collector by the relatively slow process of diffusion resulting from the gradient in the distribution of stored minority carriers in the base. In contrast, current in FETs is transported by the faster process of majority carrier drift in response to

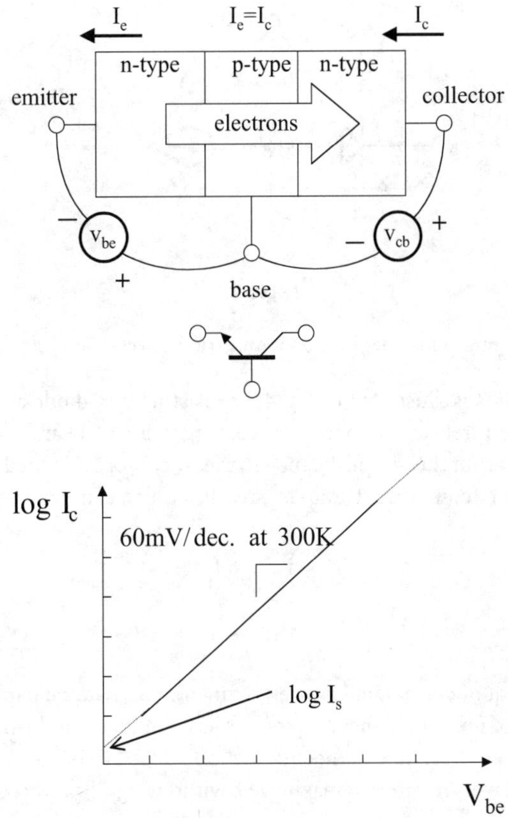

FIGURE 4.1 (a) BJT action involves carrier injection across the base from emitter to collector; (b) the ideal BJT exhibits a perfectly exponential input-output transfer function.

an electric field. By invoking the quasi-static approximation, which assumes that signals of interest change on a much longer time scale than the device time constants, carrier storage can be modeled as a lumped diffusion capacitance, C_d given by,

$$C_d = \delta Q_d / \delta V_{be} = \delta(I_c \tau_f) / \delta V_{be} = g_m \tau_f \tag{4.3}$$

where Q_d is the stored charge and τ_f is the forward transit time. The parameter τ_f can be viewed as the average time an electron spends diffusing across the base and can be expressed as a function of basic material and device parameters,

$$\tau_f = W_B^2 / 2\eta D_n \tag{4.4}$$

where W_B is the physical base width, D_n is the electron diffusion coefficient, and η is a dimensionless factor that accounts for any aiding fields in the base. Despite the limitations associated with carrier storage and diffusion, BJTs have demonstrated excellent high frequency performance. This is partly due to their vertical nature where dimensions such as W_B can be significantly less than 100 nm. In contrast, FETs are laterally arranged devices where the critical dimension through which carriers drift is determined by lithography and is typically larger than 100 nm even in the most advanced technologies. Furthermore, the fundamental limit contained in τ_f is never achieved in practice since other factors, both device and circuit-related, conspire to reduce the actual bandwidth.

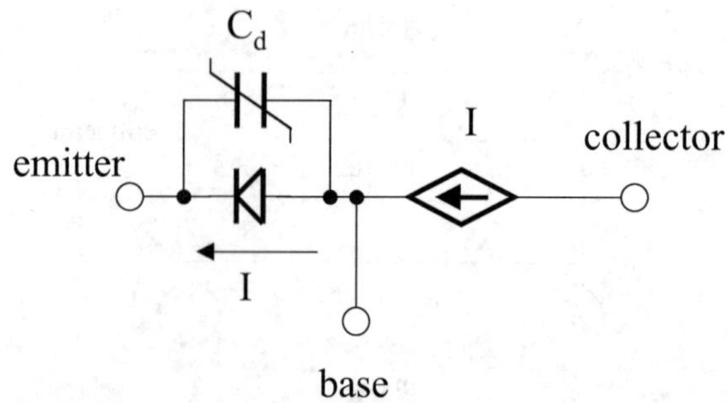

FIGURE 4.2 First order BJT model includes the nonlinear carrier injection and storage.

A first order model of BJTs is illustrated in Fig. 4.2 consisting of a diode at the base-emitter junction whose current is conveyed entirely to the collector according to Eq. (4.1) and a diffusion capacitance that models the carrier storage. From this simplistic model, the transistor f_t, defined as the frequency at which the common-emitter short-circuit current gain, h_{21}, reaches unity, can be calculated from,

$$h_{21} = i_c/i_b = f_t/f. \tag{4.5a}$$

$$f_t = g_m/2\pi C_d = 1/2\pi\tau_f. \tag{4.5b}$$

Note that f_t is independent of current and is solely a function of material parameters and device design details. Pushing this figure of merit to higher frequencies has dominated industrial and academic research for years. Polysilicon emitter BJTs, Si/SiGe heterojunction BJTs (HBTs), and III-V-based HBTs are all attempts to craft the active layers in order to maximize f_t without significantly compromising other device aspects.

4.2.1 Parasitics and Refinements

The first order model captures the essential operation of the BJT. However, actual devices deviate from that ideal in many aspects. As electrons diffuse across the base, some are lost to the process of recombination with holes as illustrated in Fig. 4.3a. Excess holes are also parasitically injected into the emitter. This overall loss of holes must originate at the base terminal and represents a finite base current, I_b, which is related to I_c via the parameter, β, known as the common emitter current gain,

$$\beta = I_c/I_b. \tag{4.6}$$

The effect of finite β is to modify the emitter-base diode current by a correction factor, α, known as the common-base current gain, that accounts for the extra base current. Since $I_e = I_b + I_c$, it is easy to show that,

$$\alpha = \beta/\beta + 1 < 1 \tag{4.7}$$

so that the emitter-base diode current now becomes I_s/α. Many factors figure into determining β, including W_B and the highly variable recombination lifetime τ_r. Since β is a poorly controlled parameter with potential variations of ±50%, it is inadvisable to design circuits that depend on its exact value. Fortunately, it is usually possible to design ICs that are insensitive to β as long as β is large, say >50, thanks to matching of co-integrated devices. Figure 4.3b illustrates the well-known BJT Gummel plot, which shows the translinearity of I_c as well as the finite I_b in response to V_{be}. The separation between the two curves represents β. As shown in Fig. 4.4, β represents a low frequency asymptote of h_{21} that exhibits

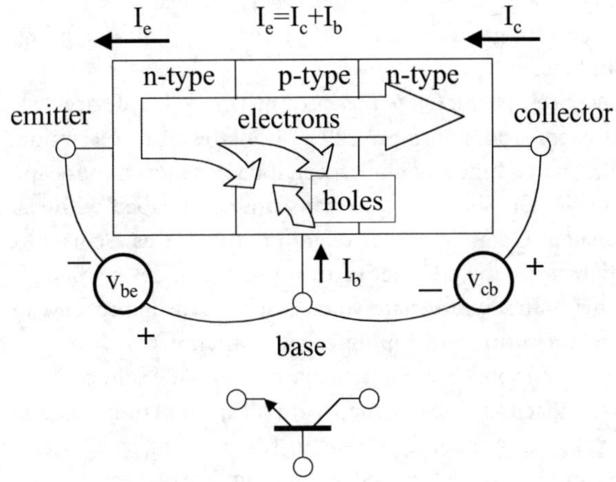

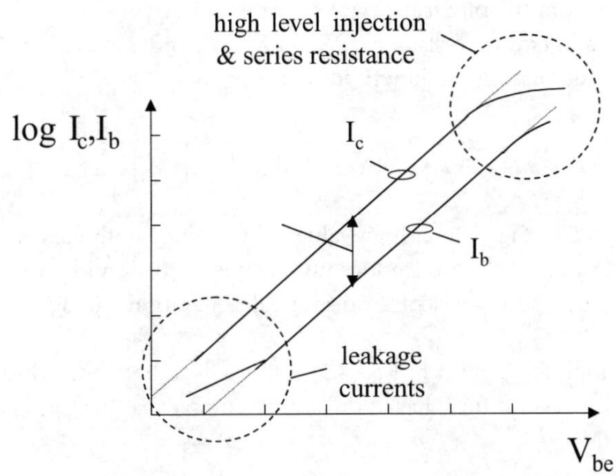

FIGURE 4.3 (a) Recombination of electrons and holes leads to finite current gain, β; (b) the Gummel plot illustrates the effects of finite β and other nonidealities.

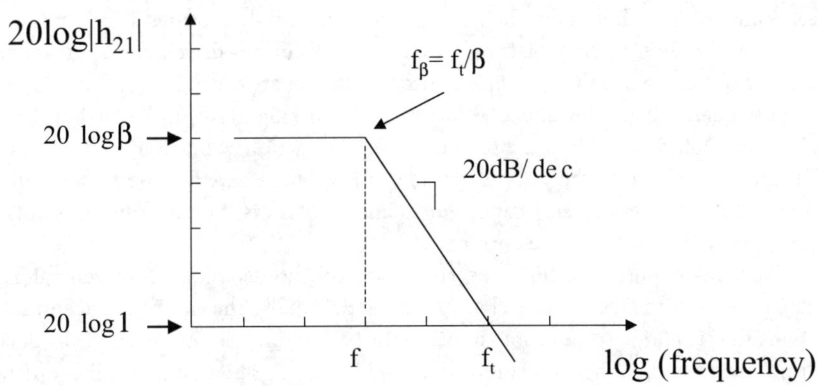

FIGURE 4.4 The frequency dependence of h_{21} captures finite β and f_t.

a single pole roll off toward f_t at a frequency, f_β given by f_t/β. Since RFICs most often operate well above f_β, the useful measure of current gain is actually f_t/f rather than β, although high DC β is still important for low noise and ease of biasing.

Traditionally, BJTs have been characterized as current-controlled devices where a forced I_b drives an I_c into an load impedance, consistent with Shockley's trans-resistor description. Now if I_b is considered a parasitic nuisance rather than a fundamental aspect, it becomes even more appropriate to view the BJT as a voltage-controlled device that behaves as a transconductor, albeit with exponential characteristics. In fact, contemporary analog IC design avoids operating the BJT as a current-controlled device due to the unpredictability of β. Instead, the designer takes advantage of device matching in an IC environment and translinearity to provide the appropriate voltage drive. This approach can be shown to be robust against device, process, temperature, and supply voltage variations.

Superimposed on the basic model are parasitic ohmic resistances in series with each active terminal (R_b, R_e, R_c) and parasitic capacitances associated with all pn junction depletion regions (C_{jc}, C_{je}, C_{js}), including the collector-substrate junction present to some extent in all technologies. Since parasitics degrade the idealized DC and RF characteristics captured in Eqs. (4.1) and (4.5), a major effort is focused on minimizing them through aggressive scaling and process engineering. Their values and bias dependencies can be estimated from the physical device structure and layout.

In particular, the base and emitter resistance, R_b and R_e, soften the elegant exponential characteristics in Eq. (4.1) by essentially de-biasing the junction since

$$I_c = I_s \exp\left(\left[V_{be} - I_b R_b - I_e R_e\right] / V_t\right). \tag{4.8}$$

This effect is also illustrated in Fig. 4.3b at high values of V_{be} where both curves appear to saturate. This departure from ideal translinearity can introduce unwelcome distortion into many otherwise linear ICs. Furthermore, these resistances add unwelcome noise and degeneration (voltage drops) as will be discussed later.

The idealized formulation for f_t given in Eq. (4.5) also needs to be modified to account for parasitics. A more comprehensive expression for f_t based on a more complex equivalent circuit results in,

$$f_t = \left(2\pi^*\left[\tau_f + \left(C_{je} + C_{jc}\right)/g_m + C_{jc}\left(R_c + R_e\right)\right]\right)^{-1} \tag{4.9}$$

where now f_t is dependent on current through g_m charging of C_{je} and C_{jc}. To achieve peak f_t, high currents are required to overcome the capacitances. As the intrinsic device τ_f has been reduced, the parasitics have become a dominant part of the BJTs' high frequency performance requiring even higher currents to reach the lofty peak values of f_t. It should also be kept in mind that f_t only captures a snapshot of the device high frequency performance. In circuits, transistors are rarely current driven and short circuited at the output. The base resistance and substrate capacitance that do not appear in Eq. (4.9) can have significant impact on high frequency IC performance. While various other figures of merit such as f_{max} have been proposed, none can capture the complex effects of device interactions with source and load impedances in a compact form. The moral of the story is that ideal BJTs should have low inertia all around, i.e., not only high peak f_t but also low parasitic capacitances and resistances, so that time constants in general can be minimized at the lowest possible currents.

At the high currents required for high f_t, second order phenomena known in general as high level injection begin to corrupt the DC and RF characteristics. Essentially, the electron concentration responsible for carrying the current becomes comparable to the background doping levels in the device causing deviations from the basic theory that assumes low level injection. The dominant phenomenon known as the Kirk effect or base-pushout manifests itself as a sudden widening of the base width at the expense of the collector. This translates into a departure from translinearity with dramatic drops in both β and f_t. A number of other effects have been identified over the years such as Webster effect, base conductivity

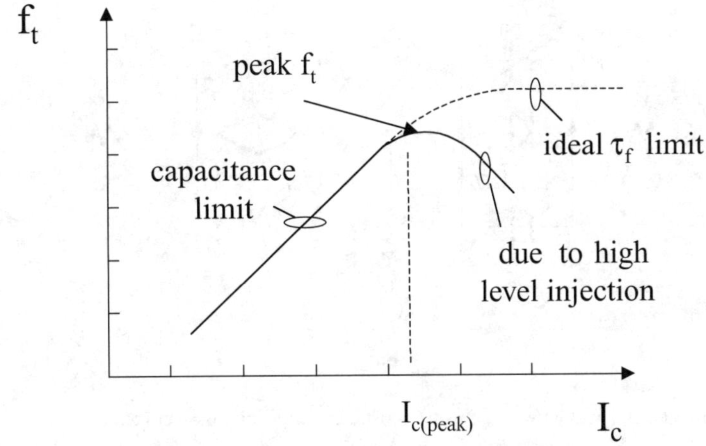

FIGURE 4.5 The f_t peaks at a particular current when high level injection occurs.

modulation, and current crowding. High level injection sets a practical maximum current at which peak f_t can be realized. Figure 4.5 illustrates the typical behavior of f_t with I_c. To counteract high level injection, doping levels throughout the device have increased at a cost of higher depletion capacitances and lower breakdown voltages. Since modeling of these effects is very complex and not necessarily included in many models, it is dangerous to design in this regime.

So far, the transistor output has been considered a perfect current source with infinite output resistance. In reality, as the output voltage swings, the base width is modulated, causing I_s and thus I_c to vary for a fixed V_{be}. This is exactly the effect of an output resistance and is modeled by a parameter V_a, the Early voltage,

$$r_o = \delta V_{ce} / \delta I_c = V_a / I_c . \tag{4.10}$$

As illustrated in the output common-emitter characteristics, I_c vs. V_{ce}, shown in Fig. 4.6a, V_a represents the common extrapolation point where all tangents intersect the V_{ce} axis. The effect of r_o is to set the maximum small-signal unloaded voltage gain since

$$A_v = \delta V_o / \delta V_i = \left(\delta I_c / \delta V_i\right)\left(\delta V_o / \delta I_c\right) = -g_m r_o = -V_a / V_t \tag{4.11}$$

For typical values of $V_a = 50$ V, $A_v = 2000$ (66 dB) at room temperature. Note that this gain is independent of I_c and is quite high for a single device, representing one of the main advantages of BJTs over FETs where the maximum gain is usually limited to < 40 dB for reasonable gate lengths. In modern devices,

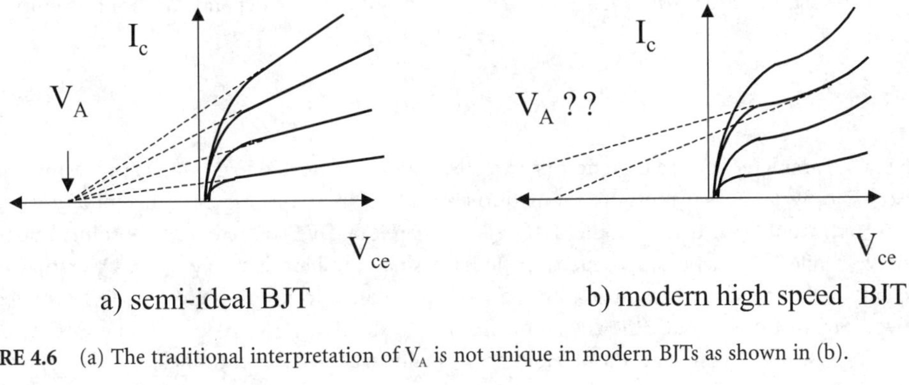

FIGURE 4.6 (a) The traditional interpretation of V_A is not unique in modern BJTs as shown in (b).

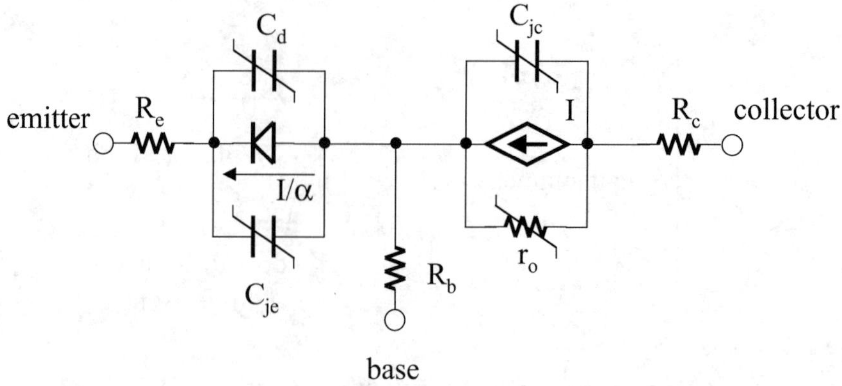

FIGURE 4.7 Augmented BJT model with parasitic resistances and finite base current.

however, V_a is not constant due to complex interactions between flowing electrons and internal electric fields. The net effect illustrated in Fig. 4.6b indicates a varying V_a depending on I_c and V_{ce}. This is often termed soft breakdown or weak avalanche in contrast to actual breakdown which will be discussed shortly. The effect of a varying V_a is to introduce another form of distortion since the gain will vary according to bias point. Figure 4.7 shows a more complete model that includes the fundamental parameters as well as the parasitics and refinements considered so far. It resembles a simplified version of the popular Gummel-Poon model found in many simulators.

Another form of pseudo-breakdown occurs in ultra-narrow base BJTs operating at high voltages. If the Early effect or base-width modulation is taken to an extreme, the base eventually becomes completely depleted. After this point, a further change in collector voltage directly modulates the emitter-base junction leading to an exponential increase in current flow. This phenomenon known as punchthrough fortunately has been mitigated by the fact that as base widths have narrowed, the base doping has been forced to increase so as to maintain a reasonable base resistance. Furthermore, since higher collector doping levels have been necessary to fight high level injection, true breakdown has become the voltage limiting mechanism rather than punchthrough.

Just as high level injection limits the maximum operating current, junction breakdown restricts the maximum operating voltage. When the collector voltage is raised, the collector base junction is reverse biased. The resulting electric field reaches a critical point where valence electrons are literally ripped out of their energy band and promoted to the conduction band while leaving holes in the valence band. The observed effect known as avalanche breakdown is a dramatic increase in current. The breakdown of the collector-base junction in isolation, i.e., with the emitter open circuited, is termed the BV_{cbo} where the "o" refers to the emitter open. Another limiting case of interest is when the base is open while a collector-emitter voltage is applied. In this case, an initial avalanche current acts as base current that induces more current flow, which further drives the avalanche process. The resulting positive feedback process causes the BV_{ceo} to be significantly lower than BV_{cbo}. The relationship clearly depends on β and is empirically modeled as,

$$BV_{ceo} = BV_{cbo} \Big/ \beta^{1/n} \tag{4.12}$$

where n is a fit parameter. A third limit occurs when the base is AC shorted to ground via a low impedance. In this scenario, avalanche currents are shunted to ground before becoming base current and BV_{ces} ("s" means shorted) would be expected to be $BV_{cbo} + V_{be}$. Therefore, BV_{ces} represents an absolute maximum value for V_{ce} while BV_{ceo} represents a pessimistic limit since the base is rarely open. Operating in the intermediate region requires care in setting the base impedance to ground and knowing its effect on breakdown. Figure 4.8 illustrates the transition from BV_{ceo} to BV_{ces}. The base-emitter junction is also sensitive to reverse bias. In this case, the breakdown is usually related to Zener tunneling and is represented

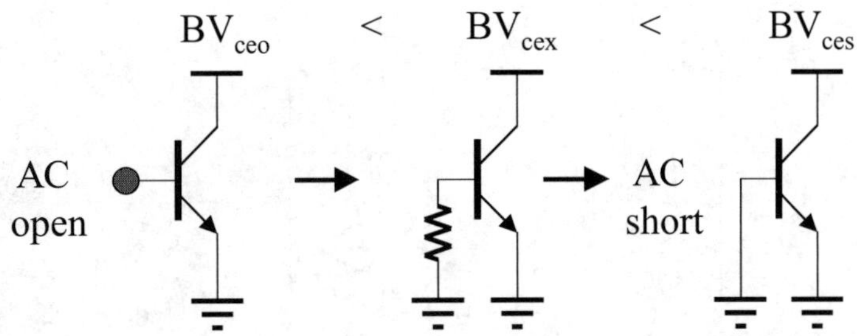

FIGURE 4.8 The AC base impedance helps determine the effective breakdown voltage.

by BV_{ebo}. Excessive excursions toward BV_{ebo} can cause injection of energetic carriers into surrounding dielectrics, leading to leakage currents, degraded reliability, and possibly device failure. Since these breakdown mechanisms in general are poorly modeled, it is recommended to operate well below BV_{ceo} and BV_{ebo}.

Even when these guidelines are followed, there are still situations where peak voltages might exceed breakdown limits. For instance, modern ICs in portable products often require a disable mode in which the part consumes virtually no current so as to conserve battery life. When disabled, it is common for transistors that are normally in a safe operating condition to find themselves with reverse-biased junctions in excess of BV_{ceo} and BV_{ebo}. The resulting leakage currents can lead to expectedly high disable currents, accidental activation of certain circuit blocks, and again to potential reliability failures. A second situation is when output stages drive variable reactive loads. It is possible (and likely) that certain conditions will cause 2 to 3 times the supply voltage to appear directly across a transistor leading to certain catastrophic breakdown.

As devices have been scaled down in size, isolated in dielectrics, and driven with higher current densities, self-heating has become increasingly problematic. The thermal resistance, R_{th}, and capacitance, C_{th}, model the change in junction temperature with power and the time constant of its response. The DC and AC variations in the junction temperature induce global variations in nearly all model parameters causing bias shifts, distortion, and even potentially catastrophic thermal runaway. Furthermore, local heating in one part of a circuit, say an output power stage, can start affecting neighboring areas such as bias circuits creating havoc. It is essential to selectively model thermal hotspots and ensure that they are within maximum temperature limits and that their effect on other circuit blocks is minimized by proper layout.

4.2.2 Dynamic Range

The limits on signal integrity are bounded on the low end by noise and on the high end by distortion. It is not appropriate to claim a certain dynamic range for BJTs since the circuit topology has a large impact on it. However, the fundamental noise sources and nonlinearities in BJTs can be quantified and then manipulated by circuit design. It is customary to refer the effects of noise and nonlinearity to either the input or the output of the device or circuit to allow fair comparisons and to facilitate cascaded system-level analyses. Here, all effects will be referred to the input, noting that the output quantity is simply scaled up by the appropriate gain.

There are three fundamental noise mechanisms that are manifested in BJTs. Figure 4.9 presents a linearized small-signal model that includes the noise sources. The objective will be to refer all noise sources to equivalent voltage and current noise sources, e_n and i_n, respectively, at the input. Shot noise is associated with fluctuations in current caused by the discrete nature of electronic charge overcoming the junction potential. The presence of the forward biased base-emitter junction results in collector and base shot noise, i_{cn}, and i_{bn}, which are given by,

$$i_{cn}^2 = 2qI_c;\ i_{bn}^2 = 2qI_b \tag{4.13}$$

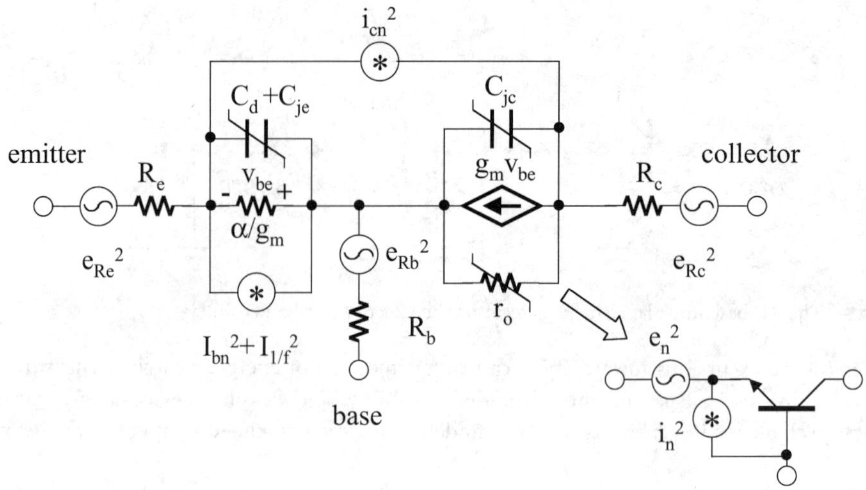

FIGURE 4.9 The fundamental noise sources can be transformed to equivalent input-referred sources.

where q is the electronic charge. This is a major disadvantage of BJTs with respect to FETs, which are free of shot noise due to the absence of potential barriers that determine current flow. Each physical resistance carries an associated Johnson noise represented as

$$e_{Rx}^2 = 4kTR_x \tag{4.14}$$

where R_x is the appropriate resistance and k is Boltzmann's constant. Finally, a 1/f noise source, $i_{1/f}^2$ associated primarily with recombination at semiconductor surfaces appears at the input. The 1/f noise component decreases with increasing frequency, as expected. A corner frequency is often quoted as the point where the 1/f noise and the white thermal and shot noise cross over. It is an empirical value that varies from device to device and depends on bias conditions and source impedance. BJTs typically offer lower 1/f corner frequency than FETs since surfaces play a lesser role in BJTs. Except for broadband DC-coupled circuits and strongly nonlinear circuits such as oscillators and mixers that perform frequency conversion, 1/f noise is insignificant at high frequencies and will be ignored for simplicity. Note that treating noise in nonlinear systems is quite complex and remains an active research topic.

Once the noise sources are input referred, e_n and i_n are approximately given by

$$e_n^2 = 4kT\left(R_b + R_e + 1/2g_m\right) \tag{4.15a}$$

$$i_n^2 = 2q\left(I_b + I_c \Big/ \left|h_{21}(f)\right|^2\right). \tag{4.15b}$$

An arbitrarily higher I_c reduces e_n through the larger g_m but has mixed effects on i_n depending on how $h_{21}(f)$ varies with current. An arbitrarily larger device reduces e_n by lowering R_b and R_e but the commensurate increase in capacitance might degrade $h_{21}(f)$ and thus i_n. The moral here is that minimization of noise requires a trade-off in I_c and device size given an f_t equation and parasitic scaling rules with device geometry.

Distortion originates in nonlinearities within the device. For simplicity, distortion will be described in terms of a power series to derive expressions for harmonic distortion, intermodulation distortion, and compression. It is assumed that linearities up to only third order are considered, limiting the accuracy of the results as compression is approached. While this approach is valid at low frequencies, capacitive

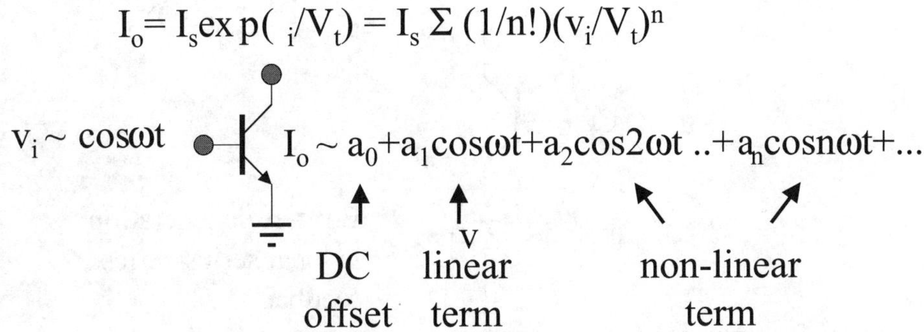

FIGURE 4.10 A power series expansion of the exponential junction nonlinearity can be used to quantify distortion.

effects at higher frequencies greatly complicate nonlinear analysis and more generalized Volterra series must be invoked. However, the general idea is still valid. Note that not all systems can be mathematically modeled by power or Volterra series; for example, a hard limiting amplifier that shows sudden changes in characteristics.

In BJTs, the fundamental transfer function given in Eq. (4.1) represents one of strongest nonlinearities. As shown in Fig. 4.10, by expanding Eq. (4.1) in its power series up to 3rd order, the following measures of input-referred figures of merit are derived in dBV (mV),

$$P_{1db} = 0.933\ V_t = 24.3\ \text{mV} = -32.3\ \text{dBV} \tag{4.16a}$$

$$\text{IIP3} = 2.83\ V_t = 73.5\ \text{mV} = -22.7\ \text{dBV} \tag{4.16b}$$

$$\text{P3OI} = 4.89\ V_t = 127\ \text{mV} = -17.9\ \text{dBV} \tag{4.16c}$$

where P_{1db} is the 1-dB gain compression point, IIP3 is the third order intercept point, and P3OI is the third harmonic intercept point. Note that the conversion from dBV to dBm referenced to 50 Ω is + 10 dB. From the analysis, it can be inferred that P_{1db} occurs for peak input swings of about V_t while a third order intermodulation level of 40 dBc is achieved for an input swing of only 0.0283 V_t or 0.73 mV at room temperature. These values are quite low, indicating that circuit linearization techniques must be invoked to exceed these limits. FETs on the other hand possess a more benign input nonlinearity that ranges from square law in long channel devices to nearly linear in short channels. While this is not the sole source of nonlinear behavior, FETs are expected to have superior linearity as compared to BJTs from this perspective. It is interesting to observe that the commonly used rule of thumb stating that IIP3 is approximately 10 dB higher than P_{1db} is true in this case. As mentioned earlier, higher order terms, parasitics, and reactive effects modify the results in Eq. (4.16) but they do provide a useful order of magnitude.

A common technique for extending overall dynamic range is emitter degeneration, which as illustrated in Fig. 4.11, essentially amounts to series-series feedback. The impedance in series with the emitter reduces the signal that appears directly across the emitter base leading to lower input-referred distortion. Note that the gain is proportionately reduced so that output-referred distortion is not improved to the same degree. The potentially superior distortion performance of FETs is analogous to a strongly degenerated BJT with the accompanying lower gain. Since this degenerating impedance is directly in the signal path, it also affects noise. If the impedance is a resistor, it will contribute noise as if it were an extra R_e in Eq. (4.15a). If R_e is not a dominant noise contributor, this approach can extend dynamic range significantly. Another option is to use a degenerating inductor, which at high frequencies provides adequate degeneration while only contributing noise associated with its finite quality factor, Q. Other feedback

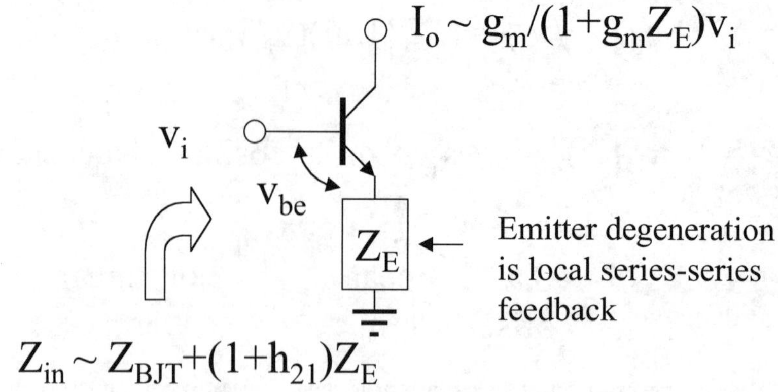

FIGURE 4.11 Emitter degeneration can improve dynamic range and change impedance levels.

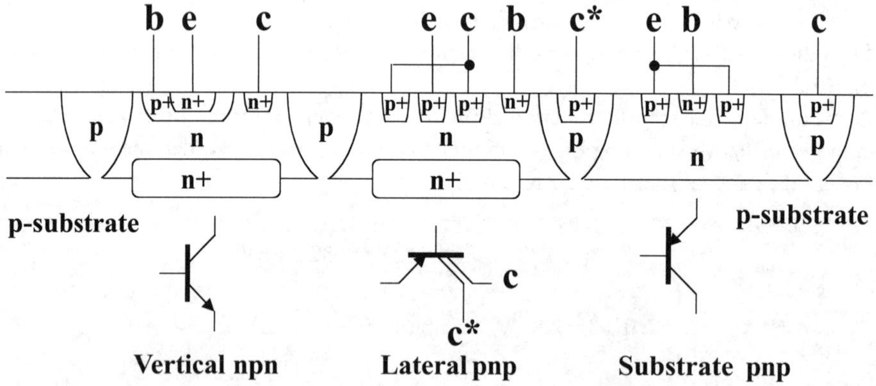

FIGURE 4.12 Both lateral and substrate pnp BJTs can be generated in a typical npn process.

techniques are also possible to improve dynamic range, but care must be taken to ensure stability particularly around longer feedback loops. Additional benefits of using local feedback are desensitization of input impedance and gain to device parameters.

4.2.3 Complementary Pnp

The availability of a pnp device to complement the npn offers numerous advantages, particularly at lower supply voltages. For example, pnp active loads can provide more voltage gain at a given supply voltage; signals can be "folded" up and down to avoid hitting the rails and balanced complementary push-pull topologies can operate near rail-to-rail. In general, the p-type device is slower than the n-type device across most technologies due to the fundamentally lower hole mobility with respect to electron mobility. Furthermore, in practice, the pnp transistor is often implemented as a parasitic device, i.e., the fabrication process is optimized for the npn and the pnp shares the available layers. Figure 4.12 illustrates how two different pnp's can be constructed in an npn process (p-type substrate). The so-called substrate pnp has the substrate double as its collector and thus only appears in a grounded collector arrangement. The lateral pnp frees the collector terminal, but carries along a parasitic substrate pnp of its own that injects large currents into the substrate and degrades f_t and β. In some cases, true complementary technologies are available in which both the npn's and pnp's are synthesized and optimized separately.

When only lateral and substrate pnp's are available, they are usually used outside of the high-speed signal path due to their compromised RF performance. They appear prominently in bias generation and distribution as well as in low and intermediate frequency blocks. When using either pnp, care must be taken to account for current injection into the substrate, which acts locally as a collector. This current can induce noise and signal coupling as well as unintentional voltage drops along the substrate. It is customary to add substrate contacts around the pnp's to bring this current to the surface before it disperses throughout the IC. If true high-quality vertical pnp's are available that are balanced in performance with respect to the npn's, they can be used for high-frequency signal processing, enabling a number of circuit concepts commonly used at lower frequencies to be applied at RF.

4.2.4 Topologies

Several basic single-transistor and composite transistor configurations are commonly encountered in RFICs. Figure 4.13 shows the three basic single-transistor connections known as common-emitter (CE), common-base (CB), and common-collector (CC). The name refers to the fact that the common terminal is AC grounded. Their properties are well covered in numerous texts and only a brief discussion on their general properties and applications will be presented here.

The CE stage is the most commonly considered configuration since it provides a fairly high input impedance, high output impedance, and both current gain, $h_{21}(f)$, and voltage gain, $g_m R_L$. The noise and distortion properties are the same as considered earlier for the BJT. It is commonly used in single-ended, low noise amplifiers (LNA) and power amplifiers (PA). A well-known drawback of the CE stage at high frequencies is the feedback C_{jc}, which degrades gain via Miller multiplication and couples input and output networks leading to detuning of matching networks and possible instability.

As noted earlier, emitter degeneration can be used not only to improve dynamic range but also to help set impedance levels. It is sometimes necessary for the IC to present a standard impedance level at its input and output terminals to provide proper terminations. The use of emitter inductive degeneration has the fortuitous property of generating a synthetic resistive input impedance at high frequencies without contributing noise. As illustrated in Fig. 4.11 earlier, the transformed value is given by

$$R_{eff} = h_{21}(f) Z_E = (f_t / f)(2\pi f L_E) = 2\pi\, f_t\, L_E \tag{4.17}$$

which now appears in series with the input terminal.

The CB stage provides a well-predicted low input impedance given by $1/g_m$, high output impedance and voltage gain, $g_m R_L$, but has near unity current gain, α. It acts as an excellent current buffer and is often used to isolate circuit nodes. The CB appears most often in tight synergy with other transistors arranged in composite configurations. Although it lacks the capacitive feedback present in the CE that

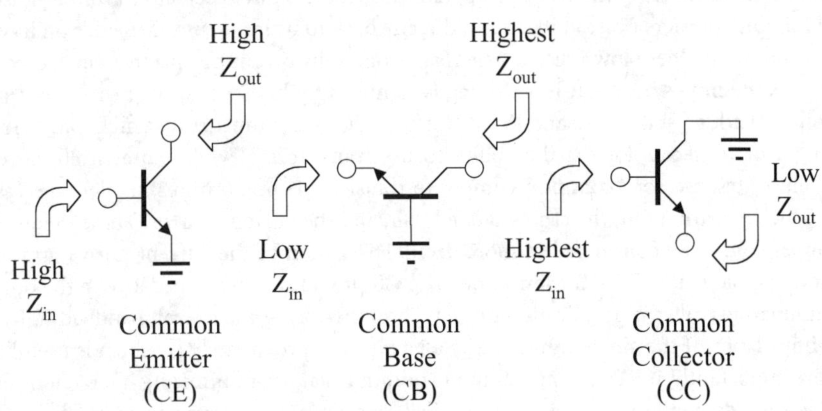

FIGURE 4.13 The three basic single BJT stages offer different features.

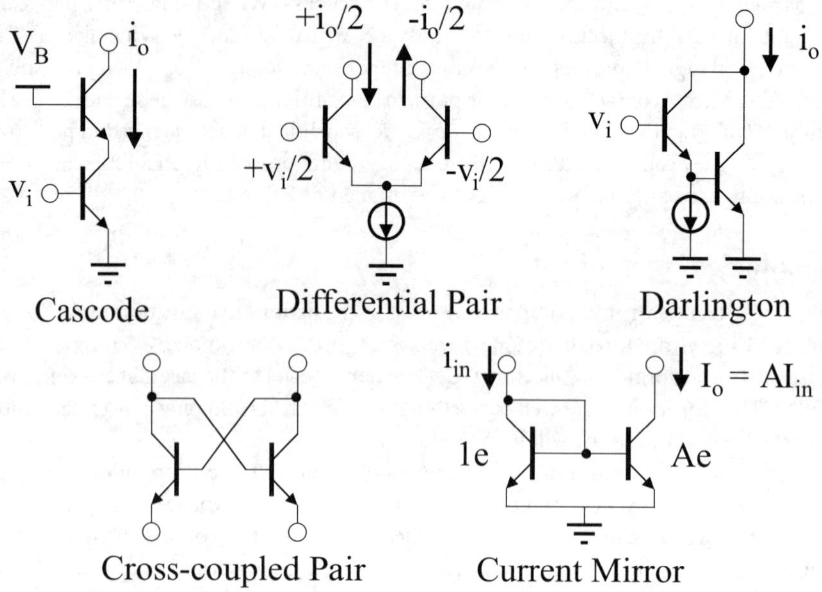

FIGURE 4.14 These common two-transistor configurations are used extensively in BJT ICs.

degrades gain and destabilizes operation, any series feedback at the grounded base node can lead to instability. The feedback element might be due to metallization, package parastics, or even the actual device base resistance. Extreme care must be exercised when grounding the CB stage. Another difference is that with a grounded base, the larger BV_{cbo} sets the voltage limit. The noise and distortion properties are identical to the CE.

The CC stage is often known as an emitter follower. It is predominantly used as a level shifter and as a voltage buffer to isolate stages and provide extra drive. It offers high input impedance, low output impedance, and near unity voltage gain. The impedance buffering factor is roughly $h_{21}(f)$ and at frequencies approaching f_t, its effectiveness is diminished as $h_{21}(f)$ nears unity. In this case it is common to see several stages of cascaded followers to provide adequate drive and buffering. The CC stage is a very wideband stage since direct capacitive feedthrough via C_d and C_{je} cancels the dominant pole to first order. However, this same capacitive coupling from input to output can cause destabilization, particularly with capacitive loads. In fact, this is the basis of operation of the Colpitts oscillator. From a noise point of view, the CC simply transfers noise at its output directly back to its input and then adds on its own noise; therefore it is not used where low noise is essential. From a distortion perspective, the CC can be fairly linear as long as the impedance that it is driving is significantly higher than its own output impedance $1/g_m$. For minimal added shot noise and low distortion, CC stages must run at high bias currents.

The strength of IC technology is the ability to use transistors at will at practically no extra cost. Figure 4.14 illustrates five very common composite transistor configurations, namely the cascode, the differential pair, the Darlington, the cross-coupled pair, and the current mirror. These ubiquitous forms are common enough to be considered in more detail. Discussion of the current mirror appears later.

The cascode topology is a CE-CB connection. The CB provides a low impedance to the output of the CE stage, eliminating Miller multiplication of its C_{jc} and increasing the overall bandwidth. It also raises the output impedance of the single transistor by a factor of approximately β, which is useful in current sources. More importantly at RF, it improves input-output isolation, minimizing interactions that reduce the gain. The cascode configuration is common in LNAs since higher tuned power gain is achievable with minimal degradation in noise. Note that the CB transistor can be significantly smaller than the CE transistor, minimizing the parasitic capacitance at its output. The cascode requires a higher supply voltage than a stand-alone CE stage by at least a $V_{ce(sat)}$ to keep the lower BJT in its active region.

The differential pair (also known as a long-tailed pair) can be thought of as a CC-CB connection when driven in a single-ended fashion or as parallel CE stages when driven differentially. A tail current source or a current setting resistance is required to establish the operating point. This ubiquitous, cannonical form has a long, distinguished history in operational amplifiers, mixers, IF/RF amplifiers, digital ECL/CML gates, and even oscillators and latches when cross-coupled. The basic operation relies on the controlled steering of current from one branch to the other. Notice that the dynamic range is modified by the fact that noise sources from two transistors contribute to the total noise while the signal is divided across two junctions. The noise associated with the tail current source appears as common mode noise and does not affect the input noise if differential signaling is used. The structure can be enhanced with the addition of emitter degeneration, cascoding CB stages, and buffering CC stages.

The Darlington connection is a CE stage buffered at the input by CC stage. This configuration behaves essentially like a super-transistor with greater impedance buffering capability and higher $h_{21}(f)$. It typically appears at input and output interfaces where drive and buffering requirements are most severe. The Darlington and the differential pair are sometimes called f_t multipliers since the input signal appears across two transistor junctions in series while the output signal is a parallel combination of currents. For a fixed input bias current per device, the effective input capacitance is $C_d/2$ while the effective transconductance is still g_m, giving from Eq. (4.5) an $f_t^{eff} = 2f_{tBJT}$. This of course ignores additional higher order poles and requires twice the current of a single BJT. In analogy to the differential pair, noise and distortion occur across two junctions in series.

The cross-coupled connection is derived from the differential pair. Following the signals around the loops leads to a positive feedback condition that can be used to induce regeneration for Schmidt triggers, multivibrators, and latches or negative resistance for tuned VCOs and gain peaking. It has also been used to synthesize voltage-to-current converters, bias cells, multipliers, and other functions. Since positive feedback is inevitably invoked in cross-coupled devices, care must be taken to avoid instability when it is undesired.

Note that npn-pnp composite structures are also possible creating many permutations that might offer certain advantages. These are illustrated in Fig. 4.15 in the form of the double up-down emitter-follower, the folded cascode, the composite pnp, and complementary mirror. The first two examples reduce the supply voltage requirements over their npn-only embodiment, the third transforms a poor quality pnp into a nearly ideal device, and the fourth provides a voltage controlled mirror ratio.

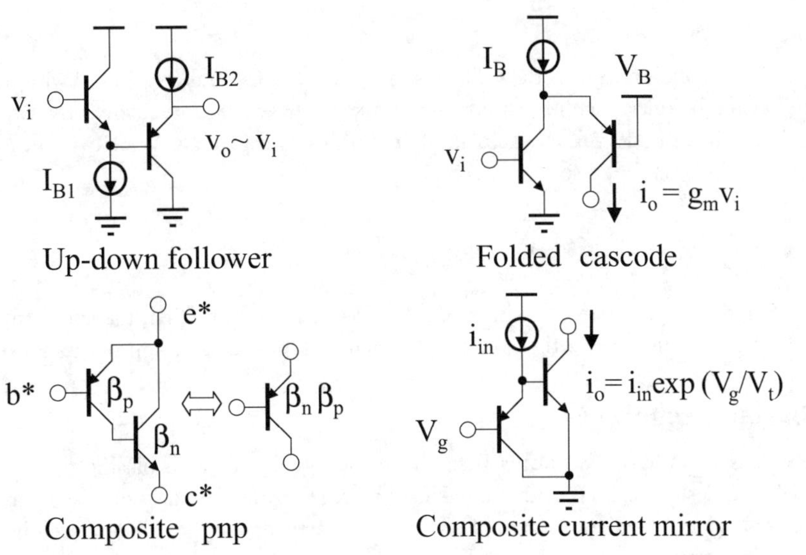

FIGURE 4.15 Composite npn-pnp combinations add functionality at lower supply voltages.

4.2.5 Translinear Circuits

The concept of translinearity was observed in deriving the basic properties of BJTs and its usefulness was alluded to in general terms. In particular, if the signals of interest are conveyed as currents rather than voltages, a large class of linear and nonlinear functions can be synthesized. When current-mode signals are used, extremely large dynamic ranges are possible since the junction essentially compresses the current signal logarithmically to a voltage; i.e., a 60 dB (1000x) signal current range corresponds to 180 mV of voltage swing at room temperature. The relatively small voltage swings represent a low impedance through which capacitances charge and discharge and thus the bandwidth of these circuits achieves broadband operation out to near the device limits represented by f_t. Furthermore, reduced internal voltage swings are consistent with lower supply voltages.

The simplest example of a translinear circuit is the well-known current mirror shown in Fig. 4.14. The input current, I_{in}, is mirrored to the output, I_{out} according to a scaling factor associated with the ratio of device sizes. Inherent in this process is a nonlinear conversion from I_{in} to the common V_{be} and then a second related nonlinear conversion from V_{be} to I_{out}. The impedance at the common base node is nominally a parallel combination of $1/g_m$ and $(1 + A)C_d$, which results in a low time constant on the order of τ_f.

A large family of translinear circuits has been synthesized by applying the translinear principle stated as: *In a closed loop containing only junctions, the product of current densities flowing in the clockwise direction is equal to the product of current densities flowing in the counterclockwise direction.*

The principle has a corollary when voltage sources are inserted into the loop, namely that the products are equal to within a factor $\exp(V_a/V_t)$, where V_a is the applied voltage. It is merely a restatement that logarithmic multiplication and division, i.e., addition and subtraction of V_{be}'s, is tantamount to products and quotients of currents. In this sense, the BJT can be thought of as the mathematical transistor. Note that since the signals are in current form, the distortion is not caused by the exponential characteristics but is actually due to *departure* from it, i.e., series resistances, high level injection, and Early voltage effects conspire to distort the signals.

Fig. 4.16 illustrates three examples of translinear IC design. The first example is an analog squarer/divider with input signals I_x, I_y and output, I_o. By following the translinear principle, it can be shown that the mathematical operation

$$I_o = I_x^2 / I_y \tag{4.18a}$$

is performed. The second example is the well-known Gilbert-cell multiplier/gain cell with predistortion. Again, by following the rules, the differential output currents have the same form as the two inputs. The third example is a linear in dB variable gain amplifier. In this case a voltage $V_g = R_g I_g$ is inserted into the loop and Eq. (4.18a) is modified so that

$$i_o = I_x / V_t \exp\left(V_g / V_t\right) V_i = I_x / V_t G_o 10^{Vg/Vt} \tag{4.18b}$$

Note that if V_g and I_x are engineered to be proportional to temperature, then the gain control becomes stable with temperature. In all cases, the operating frequency of these circuits approaches the technology f_t.

4.2.6 Biasing Techniques

Microwave IC designers have historically focused on the signal path with biasing treated as an afterthought. Analog IC designers, on the other hand, have placed a strong emphasis on biasing since it usually determines the sensitivity and robustness of a circuit. In particular, as was already seen in the variable gain amplifier, it is often necessary to generate a current that is proportional to absolute temperature, PTAT. Other times it is necessary to generate a current that is complementary to absolute temperature, CTAT. Finally sometimes a zero temperature coefficient current, ZTAT, is desired. The choice on tem-

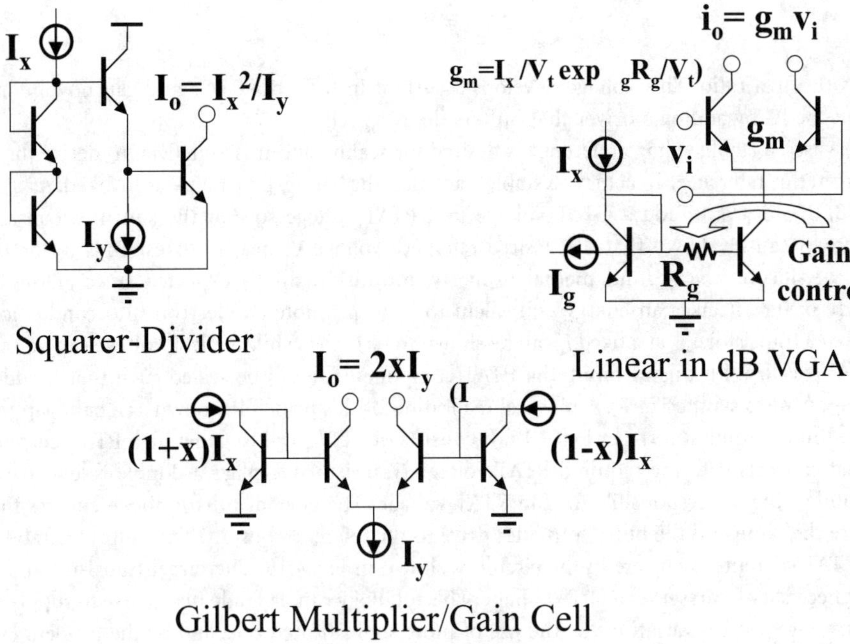

FIGURE 4.16 Numerous functions can be implemented using translinear principles.

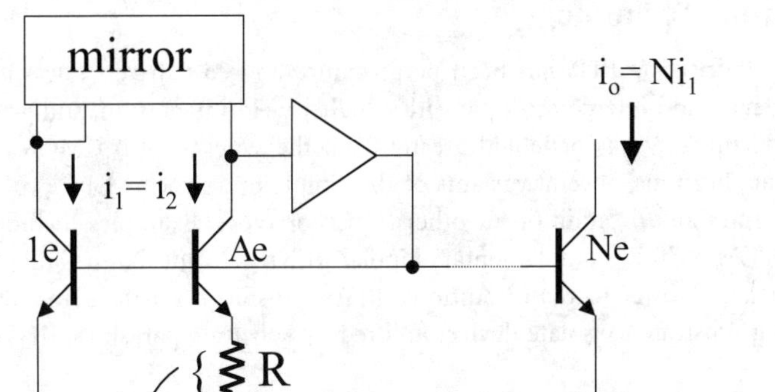

FIGURE 4.17 The ΔV_{be} cell that generates a PTAT current is ubiquitous in BJT biasing.

perature shaping depends on what is needed. In differential amplifiers where the gain is given by $g_m R_L = I_o R_L / 2V_t$, it is appropriate to choose a PTAT bias current for a stable gain; however, if the amplifier is actually a limiter, then a stable limited voltage swing requires a ZTAT shaping. By far the most common and fundamental bias current shape for BJTs turns out to be PTAT. From the PTAT form, other shapes can be derived.

The fundamental way to synthesize a bias sub-circuit is to start with a ΔV_{be} cell that generates a PTAT current. This cell, illustrated in Fig. 4.17, is a modified translinear loop. Two area ratioed BJTs each forced to carry the same current generate a difference in V_{be} given by

$$\Delta V_{be} = V_t \ln(A) \tag{4.19}$$

where A is the area ratio. This voltage develops a current in R that is PTAT and can now be mirrored throughout the IC by adding a driver that buffers the ΔV_{be} cell.

In some cases, a stable voltage reference is desired for scaling and it is beneficial to derive the various currents from this reference. To achieve a stable reference, the bandgap principle is invoked. As illustrated in Fig. 4.18, the idea is to add a CTAT voltage to a PTAT voltage so that the sum is a constant with temperature. It can be shown that the desired reference voltage V_g nearly corresponds to the bandgap energy, E_g of silicon, a very fundamental property. Intuitively, this is expected since at the extreme temperature of 0 K, it takes an energy equivalent to E_g to promote an electron into conduction. It so happens that a transistor V_{be} at a fixed I_c can be shown to be CTAT while a PTAT voltage can be generated from the ΔV_{be} cell. With this in mind, the PTAT component should be scaled such that it adds to the CTAT component to synthesize V_g. A physical realization of this principle known as a bandgap reference is illustrated in its simplest form in Fig. 4.19. It consists of a ΔV_{be} cell to generate a PTAT current and a resistor that converts it to the required PTAT voltage. Transistor Q1 plays a double role as part of the ΔV_{be} cell and as the V_{be} responsible for the CTAT voltage. The current mirror above ensures that both currents are the same and the buffer provides drive to the reference output. This output can be used to generate PTAT currents elsewhere in the circuit as shown in Fig. 4.19. The target transistor and resistor scaling are necessary to preserve the PTAT shape. This topology can be made insensitive to supply voltage, temperature, and process variations by the use of more sophisticated cells and by the prudent choice of component sizes and layout.

4.2.7 Fabrication Technology

The technology for fabricating BJTs has been performance and cost driven toward smaller transistor geometries, lower device and interconnect parasitics, higher yield/integration, and greater functionality. Silicon-based BJT technology has benefited greatly from the synergy with CMOS processes targeting VLSI/ULSI digital applications. Several variants of the simple npn process have evolved over the years that feature better npn's and/or addition of other transistor types. Examples include BiCMOS, which integrates BJTs with CMOS; fully complementary bipolar with true vertical npn's and pnp's; SiGe/Si HBT, which offers higher f_t and lower R_b than traditional BJTs; and silicon-on-insulator (SOI) processes that rely on an insulating substrate to isolate devices and reduce substrate parasitics. III-V based HBTs have

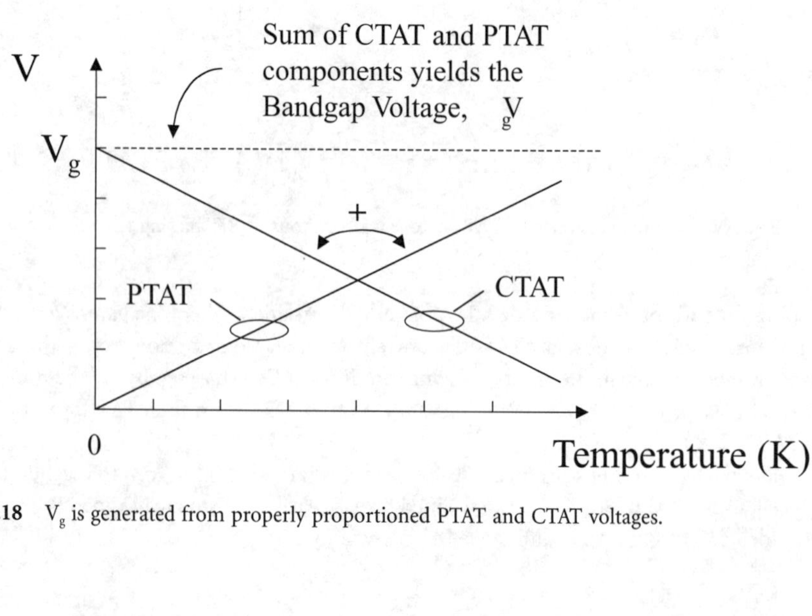

FIGURE 4.18 V_g is generated from properly proportioned PTAT and CTAT voltages.

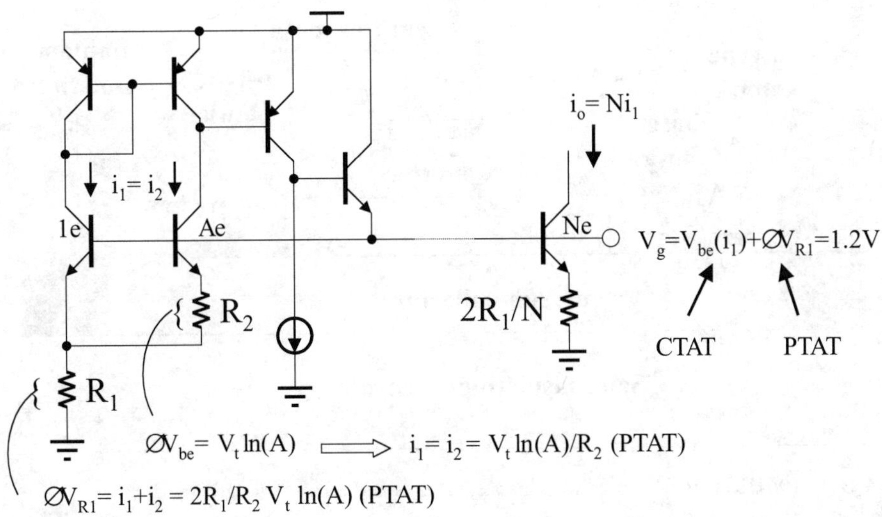

FIGURE 4.19 The Brokaw cell provides PTAT currents and a stable voltage reference.

developed in a different direction since raw device speed has been more than adequate, but integration levels and process complexity have limited their availability. Process features can be grouped into two general categories: the active device, which determines the transistor performance, and the back-end process, which defines the interconnect metallization, dielectric isolation, passive components, and through-vias. The back-end process is particularly critical in RFICs since they set limits to IC parasitics and are responsible for the quality of resistors, capacitors, and inductors.

The silicon BJT process has evolved from the junction isolated buried collector process shown in Fig. 4.12 to the oxide trench isolated double-poly (emitter and base) process diagrammed in Fig. 4.20. The double poly structure with self-aligned spacers allows ultra-small emitter-base structures to be defined while simultaneously providing low extrinsic base resistance and low emitter charge storage. A selective collector implant under the active emitter helps delay the Kirk effect without significantly increasing the depletion capacitance. In general, the devices are defined by diffusions into the substrate with interconnects and passives constructed over dielectrics above the substrate. Standard aluminum metals have been silicided for improved reliability. Overall, this has led to significantly reduced device footprints and feature sizes with emitter geometries down to < 0.35 μm and f_t ranging from 30 GHz for standard BJTs to > 100 GHz for SiGe HBTs. Refinements such as silicided base handles and optimized epitaxially deposited bases promise even further improvements in device performance. On the back-end side, thin-film resistors and metal-metal capacitors are replacing their polysilicon and diffused versions bringing lower temperature coefficients, reduced parasitics and coupling to the substrate, trimming

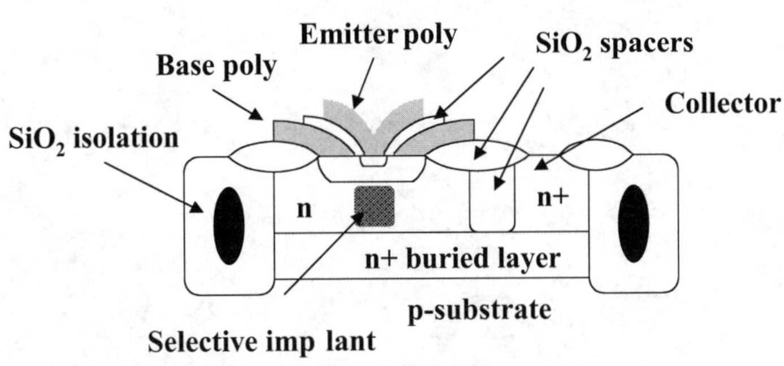

FIGURE 4.20 Modern processes feature double-poly, SiO_2 trenches and sub-μm lithography.

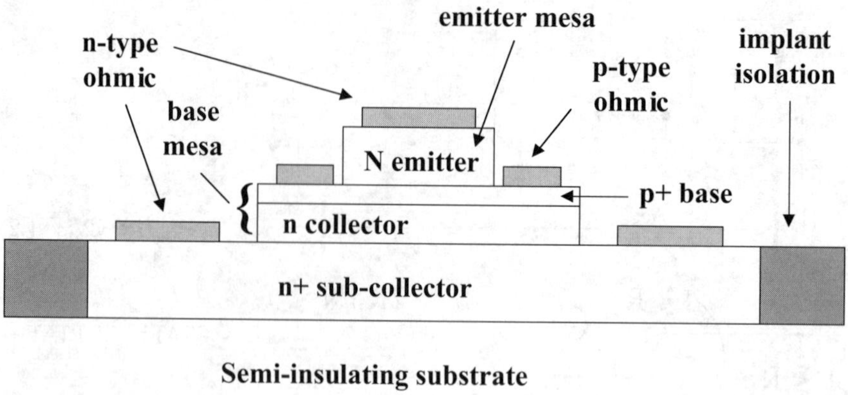

FIGURE 4.21 III-V HBTs are fabricated with mesa structures and relaxed lithography.

capability, and bias insensitivity. Furthermore, advanced multiple level interconnect modules and planarization techniques have been adopted from CMOS processes in the form of copper metalization, chemical-mechanical polishing, tungsten plugs, and low dielectric insulators. For RFICs, advanced interconnects are desirable not for high packing density, but for defining high quality passives away from the lossy substrate, particularly inductors, which are essential in certain applications.

The fabrication technology for the III-V HBTs is significantly different from that of silicon BJTs. The main difference stems from the fact that the starting material must be grown by advanced epitaxial techniques such as molecular beam epitaxy (MBE) or metallo-organic chemical vapor deposition (MOCVD). Each layer in the stack is individually optimized as needed. As a result of the predefined layers as well as thermal limits that disallow high temperature diffusion, the device is literally carved into the substrate leading to the wedding-cake triple mesa structure depicted in Fig. 4.21. The nature of the process limits minimum emitter geometries to about 1 μm, although self-aligned structures are used to minimize extrinsic areas. Values of f_t range from 30 GHz for power devices to 200 GHz for the highest speed structures. Interestingly, the improvements being made to silicon BJTs such as epitaxial bases, polysilicon emitters, and silicided bases, make use of deposited active layers, which is reminiscent of III-V technologies based on epitaxy. The back-end process inevitably consists of thin film resistors and metal-metal capacitors. Since the substrate is semi-insulating by definition, high-Q monolithic inductors are already available. Some processes offer backside vias that literally provide a direct connection from the front-side to the back-side of the wafer enabling easy and effective grounding at any point on the circuit.

5

Heterostructure Bipolar Transistors

William Liu
Texas Instruments

5.1 Basic Device Principle

Heterojunction bipolar transistors (HBTs) differ from conventional bipolar junction transistors (BJTs) in their use of hetero-structures, particularly in the base-emitter junction. In a uniform material, an electric field exerts the same amount of force on an electron and a hole, producing movements in opposite directions as shown in Fig. 5.1a. With appropriate modifications in the semiconductor energy gap, the forces on an electron and a hole may differ, and at an extreme, drive the carriers along the same direction as shown in Fig. 5.1b.[1] The ability to modify the material composition to independently control the movement of carriers is the key advantage of adopting hetero-structures for transistor design.

Figure Fig. 5.2 illustrates the band diagram of a *npn* BJT under normal operation, wherein the base-emitter bias (V_{BE}) is positive and the base-collector bias (V_{BC}) is negative. The bipolar transistor was detailed in the previous chapter. Here we use Fig. 5.2 to emphasize that the energy gap is the same throughout the entire transistor structure. The Fermi levels and the depletion regions of the band diagram reflect the BJT design constraint that the doping level is highest in the emitter, and lowest in the collector. The Fermi level in the emitter is above the conduction band, signifying that the emitter is degenerately doped.

A BJT has two key current components. The first is the forward-injection current conducted by the electrons. These electrons are emitted at the emitter, crossing the base-emitter potential barrier with the help of the V_{BE} bias, diffusing through the thin base layer as minority carriers, and then are swept by the large electric field in the reverse-biased base-collector junction. The carriers finally leave the collector, forming the collector current (I_C). In their journey, only a few electrons are lost through recombinations with the majority holes in the base. We therefore denote the electron current as I_C in Fig. 5.2, even though the current magnitude in the emitter is slightly larger than that which finally exits the collector terminal. Of the three phases of the journey, the diffusion through the base layer is the rate-limiting step. Hence, a quantitative expression of I_C, to be developed shortly, relates intimately to the base layer parameters.

By varying V_{BE}, we control the amount of the electron injection from the emitter to the base, and subsequently, the amount of electrons collected at the collector. This current flow, however, is independent of V_{BC}, as long as the base-collector junction is reverse biased. The property that the desired signal (I_C) is modified by the input bias (V_{BE}) but unaffected by the output bias (V_{BC}) fulfills the requirement of a

0-8493-1562-X/03/$0.00+$1.50

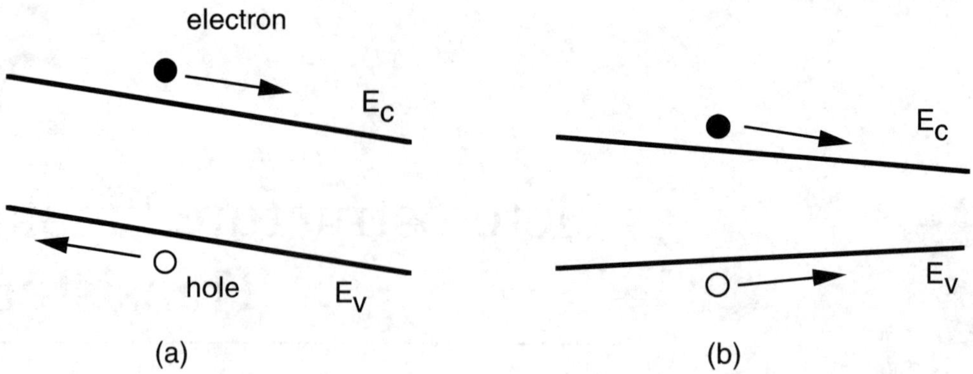

FIGURE 5.1 (a) An electric field exerts the same amount of force (but in opposite directions) on an electron and a hole. (b) Electron and hole can move in the same direction in a hetero-structure. (After Ref. [1], with permission.)

sound three-terminal device. As the input and output ports are decoupled from one another, complicated circuits based on the transistors can be easily designed.

The second current component is composed of holes that are back-injected from the base to the emitter. This current does not get collected at the collector terminal and does not contribute to the desired output signal. However, the moment a V_{BE} is applied so that the desired electrons are emitted from the emitter to the base, these holes are back-injected from the base to the emitter. The bipolar transistor is so named to emphasize that both electron and hole play significant roles in the device operation. A good design of a bipolar transistor maximizes the electron current transfer while minimizing the hole current. As indicated on Fig. 5.2, we refer the hole current as $I_{B,\text{back-inject}}$. It is a base current because the holes come from the base layer where they are the majority carriers.

We mentioned that I_C is limited by the diffusion process in the base, wherein the electrons are minority carriers. Likewise, $I_{B,\text{back-inject}}$ is limited by the diffusion in the emitter, wherein the holes are minority carriers. Fisk's law states that a diffusion current density across a layer is equal to the diffusion coefficient times the carrier concentration gradient.[2] (From the Einstein relationship, the diffusion coefficient can be taken to be kT/q times the carrier mobility.) The carrier concentrations at one end of the base layer (for the calculation of I_C) and the emitter layer (for the calculation of $I_{B,\text{back-inject}}$) are both proportional to $\exp(qV_{BE}/kT)$, and are 0 at the other end. An application of Fisk's Law leads to:[3,4]

$$I_C = \frac{qA_{\text{emit}}D_{n,\text{base}}}{X_{\text{base}}}\frac{n_{i,\text{base}}^2}{N_{\text{base}}}\exp\left(\frac{qV_{BE}}{kT}\right) \tag{5.1}$$

$$I_{B,\text{back-inject}} = \frac{qA_{\text{emit}}D_{p,\text{emit}}}{X_{\text{emit}}}\frac{n_{i,\text{emit}}^2}{N_{\text{emit}}}\exp\left(\frac{qV_{BE}}{kT}\right) \tag{5.2}$$

where, for example, A_{emit} is the emitter area; $D_{n,\text{base}}$ is the electron diffusion coefficient in the base layer; N_{base} is the base doping; X_{base} is the base thickness; and $n_{i,\text{base}}$ is the intrinsic carrier concentration in the base layer. I_C is proportional to the emitter area rather than the collector area because I_C is composed of the electrons injected from the emitter. For homojunction BJTs, $n_{i,\text{base}}$ is identical to $n_{i,\text{emit}}$. The ratio of the desired I_C to the undesired $I_{B,\text{back-inject}}$ is,

$$\frac{I_C}{I_{B,\text{back-inject}}} = \frac{X_{\text{emit}}}{X_{\text{base}}}\frac{D_{n,\text{base}}}{D_{p,\text{emit}}} \times \frac{N_{\text{emit}}}{N_{\text{base}}} \quad \text{(homojunction BJT)} \tag{5.3}$$

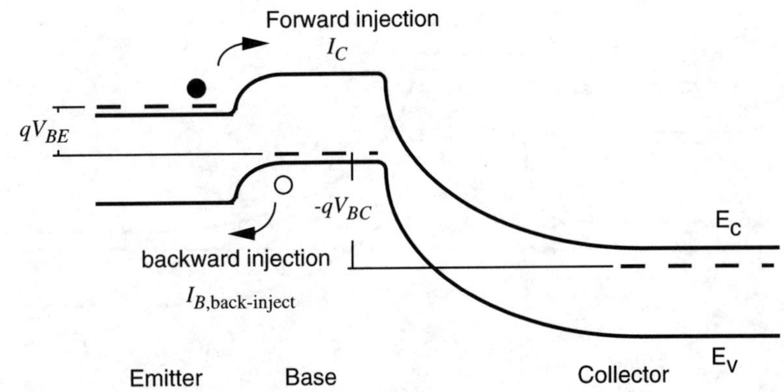

FIGURE 5.2 Band diagram of a *npn* bipolar junction transistor.

Because the diffusion coefficients and layer thicknesses are roughly equal on the first order, Eq. (5.3) demonstrates that the emitter doping of a BJT must exceed the base doping in order for I_C to exceed $I_{B,\text{back-inject}}$. For most applications, we would actually prefer the base doping be high and the emitter doping, low. A high base doping results in a small base resistance and a low emitter doping reduces the base-emitter junction capacitance, both factors leading to improved high frequency performance. Unfortunately, these advantages must be compromised in the interest of minimizing $I_{B,\text{back-inject}}$ in homojunction transistors.

A heterojunction bipolar transistor (HBT), formed by a replacement of the homojunction emitter by a larger energy gap material, enables the design freedom of independently optimizing the $I_C/I_{B,\text{back-inject}}$ ratio and the doping levels. Figure 5.3 illustrates the band diagrams of *Npn* HBTs. The capital *N* in "*Npn*" rather than a small letter *n* emphasizes that the emitter is made of a larger energy gap material than the rest. It is implicit that the base-collector junction of an HBT is a homojunction. An HBT whose base-collector junction is also a heterojunction is called a double heterojunction bipolar transistor (DHBT), typically found in the InP/InGaAs material system.

The energy gap difference in the base-emitter heterojunction of an HBT burdens the holes from the base to experience a much larger energy barrier than the electrons from the emitter. With the same application of V_{BE}, the forces acting on the electrons and holes differ, favoring the electron injection from the emitter into the base to the hole back-injection from the base into the emitter. Figure 5.3 shows that the difference in the electron and hole barriers is ΔE_v, the valence band discontinuity. The $I_C/I_{B,\text{back-inject}}$ ratio of Eq. (5.3) can be extended to the HBT as:[1]

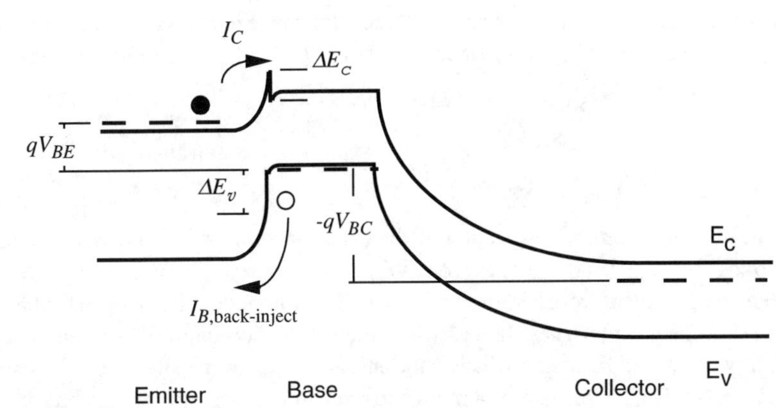

FIGURE 5.3 Band diagram of a *Npn* abrupt heterojunction bipolar transistor.

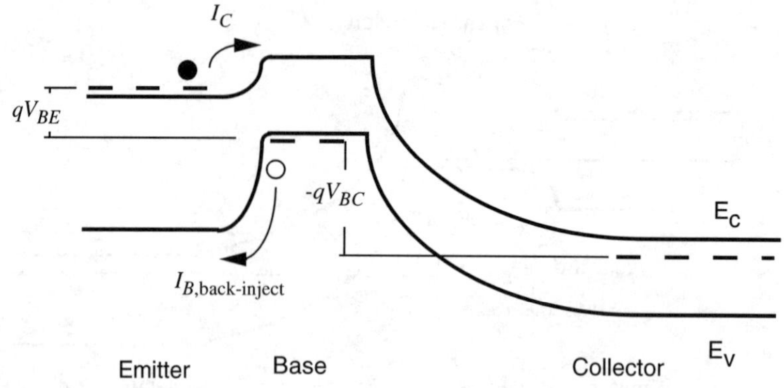

FIGURE 5.4 Band diagram of a *Npn* graded heterojunction bipolar transistor.

$$\frac{I_C}{I_{B,\text{back-inject}}} = \frac{X_{\text{emit}}}{X_{\text{base}}} \frac{D_{n,\text{base}}}{D_{p,\text{emit}}} \times \frac{N_{\text{emit}}}{N_{\text{base}}} \exp\left(\frac{\Delta E_v}{kT}\right) \quad \left(\text{abrupt HBT}\right) \tag{5.4}$$

The exponential factor is the key to the fact that base doping can be made larger than the emitter doping without adversely affecting the HBT performance.

We qualify the expression in Eq. (5.4): that is for an *abrupt* HBT. This means that an abrupt change of material composition exists between the emitter and the base. An example is an $InP/In_{0.53}Ga_{0.47}As$ HBT. The indium and gallium compositions form a fixed ratio so that the resulting $In_{0.53}Ga_{0.47}As$ layer has the same lattice constant as InP, the starting substrate material. Otherwise, the dislocations due to lattice mismatch prevent the device from being functional. In fact, although the idea of HBT is as old as the homojunction transistor itself,[5] HBTs have emerged as practical transistors only after molecular beam epitaxy (MBE) and metal-organic chemical vapor deposition (MOCVD) were developed to grow high-quality epitaxial layers in the 1980s.[6] Another example of an abrupt HBT is the $Al_{0.3}Ga_{0.7}As/GaAs$ HBT, in which the entire emitter layer consists of $Al_{0.3}Ga_{0.7}As$, while the base and collector are GaAs. When the AlGaAs/GaAs material system is used, we can take advantage of the material property that AlAs and GaAs have nearly the same lattice constants. It is therefore possible to grow an $Al_xGa_{1-x}As$ layer with any composition x and still be lattice-matched to GaAs. When the aluminum composition of an intermediate AlGaAs layer is graded from 0 at the GaAs base to 30% at the $Al_{0.3}Ga_{0.7}As$ emitter, then the resulting structure is called a *graded* HBT.

The band diagram of a graded HBT is shown in Fig. 5.4. A graded HBT has the advantage that the hole barrier is larger than the electron barrier by ΔE_g, the energy-gap discontinuity between the emitter and the base materials. Previously, in an abrupt HBT, the difference in the hole and electron barriers was only ΔE_v. The additional barrier brings forth an even larger $I_C/I_{B,\text{back-inject}}$ ratio for the graded HBT:

$$\frac{I_C}{I_{B,\text{back-inject}}} = \frac{X_{\text{emit}}}{X_{\text{base}}} \frac{D_{n,\text{base}}}{D_{p,\text{emit}}} \times \frac{N_{\text{emit}}}{N_{\text{base}}} \exp\left(\frac{\Delta E_g}{kT}\right) \quad \left(\text{Graded HBT}\right) \tag{5.5}$$

Equation (5.5) is obtained from the inspection of the band diagram. It is also derivable from Eqs. (5.1) and (5.2) by noting that $n_{i,\text{base}}^2 = n_{i,\text{emit}}^2 \times \exp(\Delta E_g/kT)$.

The amount of improvement from a homojunction to a heterojunction depends on the band alignment properties of the two hetero-materials. Table 5.1 lists the conduction band discontinuity (ΔE_c), valence band discontinuity (ΔE_v), and their sum which is the energy-gap discontinuity (ΔE_g), for several material systems used for HBTs. The energy gap of a semiconductor in the system is also given for convenience. The values shown in Table 5.1 are room temperature values.

TABLE 5.1 Heterojunction Parameters at Room Temperature

	$Al_xGa_{1-x}As/GaAs$ ($x \leq 0.45$)	$Ga_{0.51}In_{0.49}P/GaAs$	$InP/In_{0.53}Ga_{0.47}As$	$In_{0.52}Al_{0.48}As/In_{0.53}Ga_{0.47}As$	$Si/Si_{0.8}Ge_{0.2}$
ΔE_g (eV)	$1.247 \cdot x$	Ordered: 0.43 Disordered: 0.46	0.60	0.71	0.165
ΔE_c (eV)	$0.697 \cdot x$	Ordered: 0.03 Disordered: 0.22	0.23	0.50	
ΔE_v (eV)	$0.55 \cdot x$	Ordered: 0.40 Disordered: 0.24	0.37	0.21	
E_g (eV)	GaAs: 1.424	Ordered: 1.424	InP: 1.35	InGaAs: 0.75	Si: 1.12

The most popular III-V material system is $Al_xGa_{1-x}As/GaAs$. The table lists its band alignment parameters for $x \leq 0.45$. At these aluminum mole fractions, the AlGaAs layer is a direct energy-gap material, in which the electron wave function's wave vectors **k** at both the conduction band minimum and the valence band maximum are along the same crystal direction.[2] Since nearly all semiconductor valence band maximums take place when **k** is [000], the conduction band minimum in a direct energy-gap material is also at **k** = [000]. The conduction band structure surrounding this **k** direction is called the Γ valley. There are two other **k** directions of interest because the energy band structures in such directions can have either the minimum or a local minimum in the conduction band. When **k** is in the [100] direction, the band structure surrounding it is called the X valley, and when **k** is in the [111] direction, the L valley. As the aluminum mole fraction of $Al_xGa_{1-x}As$ exceeds 0.45, it becomes an indirect energy-gap material, with its conduction band minimum residing in the X valley. The band alignment parameters at $x > 0.45$ are found elsewhere.[3] SiGe is an indirect energy-gap material; its conduction band minimum is also located in the X valley. Electron-hole generation or recombination in indirect energy-gap materials requires a change of momentum ($\mathbf{p} = \hbar\mathbf{k}$), a condition that generally precludes them from being useful for laser or light-emitting applications.

Particularly due to the perceived advantage in improved reliability, HBTs made with the GaInP/GaAs system have gained considerable interest.[7] The band alignment of GaInP/GaAs depends on whether the grown GaInP layer is ordered or disordered, as noted in Table 5.1. The crystalline structure in an *ordered* GaInP layer is such that sheets of pure Ga, P, In, and P atoms alternate on the (001) planes of the basic unit cell, without the intermixing of the Ga and In atoms on the same lattice plane.[8] When the Ga, In and P atoms randomly distribute themselves on a plane, the GaInP layer is termed *disordered*.

The processing of AlGaAs/GaAs and GaInP/GaAs HBTs is fairly simple. Both wet and dry etching are used in production environments, and ion implantation is an effective technique to isolate the active devices from the rest. The processing of InP/InGaAs and InAlAs/InGaAs materials, in contrast, is not straightforward. Because of the narrow energy-gap of the InGaAs layer, achieving an effective device isolation often requires a complete removal of the inactive area surrounding the device, literally digging out trenches to form islands of devices. Further, the dry-etching, and its associated advantages such as directionality of etching, is not readily/easily available for InGaAs.[9] However, the material advantages intrinsic to InP/InGaAs and InAlAs/InGaAs make them the choice for applications above 40 GHz. In addition, the turn-on voltage, the applied V_{BE} giving rise to a certain collector current, is smaller for HBTs formed with InGaAs base compared to GaAs base (due to the energy-gap difference). The turn-on characteristics of various HBTs are shown in Fig. 5.5.

A calculation illustrates the advantage of an HBT. Consider an AlGaAs/GaAs HBT structure designed for power amplifier applications, as shown in Fig. 5.6.[10] The emitter and the base layers of the transistor are: $N_{emit} = 5 \times 10^{17}$ cm^{-3}; $N_{base} = 4 \times 10^{19}$ cm^{-3}; $X_{emit} \approx 1300$ Å; and $X_{base} = 1000$ Å. We shall use the following diffusion coefficients for the calculation: $D_{n,base} = 20$ and $D_{p,emit} = 2.0$ cm^2/V-s. For a graded $Al_{0.3}Ga_{0.7}As/GaAs$ heterojunction, ΔE_g is calculated from Table 5.1 to be $0.3 \times 1.247 = 0.374$ eV. The ratio for the graded HBT, according to Eq. (5.5), is,

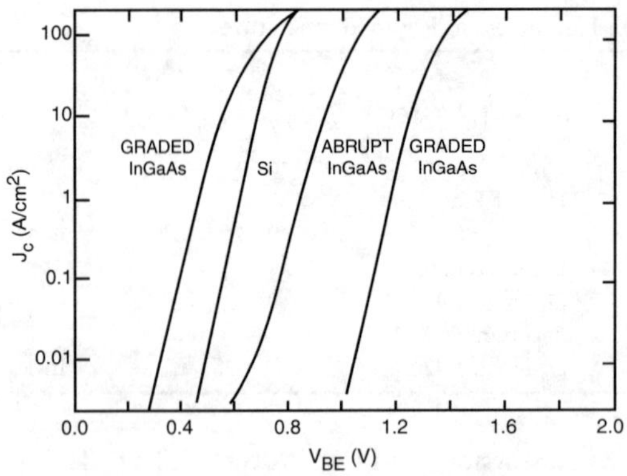

FIGURE 5.5 Turn-on characteristics of bipolar transistors based on various material systems. (After Ref. [6], with permission.)

Material	Thickness (Å)	Composition x	Doping (cm^{-3})	
n-$In_xGa_{1-x}As$	800	$0 \to 0.6$	$> 3 \times 10^{19}$	
n-GaAs	2000		5×10^{18}	
N-$Al_xGa_{1-x}As$	300	$0.3 \to 0$	1×10^{18}	
N-$Al_xGa_{1-x}As$	1000	0.3	5×10^{17}	Emitter
N-$Al_xGa_{1-x}As$	300	$0 \to 0.3$	5×10^{17}	
(The above grading layer is absent in abrupt HBT)				
p-GaAs	1000		4×10^{19}	Base
n-GaAs	7000		3×10^{16}	Collector
n-GaAs	6000		5×10^{18}	Subcollector
Semi-insulating GaAs Substrate				

FIGURE 5.6 Typical HBT epitaxial structure designed for power amplifier applications. (After Ref. [10], with permission.)

$$\frac{I_C}{I_{B,\text{back-inject}}} = \frac{1300}{1000}\frac{20}{2.0} \times \frac{5 \times 10^{17}}{4 \times 10^{19}} \exp\left(\frac{0.374}{0.0258}\right) = 2.5 \times 10^5$$

For an abrupt $Al_{0.3}Ga_{0.7}As$/GaAs HBT, ΔE_v is the parameter of interest. According to the table, ΔE_v at $x =$ 0.3 is $0.55 \times 0.3 = 0.165$ eV. Therefore, Eq. (5.4) leads to:

$$\frac{I_C}{I_{B,\text{back-inject}}} = \frac{1300}{1000}\frac{20}{2.0} \times \frac{5 \times 10^{17}}{4 \times 10^{19}} \exp\left(\frac{0.165}{0.0258}\right) = 97$$

Consider a Si BJT with identical doping levels, layer thicknesses, and diffusion coefficients, except that it is a homojunction transistor so that $\Delta E_g = 0$. Using Eq. (5.3), we find the $I_C/I_{B,\text{back-inject}}$ ratio to be:

$$\frac{I_C}{I_{B,\text{back-inject}}} = \frac{1300}{1000}\frac{20}{2.0} \times \frac{5 \times 10^{17}}{4 \times 10^{19}} = 0.16.$$

The useful collector current in the homojunction transistor is only 1/6 of the undesirable back-injection current. This means the device is useless. In contrast, both the graded and the abrupt HBTs remain functional, despite the higher base doping in comparison to the emitter.

5.2 Base Current Components

$I_{B,\text{back-inject}}$ is only one of the five dominant base current components in a bipolar transistor. We have thus far considered only $I_{B,\text{back-inject}}$ because it is the distinguishing component between a HBT and a BJT. Once $I_{B,\text{back-inject}}$ is made small in a HBT through the use of a heterojunction, the remaining four components become noteworthy. All of these components are recombination currents; they differ only in the locations where the recombinations take place, as shown in Fig. 5.7. They are: (1) extrinsic base surface recombination current, $I_{B,\text{surf}}$; (2) base contact surface recombination current, $I_{B,\text{cont}}$; (3) bulk recombination current in the base layer, $I_{B,\text{bulk}}$; and (4) space-charge recombination current in the base-emitter junction depletion region, $I_{B,\text{scr}}$. In the discussion of bipolar transistors, an easily measurable quantity of prime importance is the current gain (β), defined as the ratio of I_C to the total base current I_B:

$$\beta = \frac{I_C}{I_B} = \frac{I_C}{I_{B,\text{back-inject}} + I_{B,\text{surf}} + I_{B,\text{cont}} + I_{B,\text{bulk}} + I_{B,\text{scr}}} \tag{5.6}$$

Depending on the transistor geometrical layout, epitaxial layer design, and the processing details that shape each of the five base current components, the current gain can have various bias and temperature dependencies. In the following, the characteristics of each of the five base components are described, so that we can better interpret the current gain from measurement and establish some insight about the measured device.

Figure 5.8a illustrates a schematic cross-section of an HBT. Without special consideration, a conventional fabrication process results in an exposed base surface at the extrinsic base regions. (*Intrinsic region* is that underneath the emitter, and *extrinsic region* is outside the emitter mesa, as shown in Fig. 5.7.) Because the exposed surface is near the emitter mesa where the minority carrier concentration is large, and because the surface recombination velocity in GaAs is high (on the order of 10^6 cm/s), $I_{B,\text{surf}}$ is significant in these unpassivated devices. Various surface passivation techniques have been tested. The most effective method is *ledge passivation*,[11,12] formed with, for example, an AlGaAs layer on top of the GaAs base. The AlGaAs ledge must be thin enough so that it is fully depleted by a combination of the free surface Fermi level pinning above and the base-emitter junction below. If the passivation ledge is not fully depleted, the active device area would be much larger than the designed emitter. The requirement for the AlGaAs layer to be fully depleted limits the AlGaAs thickness to the order of 1000 Å, and emitter

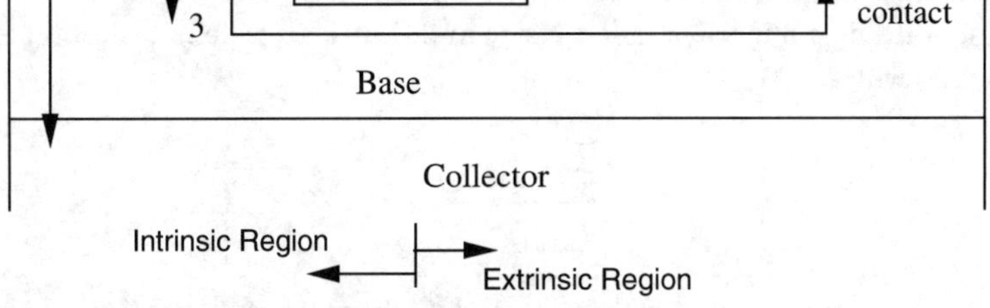

FIGURE 5.7 Locations of the four base recombination currents. The fifth base current component, not a recombination current, is $I_{B,\text{back-inject}}$ shown in Figure 5.2. (After Liu, W., Microwave and DC characterizations of Npn and Pnp HBTs, PhD. Dissertation, Stanford University, Stanford, CA, 1991.)

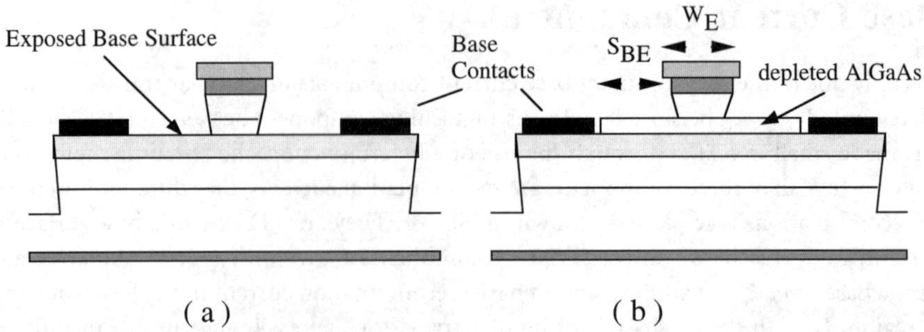

FIGURE 5.8 Schematic cross-sections of HBTs: (a) unpassivated HBTs, and (b) passivated HBTs. (After Liu, W., Microwave and DC characterizations of Npn and Pnp HBTs, PhD. Dissertation, Stanford University, Stanford, CA, 1991.)

doping to low to mid 10^{17} cm^{-3}. Although the ledge passivation was originally designed to minimize $I_{B,surf}$, it is also crucial to long-term reliability.[3,7]

Unlike I_C, $I_{B,surf}$ is proportional to the emitter periphery rather than the emitter area. For high frequency devices whose emitter is in a strip form (thus the perimeter-to-area ratio is large), $I_{B,surf}$ is a major component to the overall base current. The current gain is substantially reduced from that of a large squarish device whose perimeter-to-area ratio is small. The discrepancy in β due to emitter geometry is termed the *emitter-size effect.* Figure 5.9 displays β vs. I_C for both passivated and unpassivated devices with A_{emit} = 4 × 10 μm². The emitter area is small enough to demonstrate the benefit of the surface passivation. A large device has negligible surface recombination current and the current gain does not depend on whether the surface is passivated or not. Because the two devices are fabricated simultaneously and physically adjacent to each other, the difference between the measured βs is attributed to the additional $I_{B,surf}$ of the unpassivated device.

The second base recombination current, $I_{B,cont}$, is in principle the same as $I_{B,surf}$. Both are surface recombination currents, except $I_{B,cont}$ takes place on the base contacts whereas $I_{B,surf}$, on the extrinsic base surfaces. Because the contacts are located further away from the intrinsic emitter than the extrinsic base surface, $I_{B,cont}$ is generally smaller than $I_{B,surf}$ when the surface is unpassivated. However, it may replace $I_{B,surf}$ in significance in passivated devices (or in Si BJTs whose silicon dioxide is famous in passivating silicon). There is a characteristic distance for the minority carrier concentration to decrease exponentially from the emitter edge toward the extrinsic base region.[3] As long as the base contact is placed at roughly 3 times this characteristic length away from the intrinsic emitter, $I_{B,cont}$ can be made small. The base contact cannot be placed too far from the emitter, however. An excessively wide separation increases the resistance in the extrinsic base region and degrades the transistor's high frequency performance.

The above two recombination currents occur in the extrinsic base region. Developing analytical expressions for them requires a solution of the two-dimensional carrier profile. Although this is possible without a full-blown numerical analysis,[3] the resulting analytical equations are quite complicated. The base bulk recombination current, in contrast, can be accurately determined from a one-dimensional analysis since most of the base minority carriers reside in the intrinsic region. It is convenient to express $I_{B,bulk}$ through its ratio with I_C:

$$\frac{I_C}{I_{B,bulk}} = \frac{\tau_n}{\tau_b} \tag{5.7}$$

where τ_n is the minority carrier lifetime in the base, and τ_b, the minority carrier transit time across the base. In a typical Si BJT design, an electron spends about 10 ns diffusing through the base layer while in every 1 μs an electron is lost through recombination. The transistor then has a current gain of 1 μs/10 ns = 100. The recombination lifetime in the GaAs base material is significantly shorter than in Si, at

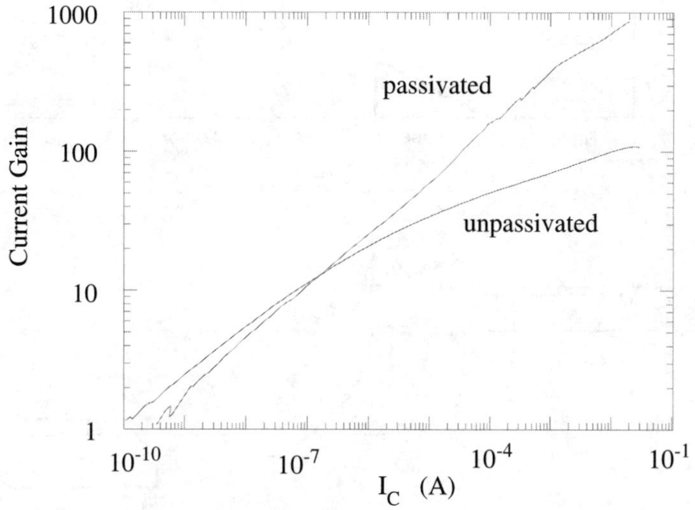

FIGURE 5.9 Measured current gain as a function of collector current for both passivated and unpassivated HBTs. (After Liu, W. et al., Diode ideality factor for surface recombination current in AlGaAs/GaAs heterojunction bipolar transistors, *IEEE Trans. Electron Devices*, 39, 2726–2732, 1992, with permission.)

about 1 ns. However, the transit time through the GaAs is also much shorter than in Si due to the higher carrier mobility in GaAs. A well-designed HBT has a τ_b of 0.01 ns; therefore, a $\beta = 100$ is also routinely obtainable in III-V HBTs.

Equation (5.7) indicates that $I_{B,\text{bulk}}$ is large if τ_n is small. The recombination lifetime of a semiconductor, a material property, is found to be inversely proportional to the doping level. When the base doping in an AlGaAs/GaAs (or GaInP/GaAs) HBT is 5×10^{18} cm^{-3}, the current gain easily exceeds 1000 when proper device passivation is made and the base contacts are placed far from the emitter. As the base doping increases to 10^{20} cm^{-3}, $I_{B,\text{bulk}}$ dominates all other base current components, and the current gain decreases to only about 10, independent of whether the extrinsic surface is passivated or not. The base doping in III-V HBTs for power applications, as shown in Fig. 5.6, is around $3 - 5 \times 10^{19}$ cm^{-3}. It is a compromise between achieving a reasonable current gain (between 40 and 200) and minimizing the intrinsic base resistance to boost the high frequency performance.

Equation (5.7) also reveals that $I_{B,\text{bulk}}$ is large when the base transit time is long. This is the time that a base minority carrier injected from the emitter takes to diffuse through the base. Unlike the carrier lifetime, which is mostly a material constant, the base transit time is a strong function of the base layer design:

$$\tau_b = \frac{X_{\text{base}}^2}{2D_{n,\text{base}}} \tag{5.8}$$

Because τ_b is proportional to the square of X_{base}, the base thickness is designed to be thin. Making the base too thin, however, degrades high frequency performance due to increased base resistance. A compromise between these two considerations results in a X_{base} at around 800 – 1000 Å, as shown in Fig. 5.6.

The derivation of Eq. (5.8) assumes that the minority carriers traverse through the base purely by diffusion. This is certainly the scenario in a bipolar transistor whose base layer is uniformly doped and of the same material composition. With energy-gap engineering, however, it is possible to shorten the base transit time (often by a factor of 3) by providing a drift field. In a Si/SiGe HBT for example, the Ge content can be linearly graded from 0 to 8% across a 300 Å base to result a *quasi-electric field* on the order of 30 to 50 kV/cm.[13] We used the term quasi-electric field to describe the electric field generated by grading the energy gap, or more specifically, the gradient of the electron affinity (χ_e). In a conventional

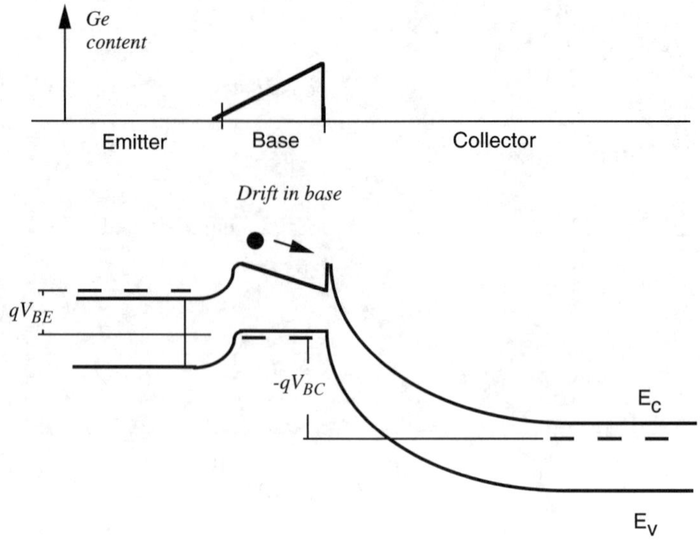

FIGURE 5.10 Band diagram of a SiGe HBT with a base quasi-electric field established by grading the germanium content.

bipolar transistor, an electric field can be established only in the presence of space charges (such as the depletion region in a *p-n* junction). In an HBT with a graded base, the overall electric field in the base layer is nonzero even though the entire base region remains charge neutral. A SiGe HBT band diagram, shown in Fig. 5.10, illustrates how a minority carrier can speed up in the presence of the band grading. The figure shows that the energy gap becomes narrower as the Ge content increases. Because the base layer is heavily doped, the quasi-Fermi level in the base is pinned to be relatively flat with respect to position. The entire energy gap difference appears in the conduction band. The base quasi-electric field, being proportional to the slope of the band bending, propels the electrons to drift from the emitter toward the collector. As the carrier movement is enhanced by the drift motion, in addition to the diffusion, the base transit time decreases.

Figure 5.10 is characteristic of the SiGe HBT pioneered by a U.S. company,[14] in which the Ge content is placed nearly entirely in the base and graded in a way to create a base quasi-electric field. The base-emitter junction is practically a homojunction; therefore, it perhaps does not strictly fit the definition of being a heterojunction bipolar transistor and the base must be doped somewhat lighter than the emitter. This type of transistor resembles a drift homojunction transistor,[15] in which a base electric field is established by grading the base doping level. A drift transistor made with dopant grading suffers from the fact that part of the base must be lightly doped (at the collector side), thus bearing a large base resistance. An alternative school of SiGe HBT places a fixed Ge content in the base, mostly promoted by European companies.[16] Transistors fabricated with the latter approach do not have a base quasi-electric field. However, the existence of the base-emitter heterojunction allows the emitter to be more lightly doped than the base, just as in III-V HBTs. In either type of SiGe HBTs, there can be a conduction band discontinuity between the base and collector layers. This base-collector junction spike (Fig. 5.10), also notable in InP-based DHBT, has been known to cause current gain fall off.[3] The spike can be eliminated by grading the Ge content from the base to inside the collector layer.

Likely due to reliability concerns or for purely historical reasons, most commercial III-V HBTs have a uniformly doped base without a base quasi-electric field. If a base electric field is desired, in AlGaAs/"GaAs" HBTs in particular, the field can be established by grading of the aluminum concentration in the AlGaAs base layer.

The fourth recombination current is the space-charge recombination current in the base-emitter depletion region. $I_{B,scr}$ differs from the other base current components in its bias dependency.

Equations (5.1) and (5.2) show that I_C and $I_{B,\text{back-inject}}$ are proportional to $\exp(qV_{BE}/nkT)$ with n, the *ideality factor*, being equal to 1. Equation (5.7) also shows that $I_{B,\text{bulk}}$ is directly proportional to I_C. Hence, $I_{B,\text{bulk}}$ has a unity ideality factor as well. Extensive measurement experiments and theoretical calculations indicate that $I_{B,\text{surf}}$ and hence, $I_{B,\text{cont}}$, have an ideality factor closer to 1 than 2.[3,17] The ideality factor of $I_{B,\text{scr}}$, in contrast, is nearly 2 because the electron and hole concentrations in the forward-biased base-emitter junction are both proportional to $\exp(qV_{BE}/2kT)$.[2] This means $I_{B,\text{scr}}$ is most significant when V_{BE} is small, at which operating region I_C is also small. A *Gummel plot*, I_B and I_C as a function of V_{BE} taken at $V_{BC} = 0$, illustrates dominance of $I_{B,\text{scr}}$ in low-current regions, as shown in Fig. 5.11. There are times when $I_{B,\text{scr}}$ dominates other base current components even at high I_C levels, particularly in graded HBTs.[3]

The previous four base current components are all recombination current. The fifth component is the back-injection current, $I_{B,\text{back-inject}}$. This component is made small in a heterojunction transistor, at least at room temperature. However, as temperature increases, the extra energy barrier provided to the hole carriers becomes less effective in impeding the back injection. When the HBT is biased with a high I_C and a certain amount of V_{BC}, the power dissipated in the device can heat up the device itself (called *self-heating*). As the HBTs junction temperature rises, $I_{B,\text{back-inject}}$ increases and the current gain decreases. This β's temperature dependency is to be contrasted with silicon BJTs current gain, which increases with temperature.[18]

5.3 Kirk Effects

Understanding the properties of the various base current components facilitates the description of the measured device characteristics, such as the β vs. I_C curve of Fig. 5.9. The previous analysis of the transistor currents implicitly assumes that the transistor operates in the normal bias condition, under which the transistor gain is seen to increase steadily with the collector current. However, I_C cannot increase indefinitely without adverse effects. Figure 5.9 reveals that, after a certain critical I_C is reached while V_{BC} is kept constant, the current gain plummets, rendering the device useless. For high-power applications, such as a power amplifier for wireless communications, HBTs are biased at a large current (with a current density on the order of 10^4 A/cm^2), not only because the output power is directly proportional to I_C, but also because the high frequency performance is superior at large I_C (but before the current gain falls). Therefore, it is imperative to understand the factor setting the maximum I_C level, below which the current gain is maintained at some finite values.

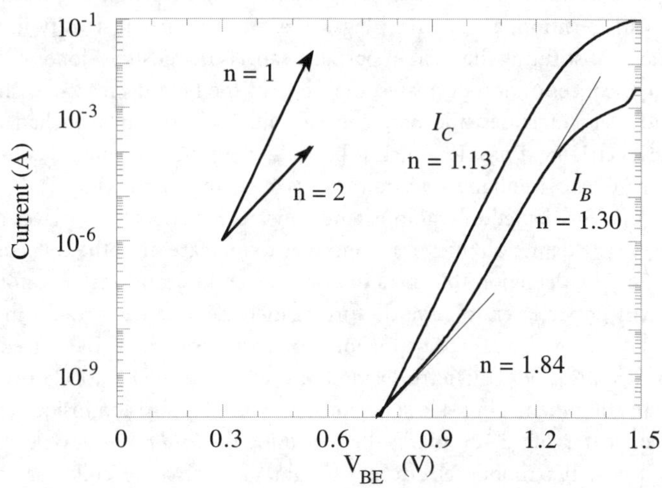

FIGURE 5.11 Measured Gummel plot of an abrupt AlGaAs/GaAs HBT. (After Liu, W., Experimental comparison of base recombination currents in abrupt and graded AlGaAs/GaAs heterojunction bipolar transistors, *Electronic Letters*, 27, 2115–2116, 1991, with permission.)

The Poisson equation relates charges to the spatial variations of the electric field (ε).[2] It is a fundamental equation, applicable to any region at any time in a semiconductor:

$$\frac{d\varepsilon}{dx} = \frac{q}{\epsilon_s}\left(p - n + N_d - N_a\right) \tag{5.9}$$

ϵ_s is the dielectric constant of the semiconductor; N_d and N_a are the donor and acceptor doping levels, respectively; and n and p are the mobile electron and hole carrier concentrations, respectively. We apply this equation to the base-collector junction of HBTs, which is typically a homojunction. Since the base doping greatly exceeds the collector doping, most of the depletion region of the base-collector junction is at the collector side. In the depleted collector region where it is doped n-type, N_d in Eq. (5.9) is the collector doping level, N_{coll}, and N_a is 0. The collector current flowing through the junction consists of electrons. If the field inside the depletion region was small, then these electrons would move at a speed equal to the product of the mobility and the electric field: $\mu_n \cdot \varepsilon$. It is a fundamental semiconductor property that, once the electric field exceeds a certain critical value ($\varepsilon_{\text{crit}} \sim 10^3$ V/cm for GaAs), the carrier travels at a constant velocity called the saturation velocity (v_{sat}). Because the electric field inside most of the depletion region exceeds 10^4 V/cm, practically all of these electrons travel at a constant speed of v_{sat}. The electron carrier concentration inside the collector can be related to the collector current density (J_C; equal to I_C/A_{emit}) as:

$$n(x) = \frac{J_C}{q v_{sat}} = \text{constant inside the base-collector junction} \tag{5.10}$$

Lastly, because there is no hole current, p in Eq. (5.9) is zero.

Equation (5.9), when applied to the base-collector junction of a HBT, is simplified to:

$$\frac{d\varepsilon}{dx} = \frac{q}{\epsilon_s}\left(-\frac{J_C}{q v_{sat}} + N_{\text{coll}}\right) \tag{5.11}$$

When J_C is small, the slope of the electric field is completely determined by the collector doping, N_{coll}. Because the doping is constant with position, solving Eq. (5.11) at negligible J_C gives rise to a field profile that varies linearly with position, as shown in Fig. 5.12a. As the current density increases, the mobile electron concentration constituting the current partially cancels the positive donor charge concentration N_{coll}. As the net charge concentration decreases, the slope of the field decreases, as shown in Fig. 5.12b. While the current density increases, the base-collector bias V_{BC} remains unchanged. Therefore, the enclosed area of the electric field profile, which is basically the junction voltage, is the same before and after the current increase. The simultaneous requirements of having a decreasing field slope and a constant enclosed area imply that the depletion region extends toward the subcollector layer and the maximum electric field decreases. The depletion thickness continues to increase until the collector is fully depleted, as shown in Fig. 5.12c. The depletion thickness does not extend beyond the collector layer because the subcollector is a heavily doped layer. Afterwards, further increase of current results in a quadrangle field profile, as shown in Fig. 5.12d, replacing the previous triangular profile. As the current density increases to a level such that $J_C = qN_{\text{coll}} \cdot v_{sat}$, the term inside the parentheses of Eq. (5.11) becomes zero. A field gradient of zero means that the field profile stays constant with the position inside the junction (slope = 0). This situation, depicted in Fig. 5.12e, marks the beginning the *field reversal*. When J_C increases further such that $J_C > qN_{\text{coll}} \cdot v_{sat}$, the mobile electrons brought about by the collector current more than compensates the fixed charges inside the collector. The net charge concentration for the first time becomes negative and the electric field takes on a negative slope (Fig. 5.12f), with a smaller magnitude at the base side of the junction than at the subcollector side. As the trend progresses, the magnitude of the field

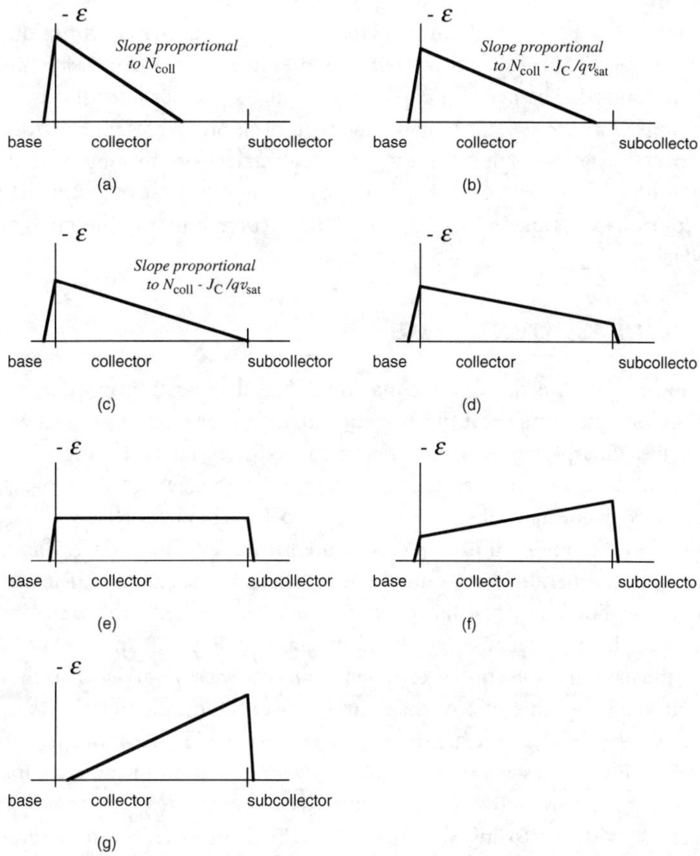

FIGURE 5.12 Electric field profile inside the base-collector junction of HBT during normal operation. (a) $J_C = 0$; (b) J_C is small; (c) J_C further increases so the entire collector is depleted; (d) further increase in J_C; (e) J_C reaches $qN_{coll} \cdot v_{sat}$; (f) $J_C > qN_{coll} \cdot v_{sat}$ and field reversal occurs; (g) further increase in J_C results in base push out as the electric field at the base-collector boundary decreases to zero.

base-collector junction eventually diminishes to zero (Fig. 5.12g). When there is no more field to block the holes from "spilling" into the collector, the *base pushout* is said to occur and the current gain falls. The device characteristics as a result of the base pushout are referred to as *Kirk effects*.[19] The above description suggests that the threshold current due to Kirk effects increases if the collector doping increases. However, in many applications where the collector doping may not be increased arbitrarily (so the operating voltage is greater than a certain value), Kirk effects then become an important mechanism affecting the current gain falloff in HBTs. For an HBT with a collector doping of 3×10^{16} cm^{-3} (Fig. 5.6), the threshold current density is roughly $J_C = qN_{coll} \cdot v_{sat} = 3.9 \times 10^4$ A/cm^2 (v_{sat} is ~ 8×10^6 cm/s). Clearly, the value of such threshold current density depends on the magnitude of the saturation velocity. Since the saturation velocity decreases with the ambient temperature, the threshold density due to Kirk effects is lower at higher temperatures.

Kirk effects confine the operating collector current to some values. Similarly, the collector-to-emitter bias (V_{CE}) has its limit, set by two physical phenomena. The first one, well analyzed in elementary device physics, is the avalanche breakdown in the base-collector junction. The base-collector junction is a reverse-biased junction with a finite electric field. The mobile electrons comprised of J_C, while moving through the junction, quickly accelerate and pick up energy from the field. When V_{CE} is small, the magnitude of the field is not too large. The energy the carriers acquired is small and is quickly dissipated in the lattice as the carriers impact upon the lattice atoms. The distance within which a carrier travels between successive impacts with the lattice atoms is called a *mean free path*. As V_{CE} increases such that the electric field approaches $10^5 - 10^6$ V/cm, the energy gained by a carrier within one mean free path can exceed the energy gap of the collector material. As the highly energetic carrier impacts the lattice atoms, the atoms are ionized. The act of a carrier impacting the lattice and thereby creating electron-hole pairs is called *impact ionization*. One single impact ionization creates an electron-hole pair, which leads to further impact ionization as the recently generated carriers also pick up enough energy to ionize the lattice. The net result of this positive feedback mechanism is a rapid rise of I_C, which puts the transistor out of useful (or controllable) range of operation. The V_{CE} corresponding the rapid rise in I_C is called the *breakdown voltage*.

5.4 Collapse of Current Gain

The breakdown voltage represents the absolute maximum bias that can be applied to a bipolar transistor. There is, in addition, one more physical phenomenon that further restricts V_{CE} to values smaller than the breakdown voltage. This phenomenon occurs in multi-finger HBTs, having roots in the thermal-electrical interaction in the device. It is termed the *collapse of current gain* (or *gain collapse*) to emphasize the abrupt decrease of current gain observed in measured I-V characteristics when V_{CE} increases past certain values. Figure 5.13 is one such example, measured from a 2-finger HBT. The figure reveals two distinct operating regions, separated by a dotted curve. When V_{CE} is small, I_C decreases gradually with V_{CE}, exhibiting a negative differential resistance (NDR). We briefly describe the cause of NDR, as it relates to the understanding of the gain collapse. The band diagrams in Figs. 5.3 and 5.4 showed that the back-injected holes from the base into the emitter experience a larger energy barrier than the emitter electrons forward injected into the base. The ratio of the desirable I_C to the undesirable $I_{B,\text{back-inject}}$ is proportional to $\exp(\Delta E_g/kT)$ in a graded HBT, and $\exp(\Delta E_v/kT)$ in an abrupt HBT. At room temperature, this ratio is large in either HBT. However, as V_{CE} increases, the power dissipation in the HBT increases, gradually elevating the device temperature above the ambient temperature. $I_C/I_{B,\text{back-inject}}$, and hence the current gain, gradually decrease with increasing V_{CE}. Since Fig. 5.13 is measured I_C for several constant I_B, the decreasing β directly translates to the gradual decrease of I_C.

As V_{CE} crosses the dotted curve, NDR gives in to the collapse phenomenon, as marked by a dramatic lowering of I_C. The *collapse locus*, the dotted curve, is the collection of I_C as a function of V_{CE} at which the gain collapse occurs. When several identical transistors are connected together to common emitter, base, and collector electrodes, we tend to expect each transistor to conduct the same amount of collector current for any biases. Contrary to our intuition, equal conduction takes place only when the power dissipation is low to moderate, such that the junction temperature rise above the ambient temperature is small. At high V_{CE} and/or high I_C operation where the transistor is operated at elevated temperatures, one transistor spontaneously starts to conduct more current than the others (even if all the transistors are ideal and identical). Eventually, one transistor conducts all the current while the others become electrically inactive. This current imbalance originates from a universal bipolar transistor property that as the junction temperature increases, the bias required to turn on some arbitrary current level decreases. Quantitatively, this property is expressed by an empirical expression relating I_C, V_{BE}, and T:

$$I_C = I_{C,\text{sat}} \cdot \exp\left[\frac{q}{kT_0} \cdot \left(V_{BE,\text{junction}} - \phi \cdot \left(T - T_0\right)\right)\right] \tag{5.12}$$

where $I_{C,\text{sat}}$ is the collector saturation current; $V_{BE,\text{junction}}$ is the bias across the base-emitter junction; T_0 is the ambient temperature; and T is the actual device temperature. The degree of the turn-on voltage change in response to a junction temperature change is characterized by ϕ, which is called the *thermal-electrical feedback coefficient*. ϕ decreases logarithmically with I_C^3, and can be approximately as 1.1 mV/°C. This means that when the junction temperature exceeds the ambient temperature by 1°C, turning on the same amount of I_C requires 1.1 mV less of $V_{BE,\text{junction}}$.

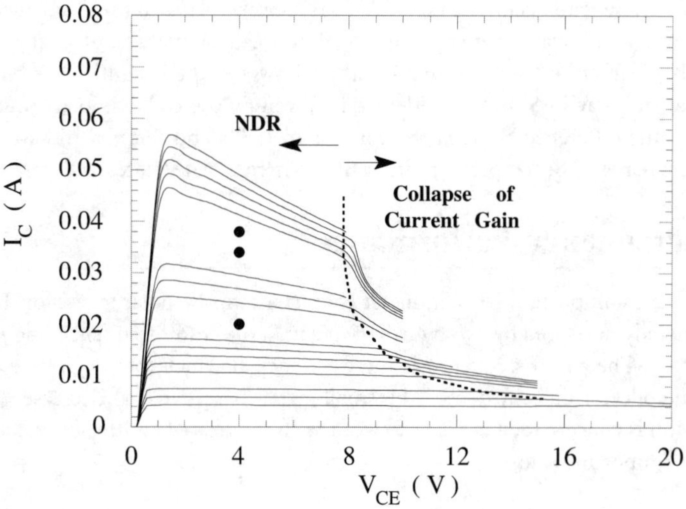

FIGURE 5.13 Measured I-V characteristics of a 2-finger AlGaAs/GaAs HBT showing two distinct regions of operation.

A multi-finger HBT can be viewed as consisting of several identical sub-HBTs, with their respective emitter, base, and collector leads connected together. If one finger (i.e., one sub-HBT) becomes slightly warmer than the others, its base-emitter junction turn-on voltage becomes slightly lower. Consequently, this particular finger conducts more current for a given fixed base-emitter voltage. The increased collector current, in turn, increases the power dissipation in the junction, raising the junction temperature even further. The gain collapse occurs when the junction temperature in one finger of the entire device becomes much hotter than the rest of the fingers, so that the feedback action of increased collector current with junction temperature quickly leads to the fact that just one particular finger conducts the entire device current. Since the transition from uniform current conduction to one finger domination occurs suddenly, the surge in junction temperature in the conducting finger quickly lowers the overall device current gain. The fundamental cause of both NDR and collapse is the current gain lowering at elevated temperatures. Their difference, however, lies in the degree of temperature increase as V_{CE} increases. In the NDR region, all fingers share relatively the same amount of current and the junction temperatures increase gradually with V_{CE}. In contrast, in the collapse region, as the device power is entirely dissipated in one finger and the junction temperature rises sharply, the current gain suddenly plummets.

The equation governing the collapse locus (per unit finger) is given by:[3,20]

$$I_{C,\text{collapse}} = \frac{kT_0}{q} \frac{1}{R_{th} \cdot \phi \cdot V_{CE} - R_E} \tag{5.13}$$

where R_{th} is the thermal resistance per finger and R_E is the emitter resistance per finger. When the individual finger current is below this critical current level (or when $I_{C,\text{collapse}}$ is negative), all fingers share the same amount of current. Above this critical current level, one finger conducts most of the current, whereas the rest of the fingers share the remaining current equally. Equation (5.13) shows that an effective method to increase $I_{C,\text{collapse}}$ is to increase R_E. The portion of the resistance that is intentionally introduced into device fabrication (such as by connecting a TaN thin-film resistor in series with the emitter electrode) is called the *ballasting resistance.* Alternatively, $I_{C,\text{collapse}}$ can be increased by reducing the thermal resistance, a goal often requiring more elaborate processes.[21]

Equation (5.13) neglects the contribution from the base resistance, R_B. For III-V HBTs, it is actually advantageous to use base ballasting; i.e., with the ballasting resistance placed in the base terminal.[22] The reason why the base ballasting approach is undesirable for Si BJT has been analyzed.[3]

The collapse of current gain occurring in III-V HBTs, closely relates to the thermal runaway in Si BJTs. HBTs suffering from gain collapse remain functional and can be biased out of the collapse region by reducing V_{CE}. Si BJTs suffering from thermal runaway, however, die instantly. The bias conditions triggering the thermal runaway in Si BJTs are identically given by the collapse locus equation [Eq. (5.13)]. The main cause of the difference between the collapse in HBTs and thermal runaway in Si BJTs is that the current gain increases with temperature in Si BJTs whereas it decreases with temperature in HBTs.[3]

5.5 High Frequency Performance

Current gain is the most important DC parameter characterizing a bipolar transistor. The high frequency properties are generally measured by two figures of merit: f_T, the cutoff frequency, and f_{max}, the maximum oscillation frequency. The cutoff frequency is the frequency at which the magnitude of the AC current gain (small-signal collector current divided by small-signal base current) decreases to unity. As far as analytical expression is concerned, it is easier to work with the related emitter-collector transit time (τ_{ec}), which is inversely proportional to f_T:

$$f_T = \frac{1}{2\pi\,\tau_{ec}}. \tag{5.14}$$

The emitter-collector transit time can be broken up into several components. The emitter charging time, τ_e, is the time required to change the base potential by charging up the capacitances through the differential base-emitter junction resistance:

$$\tau_e = \frac{kT}{q\,I_C}\cdot\left(C_{j,\mathrm{BE}} + C_{j,\mathrm{BC}}\right). \tag{5.15}$$

$C_{j,\mathrm{BE}}$ and $C_{j,\mathrm{BC}}$ denote the junction capacitances of the base-emitter and the base-collector junctions, respectively. The inverse dependence of τ_e on I_C is the primary reason why BJTs and HBTs are biased at high current levels. When the current density is below 10^4 A/cm^2, this term often dominates the overall collector-emitter transit time.

The second component is the base transit time, the time required for the forward-injected charges to diffuse/drift through base. It is given by,

$$\tau_b = \frac{X_{\mathrm{base}}^2}{\nu\cdot D_{n,\mathrm{base}}} \tag{5.16}$$

The value of ν depends on the magnitude of the base quasi-electric field. In a uniform base without a base field, ν is 2, as suggested by Eq. (5.8). Depending on the amount of energy-gap grading, ν can easily increase to 6 to 10.[3]

The space-charge transit time, τ_{sc}, is the time required for the electrons to drift through the depletion region of the base-collector junction. It is given by,

$$\tau_{sc} = \frac{X_{dep}}{2\,v_{sat}} \tag{5.17}$$

where X_{dep} is the depletion thickness of the base-collector junction. The factor of 2 results from averaging the sinusoidal of carriers current over a time period.[3] It is assumed in the derivation that, because the electric field is large throughout the entire reverse-biased base-collector junction, the carriers travel at a constant saturation velocity, v_{sat}. With a p^- collector layer placed adjacent to the p^+ base of an otherwise conventional *Npn* HBT,[23] the electric field near the base can be made smaller than $\varepsilon_{\mathrm{crit}}$. Consequently, the electrons travel at the velocity determined completely by the Γ valley. Without scattering to the L valley, the electrons continue to travel at a velocity that is much larger than v_{sat}, and τ_{sc} is significantly reduced. When carriers travel faster than v_{sat} under an off-equilibrium condition, *velocity overshoot* is said to occur.

The last term, the collector charging time, τ_c, is given by,

$$\tau_c = \left(R_E + R_C\right) \cdot C_{j,\mathrm{BC}}, \tag{5.18}$$

where R_E and R_C are the device emitter and collector resistances, respectively. The value of this charging time depends greatly on the parasitic resistances. This is the term that degrades the HBT's high frequency performance when the contacts are poorly fabricated.

The overall transit time is a sum of the four time constants:

$$\tau_{ec} = \frac{kT}{q\,I_C} \cdot \left(C_{j,\mathrm{BE}} + C_{j,\mathrm{BC}}\right) + \frac{X_{\mathrm{base}}^2}{\nu \cdot D_{n,\mathrm{base}}} + \frac{X_{dep}}{2\,v_{sat}} + \left(R_E + R_C\right) \cdot C_{j,\mathrm{BC}} \tag{5.19}$$

In most HBTs, R_E and R_C are dominated by the electrode resistances; the epitaxial resistances in the emitter and collector layers are insignificant. The cutoff frequency relates to τ_{ec} through Eq. (5.14).

The maximum oscillation frequency is the frequency at which the unilateral power gain is equal to 1. The derivation is quite involved,[3] but the final result is elegant:

$$f_{\max} = \sqrt{\frac{f_T}{8\pi\, R_B C_{j,\mathrm{BC}}}}. \tag{5.20}$$

The base resistance has three components that are roughly equal in magnitude. They are base electrode resistance ($R_{B,\mathrm{eltd}}$); intrinsic base resistance ($R_{B,\mathrm{intrinsic}}$); and extrinsic base resistance ($R_{B,\mathrm{extrinsic}}$). $R_{B,\mathrm{eltd}}$ is intimately related to processing, depending on the contact metal and the alloying recipe used to form the contact. The other two base resistance components, in contrast, depend purely on the designed base layer and the geometrical details. The HBT cross-sections illustrated in Fig. 5.8 show two base contacts placed symmetrically beside the central emitter mesa. For this popular transistor geometry, the intrinsic base resistance is given by,

$$R_{B,\mathrm{intrinsic}} = \frac{1}{12} \times R_{SH,\mathrm{base}} \frac{W_E}{L_E} \tag{5.21}$$

where W_E and L_E are the emitter width and length, respectively; and $R_{SH,\mathrm{base}}$ is the base sheet resistance in Ω/square. Where does the 1/12 factor come from? If all of the base current that goes into one of the base contacts in Fig. 5.8 leaves from the other contact, the intrinsic base resistance would follow our intuitive formula of simple resistance, equal to $R_{SH,\mathrm{base}} \times W_E/L_E$. However, during the actual transistor operation, the base current enters through both contacts, but no current flows out at the other end. The holes constituting the base current gradually decrease in number as they are recombined at the extrinsic base surface, in the base bulk layer, and in the base-emitter space-charge region. Some base current carriers remain in the base layer for a longer portion of the width before getting recombined. Other carriers get recombined sooner. The factor 1/12 accounts for the distributed nature of the current conduction, as derived elsewhere.[3] Because of the way the current flows in the base layer, the intrinsic base resistance is called a *distributed resistance*. If instead there is only one base contact, the factor 1/12 is replaced by 1/3 (not 1/6!).[24] The distributed resistance also exists in the gate of MOS transistors,[25] or III-V field-effect transistors.[4]

The extrinsic base resistance is the resistance associated with the base epitaxial layer between the emitter and the base contacts. It is given by,

$$R_{B,\text{extrinsic}} = \frac{1}{2} \times R_{SH,\text{base}} \frac{S_{BE}}{L_E} \tag{5.22}$$

where S_{BE} is the separation between the base and emitter contacts. The factor 1/2 appears in the transistor shown in Fig. 5.8, which has two base contacts. The presence of $R_{B,\text{extrinsic}}$ is the reason why most transistors are fabricated with self-aligned base-emitter contacts, and that the base contacts are deposited right next to the emitter contacts, regardless of the finite alignment tolerance between the base and emitter contact photolithographical steps. (A detailed fabrication process will be described shortly.) With self-alignment, the distance S_{BE} is minimized, to around 3000 Å.

A detailed calculation of f_T and $f_{\max}$ has been performed for a $W_E \times L_E = 2 \times 30$ µm² HBT,[4] with a S_{BE} of 0.2 µm, a base thickness of 800 Å, a base sheet resistance of 280 Ω/square, and a base diffusion coefficient of 25.5 cm²/s. Although device parameters will have different values in other HBT structures and geometries, the exemplar calculation gives a good estimation of the relative magnitudes of various terms that determine a HBTs high frequency performance. We briefly list the key results here. The HBT has: $R_E = 0.45\ \Omega$; $R_{B,\text{eltd}} = 7\ \Omega$; $R_{B,\text{extrinsic}} = 0.94\ \Omega$; $R_{B,\text{intrinsic}} = 1.56\ \Omega$; $C_{j,\text{BE}} = 0.224$ pF; $C_{j,\text{BC}} = 0.026$ pF. It was further determined at the bias condition of $J_C = 10^4$ A/cm² and $V_{BC} = -4$ V, that the collector depletion thickness $X_{dep} = 0.59$ µm. Therefore,

$$\tau_e = \frac{kT}{qI_C} \cdot \left(C_{j,\text{BE}} + C_{j,\text{BC}}\right) = \frac{0.0258\left(2.24 \times 10^{-13} + 2.59 \times 10^{-14}\right)}{1 \times 10^4 \cdot 2 \times 10^{-4} \cdot 3 \times 10^{-4}} = 1.08 \text{ ps.}$$

$$\tau_b = \frac{X^2 base}{\nu \cdot D_{n,\text{base}}} = \frac{\left(800 \times 10^{-8}\right)^2}{2 \cdot 25.5} = 1.25 \text{ ps.}$$

$$\tau_{sc} = \frac{X_{dep}}{2\,v_{sat}} = \frac{0.591 \times 10^{-4}}{2 \cdot 8 \times 10^6} = 3.69 \text{ ps.}$$

$$\tau_c = \left(R_E + R_C\right) C_{j,\text{BC}} = \left(0.453 + 4.32\right) \cdot 2.59 \times 10^{-14} = 0.124 \text{ ps.}$$

Summing up these four components, we find the emitter-collector transit time to be 6.14 ps. The cutoff frequency is therefore $1/(2\pi \cdot 6.14$ ps$) = 26$ GHz. As mentioned, although this calculation is specific to a particular power HBT design, the relative magnitudes of the four time constants are quite representative. Generally, the space-charge transit time is the major component of the overall transit time. This is unlike silicon BJTs, whose τ_b and τ_e usually dominate, because of the low base diffusion coefficient in the silicon material and the high base-emitter junction capacitance associated with high emitter doping. HBT design places a great emphasis on the collector design, while the Si BJT design concentrates on the base and emitter layers.

The total base resistance is $R_B = 7 + 0.96 + 1.95 = 9.91\ \Omega$. The maximum oscillation frequency, calculated with Eq. (5.20), is 65 GHz.

Figure 5.14 illustrates the cutoff and maximum oscillation frequencies of a state-of-art InAlAs/InGaAs HBT. The frequency responses are plotted as a function of the collector current to facilitate the circuit design at the proper bias condition. When the current is small, the cutoff frequency is small because the emitter charging time is large. As the current increases, f_T increases because the emitter charging time decreases. When the current density exceeds roughly 10^5 A/cm², Kirk effects take place and the transistor

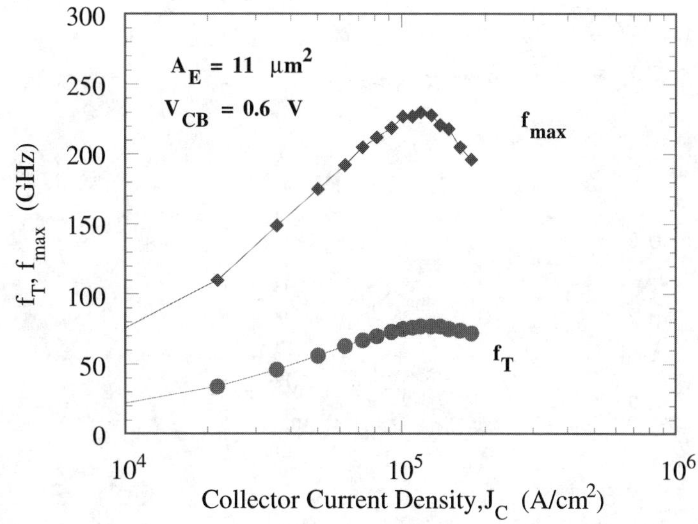

FIGURE 5.14 High-frequency performance of a state-of-the-art InAlAs/InGaAs HBT. (After Chau, H. and Kao, Y., High f_{max} InAlAs/InGaAs heterojunction bipolar transistors, IEEE IEDM, 783–786, 1993, with permission.)

performance degrades rapidly. f_{max}'s current dependence follows f_T's, as governed by the relationship of Eq. (5.20).

The high frequency transistor of the calculated example has a narrow width of 2 μm. What happens if W_E increases to a large value, such as 200 μm? A direct application of Eq. (5.19) would still lead to a cutoff frequency in the GHz range. In practice, such a 200 μm wide device will have a cutoff frequency much smaller than 1 GHz. The reason for the discrepancy is simple; the time constants appearing in Eq. (5.19), which all roughly scale with the emitter area, are based on the assumption that the current conduction is uniform within the entire emitter mesa. In HBTs, and more so in Si BJTs, the base resistance is significant. As the base current flows horizontally from the base contacts to the center of the emitter region, some finite voltage is developed, with a increasingly larger magnitude toward the center of the mesa. The effective base-emitter junction voltage is larger at the edge than at the center. Since the amount of carrier injection depends exponentially on the junction voltage at the given position, most of the injection takes place near the emitter edges, hence the term *emitter crowding*. Sometimes this phenomenon of nonuniform current conduction is referred to as the *base crowding*. Both terms are acceptable, depending on whether the focus is on the crowding of the emitter current or the base current. A figure of merit quantifying the severity of emitter crowding is the effective emitter width (W_{eff}). It is defined as the emitter width that would result in the same current level if current crowding were absent and the emitter current density were uniform at its edge value. Figure 5.15 illustrates the effective emitter width (normalized by the defined emitter width) as a function of base doping and collector current.

Because of the emitter crowding, all high-frequency III-V HBTs capable of GHz performance have a narrow emitter width on the order of 2 μm. For wireless applications in the 2 GHz range, 2 to 3 μm W_E is often used,[26] but a 6.4 μm W_E has also been reported.[27] The choice of the emitter width is a trade-off between the ease (cost) of device fabrication versus the transistor performance. We intuitively expected the W_{eff}/W_E ratio to increase as the doping increased (since the base resistance decreased). However, an increase in base doping was accompanied by shortened minority lifetime. The large increase in $I_{B,\text{bulk}}$ causes the emitter crowding to be more pronounced as N_{base} increases, as shown in Fig. 5.15. Due to differences in material properties, the emitter width in Si BJT or SiGe HBTs tends to be on the order of 0.5 μm.

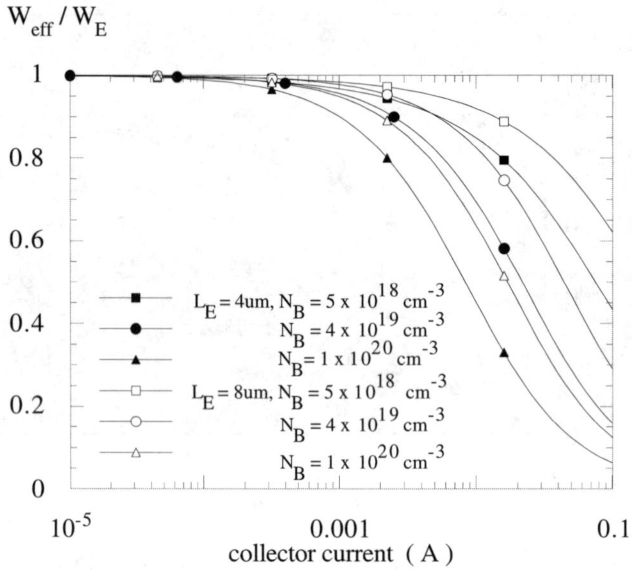

FIGURE 5.15 Calculated W_{eff}/W_E ratio. (After Liu, W. and Harris, J., Dependence of base crowding effect on base doping and thickness for Npn AlGaAs/GaAs heterojunction bipolar transistors, *Electronic Letters*, 27, 2048–2050, 1991, with permission.)

5.6 Device Fabrication

An HBT mask layout suitable for high-frequency applications is shown in Fig. 5.16, and a corresponding fabrication process is delineated in Fig. 5.17. The following discussion of processing steps assumes the AlGaAs/GaAs HBT epitaxial structure shown in Fig. 5.6. The first step is to deposit ~7000 Å of silicon dioxide, onto which the first-mask pattern "ISOL" is defined, and 1.5 μm of aluminum, evaporated and lifted off. The aluminum protects the silicon dioxide underneath during the subsequent reactive ion etch (RIE), resulting in the profile shown in Fig. 5.17a. Oxygen atoms or protons are then implanted everywhere on the wafer. This implantation is designed to make the region outside of "ISOL" pattern electrically inactive. A shallow etching is applied right after the implantation so that a trail mark of the active device area is preserved. This facilitates the alignment of future mask levels to the first mask. Afterward, both aluminum and oxide are removed with wet etching solutions, reexposing the fresh InGaAs top surface onto which a refractory metal (such as W) is sputtered. The "EMIT" mask level is used to define the emitter mesas, and Ti/Al, a conventional contact metal used in III-V technologies, is evaporated and lifted off. If a refractory metal is not inserted between the InGaAs semiconductor and the Ti/Al metal, long-term reliability problems can arise as titanium reacts with indium. During the ensuing RIE, the refractory metal not protected by the emitter metal is removed.

The resulting transistor profile is shown in Fig. 5.17b. Wet or dry etching techniques are applied to remove the exposed GaAs cap and the AlGaAs active emitter layer, and eventually reach the top of the base layer. Silicon nitride is deposited by plasma enhanced chemical vapor deposition (PECVD). This deposition is conformal, forming a nitride layer everywhere, including the vertical edges. Immediately after the nitride deposition, the whole wafer is etched by RIE. Because a vertical electric field is set up in the RIE chamber to propel the chemical species in the vertical direction, the sidewall nitride covering

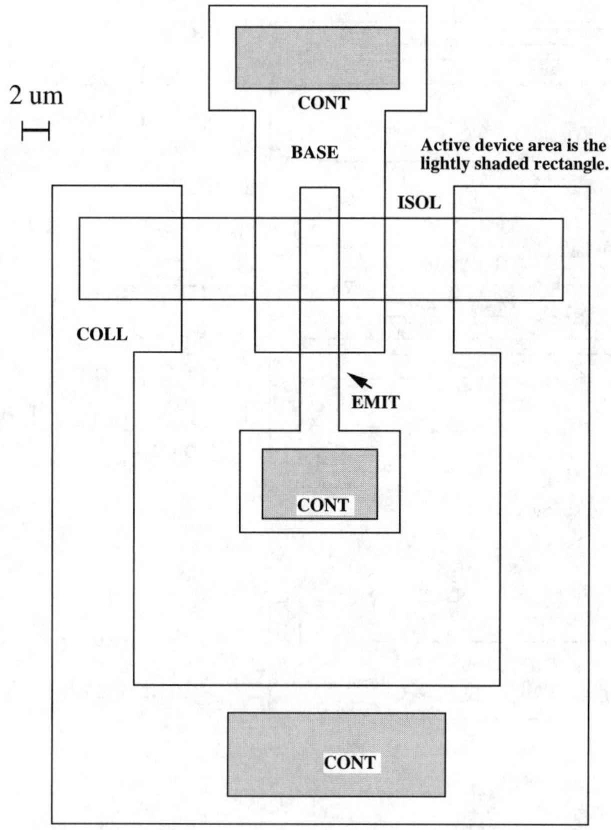

FIGURE 5.16 A HBT mask layout suitable for high-frequency applications. (After Liu, W., Microwave and DC Characterizations of Npn and Pnp HBTs, PhD. Dissertation, Stanford University, Stanford, CA, 1991.)

the sides of the emitter mesas remains untouched while the nitride layer lying on a flat surface is etched away (Fig. 5.17c). The nitride sidewall improves the device yield by cutting down the possible electrical short between the emitter contact and the soon-to-be-deposited base contacts.

The step after the nitride sidewall formation is the "BASE" lithography. As shown in the layout of Fig. 5.16, there is no separation between the "BASE" and the "EMIT" levels. The base metal during the evaporation partly lands on the emitter mesa, and some of it lands on the base layer as desired (Fig. 5.17d). The process has the desired feature that, even if the "BASE" level is somewhat misaligned with the "EMIT" level, the distance S_{BE} between the emitter and base contacts is unchanged, and is at the minimum value determined by the thickness of the nitride sidewall. In this manner, the base contact is said to be *self-aligned* to the emitter. Self-alignment reduces $R_{B,\text{extrinsic}}$ [Eq. 5.22)] and $C_{j,BC}$, hence improving the high-frequency performance. Typically, Ti/Pt/Au is the choice for the base metal.

Following the base metal formation, the collector is defined and contact metal of Au/Ge/Ni/Au is deposited (Fig. 5.17e). After the contact is alloyed at ~450°C for ~1 minute, a polyimide layer is spun to planarize the device as shown in Fig. 5.17f. The contact holes are then defined and Ti/Au is evaporated to contact various electrodes. The final device cross-section after the entire process is shown in Fig. 5.17g.

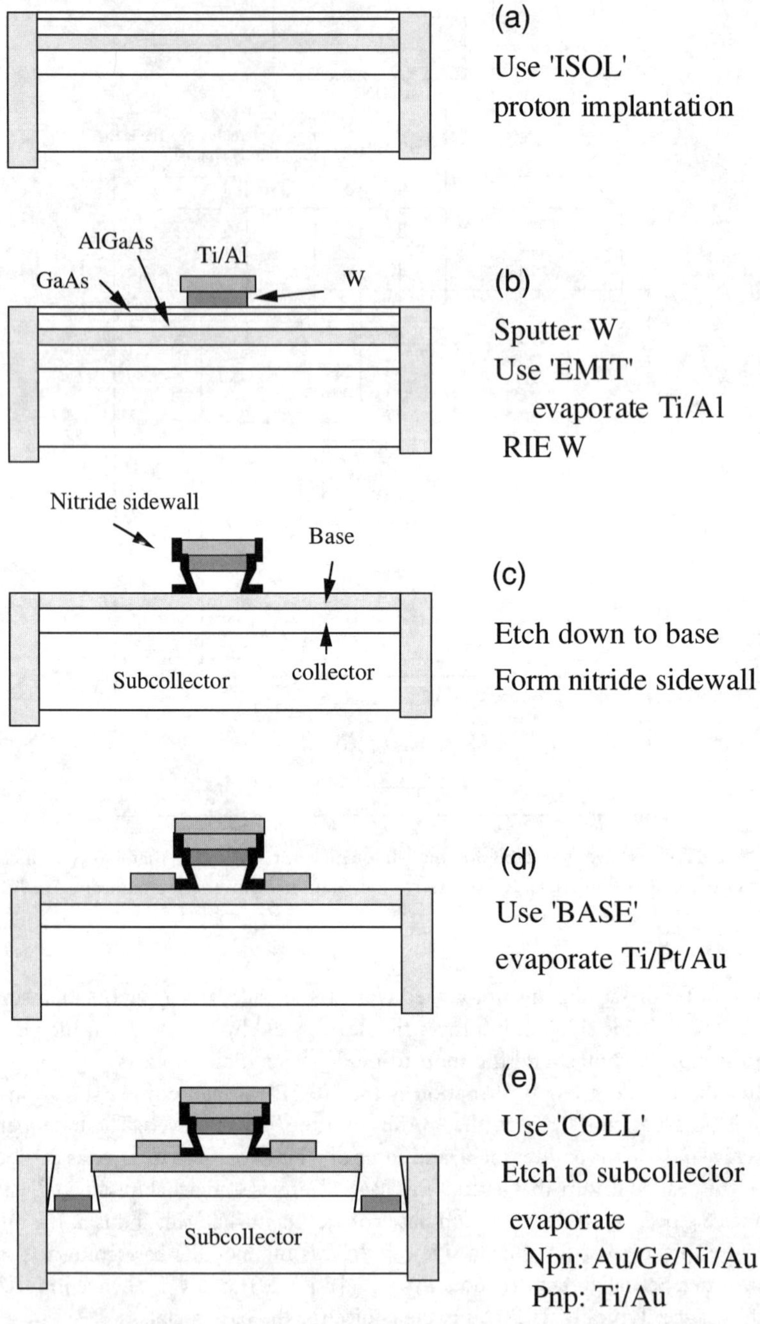

FIGURE 5.17 A high-frequency HBT process flow. (After Liu, W., Microwave and DC Characterizations of Npn and Pnp HBTs, PhD. Dissertation, Stanford University, Stanford, CA, 1991.)

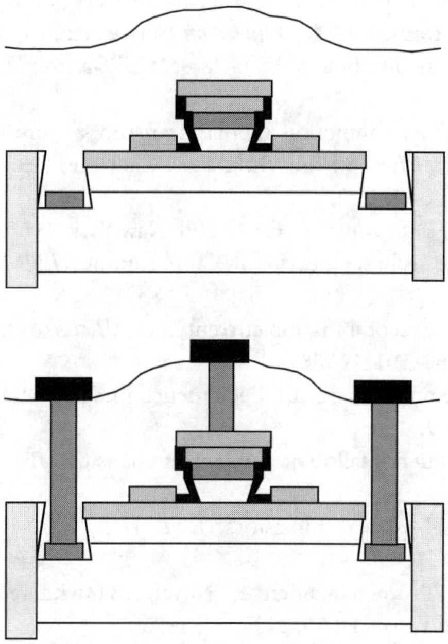

(f)

Spin on polyimide

(g)

Open contact
Form interconnect

FIGURE 5.17 (continued)

References

1. Kroemer, H, Heterostructure bipolar transistors and integrated circuits, *Proc. IEEE*, 70, 13, 1982.
2. Sah, C.T., *Fundamentals of Solid-State Electronics*, World Scientific, Singapore, 1991.
3. Liu, W., *Handbook of III-V Heterojunction Bipolar Transistors*, Wiley & Sons, New York, 1998. An in-depth discussion of several topics can be found in this handbook.
4. Liu, W., *Fundamentals of III-V Devices: HBTs, MESFETs, and HFETs/HEMTs*, Wiley & Sons, New York, 1999.
5. Kroemer, H., Theory of wide gap emitter for transistors, *Proc. IRE*, 45, 1535, 1957.
6. For pioneering HBT papers, see for example, Asbeck, P., Chang, M., Higgins, J., Sheng, N., Sullivan, G., and Wang, K., GaAlAs/GaAs heterojunction bipolar transistors: issues and prospects for application, *IEEE Trans. Electron Devices*, 36, 2032–2041, 1989. For introduction to MBE and MOCVD growth techniques, see References 3 and 4.
7. Henderson, T., Physics of degradation in GaAs-based heterojunction bipolar transistors, *Microelectronics Reliability*, 39, 1033–1042, 1999. See also, Low, T., et al., Migration from an AlGaAs to an InGaP emitter HBT IC process for improved reliability, *IEEE GaAs IC Symposium*, 153–156, 1998.
8. Liu, W. et al., Recent developments in GaInP/GaAs heterojunction bipolar transistors, in *Current Trends in Heterojunction Bipolar Transistors*, Chang, M.F., Ed., World Scientific, Singapore, 1996.
9. Chau, H. and Liu, W., Heterojunction bipolar transistors and circuit applications, in *InP-Based Material and Devices: Physics and Technology*, Wada, O. and Hasegawa, H., Eds., Wiley & Sons, New York, 1999.
10. Ali, F., Gupta, A. and Higgins, A., Advances in GaAs HBT power amplifiers for cellular phones and military applications, *IEEE Microwave and Millimeter-Wave Monolithic Circuits Symposium*, 61–66, 1996.
11. Lin, H. and Lee, S., Super-gain AlGaAs/GaAs heterojunction bipolar transistor using an emitter edge-thinning design, *Appl. Phys. Lett.*, 47, 839–841, 1985.

12. Lee, W., Ueda, D., Ma, T., Pao, Y. and Harris, J., Effect of emitter-base spacing on the current gain of AlGaAs/GaAs heterojunction bipolar transistors, *IEEE Electron Device Lett.*, 10, 200–202, 1989.
13. Patton, G., 75-GHz fr SiGe-base heterojunction bipolar transitors, *IEEE Electron Devices Lett.*, 11, 171–173, 1990.
14. Meyerson, B. et al., Silicon:Germanium heterojunction bipolar transistors; from experiment to technology, in *Current Trends in Heterojunction Bipolar Transistors*, Chang, M., Ed., World Scientific, Singapore, 1996.
15. Pritchard, R., *Electrical Characteristics of Transistors*, McGraw-Hill, New York, 1967.
16. Konig, U., SiGe & GaAs as competitive technologies for RF applications, *IEEE Bipolar Circuit Technology Meeting*, 87–92, 1998.
17. Tiwari, S., Frank, D. and Wright, S., Surface recombination current in GaAlAs/GaAs heterostructure bipolar transistors, *J. Appl. Phys.*, 64, 5009–5012, 1988.
18. Buhanan, D., Investigation of current-gain temperature dependence in silicon transistors, *IEEE Trans. Electron Devices*, 16, 117–124, 1969.
19. Kirk, C., A theory of transistor cutoff frequency falloff at high current densities, *IRE Trans. Electron Devices*, 9, 164–174, 1962.
20. Winkler, R., Thermal properties of high-power transistors, *IEEE Trans. Electron Devices*, 14, 1305–1306, 1958.
21. Hill, D., Katibzadeh, A., Liu, W., Kim, T. and Ikalainen, P., Novel HBT with reduced thermal impedance, *IEEE Microwave and Guided Wave Lett.*, 5, 373–375, 1995.
22. Khatibzadeh, A. and Liu, W., *Base Ballasting*, U.S. Patent Number 5,321,279, issued on June 14, 1994.
23. Maziar, C., Klausmeier-Brown, M. and Lundstrom, M., Proposed structure for collector transit time reduction in AlGaAs/GaAs bipolar transistors, *IEEE Electron Device Lett.*, 7, 483–385, 1986.
24. Hauser, J., The effects of distributed base potential on emitter-current injection density and effective base resistance for stripe transistor geometries, *IEEE Trans. Electron Devices*, 11, 238–242, 1964.
25. Liu, W., *MOSFET Models for SPICE Simulation, Including BSIM3v3 and BSIM4*, Wiley & Sons, New York, in press.
26. RF Microdevices Inc., A high efficiency HBT analog cellular power amplifier, *Microwave Journal*, 168–172, January 1996.
27. Yoshimasu, T., Tanba, N. and Hara, S., High-efficiency HBT MMIC linear power amplifier for L-band personal communication systems, *IEEE Microwave and Guided Wave Lett.*, 4, 65–67, 1994.

6

Metal-Oxide-Semiconductor Field-Effect Transistors

Leonard MacEachern
Carlton University

Tajinder Manku
University of Waterloo

6.1 Introduction

The insulated-gate field-effect transistor was conceived in the 1930s by Lilienfeld and Heil. An insulated-gate transistor is distinguished by the presence of an insulator between the main control terminal and the remainder of the device. Ideally, the transistor draws no current through its gate (in practice a small leakage current on the order of 10^{-18} A to 10^{-16} A exists). This is in sharp contrast to bipolar junction transistors that require a significant base current to operate. Unfortunately, the **M**etal-**O**xide-**S**emiconductor **F**ield-**E**ffect **T**ransistor (**MOSFET**) had to wait nearly 30 years until the 1960s when manufacturing advances made the device a practical reality. Since then, the explosive growth of MOSFET utilization in every aspect of electronics has been phenomenal. The use of MOSFETs in electronics became ever more prevalent when "complementary" types of MOSFET devices were combined by Wanlass in the early 1960s to produce logic that required virtually no power except when changing state. MOSFET processes that offer complementary types of transistors are known as **C**omplementary **M**etal **O**xide **S**emiconductor

0-8493-1562-X/03/$0.00+$1.50

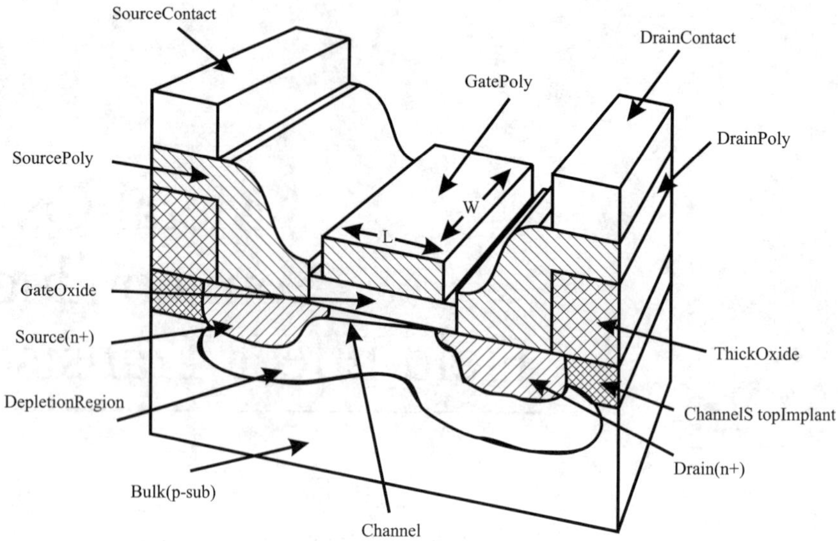

FIGURE 6.1 MOSFET physical structure.

(**CMOS**) processes, and are the foundation of the modern commodity electronics industry. The MOSFET's primary advantages over other types of integrated devices are its mature fabrication technology, its high integration levels, its mixed analog/digital compatibility, its capability for low voltage operation, its successful scaling characteristics, and the combination of complementary MOSFETs yielding low power CMOS circuits.

In this section, basic material concerning the MOSFET physical structure and operation is first presented. Nonideal effects are briefly discussed, as are important parasitics and distributed effects found in MOSFETs. Once the important parasitics and distributed effects are understood, the operation of MOSFETs at radio frequencies is examined, and important MOSFET operating parameters are given. Following this, MOSFET noise sources relevant to radio frequency designs are discussed. This section concludes with a discussion of MOSFET design and physical layout appropriate for radio frequency implementations.

6.2 MOSFET Fundamentals

Today, each of the tens of millions of MOSFETs that can occupy mere square centimetres of silicon area shares the same basic physical structure. The physical structure of the MOSFET is deceptively simple, as illustrated by the MOSFET cross-section appearing in Fig. 6.1. Visible in the figure are the various materials used to construct a MOSFET. These materials appear in layers when the MOSFET cross-section is viewed; this is a direct consequence of the processes of "doping," deposition, growth, and etching which are fundamental in conventional processing facilities.

The fabrication process of silicon MOSFET devices has evolved over the last 30 years into a reliable integrated circuit manufacturing technology. Silicon has emerged as the material of choice for MOSFETs, largely because of its stable oxide, SiO_2, which is used as a general insulator, as a surface passivation layer, and as an excellent gate dielectric.

Full appreciation of the MOSFET structure and operation requires some knowledge of silicon semiconductor properties and "doping." These topics are briefly reviewed next.

On the scale of conductivity that exists between pure insulators and perfect conductors, semiconductors fall between the extremes. The semiconductor material commonly used to make MOSFETs is silicon. Pure, or "intrinsic" silicon exists as an orderly three-dimensional array of atoms, arranged in a crystal lattice. The atoms of the lattice are bound together by covalent bonds containing silicon valence electrons. At absolute-zero temperature, all valence electrons are locked into these covalent bonds and are unavailable for current conduction, but as the temperature is increased, it is possible for an electron to gain enough thermal energy to escape its covalent bond, and in the process leave behind a covalent bond with a missing electron, or "hole." When that happens the electron that escaped is free to move about the crystal lattice. At the same time, another electron, which is still trapped in nearby covalent bonds because of its lower energy state, can move into the hole left by the escaping electron. The mechanism of current conduction in intrinsic silicon is therefore by hole-electron pair generation, and the subsequent motion of free electrons and holes throughout the lattice.

At normal temperatures intrinsic silicon behaves as an insulator because the number of free electron-hole pairs available for conducting current is very low, only about 14.5 hole-electron pairs per 1000 μm^3 of silicon. The conductivity of silicon can be adjusted by adding foreign atoms to the silicon crystal. This process is called "doping," and a "doped" semiconductor is referred to as an "extrinsic" semiconductor. Depending on what type of material is added to the pure silicon, the resulting crystal structure can either have more electrons than the normal number needed for perfect bonding within the silicon structure, or less electrons than needed for perfect bonding. When the dopant material increases the number of free electrons in the silicon crystal, the dopant is called a "donor." The donor materials commonly used to dope silicon are phosphorus, arsenic, and antimony. In a donor-doped semiconductor the number of free electrons is much larger than the number of holes, and so the free electrons are called the "majority carriers" and the holes are called the "minority carriers." Since electrons carry a negative charge and they are the majority carriers in a donor-doped silicon semiconductor; any semiconductor that is predominantly doped with donor impurities is known as "n-type." Semiconductors with extremely high donor doping concentrations are often denoted "n+ type."

Dopant atoms that accept electrons from the silicon lattice are also used to alter the electrical characteristics of silicon semiconductors. These types of dopants are known as "acceptors." The introduction of the acceptor impurity atoms creates the situation in which the dopant atoms have one less valence electron than necessary for complete bonding with neighboring silicon atoms. The number of holes in the lattice therefore increases. The holes are therefore the majority carriers and the electrons are the minority carriers. Semiconductors doped with acceptor impurities are known as "p-type," since the majority carriers effectively carry a positive charge. Semiconductors with extremely high acceptor doping concentrations are called "p+ type." Typical acceptor materials used to dope silicon are boron, gallium, and indium.

A general point that can be made concerning doping of semiconductor materials is that the greater the dopant concentration, the greater the conductivity of the doped semiconductor. A second general point that can be made about semiconductor doping is that n-type material exhibits a greater conductivity than p-type material of the same doping level. The reason for this is that electron mobility within the crystal lattice is greater than hole mobility, for the same doping concentration.

6.2.1 MOSFET Physical Structure and Operation

A cross-section through a typical n-type MOSFET, or "NFET," is shown in Fig. 6.2(a). The MOSFET consists of two highly conductive regions (the "source" and the "drain") separated by a semiconducting channel. The channel is typically rectangular, with an associated length (L) and width (W). The ratio of the channel width to the channel length, W/L, is an important determining factor for MOSFET performance.

As shown in Fig. 6.2(a), the MOSFET is considered a four-terminal device. These terminals are known as the gate (G), the bulk (B), the drain (D), and the source (S), and the voltages present at these terminals collectively control the current that flows within the device. For most circuit designs, the current flow from drain to source is the desired controlled quantity.

The operation of field-effect transistors (FETs) is based upon the principal of capacitively controlled conductivity in a channel. The MOSFET gate terminal sits on top of the channel, and is separated from

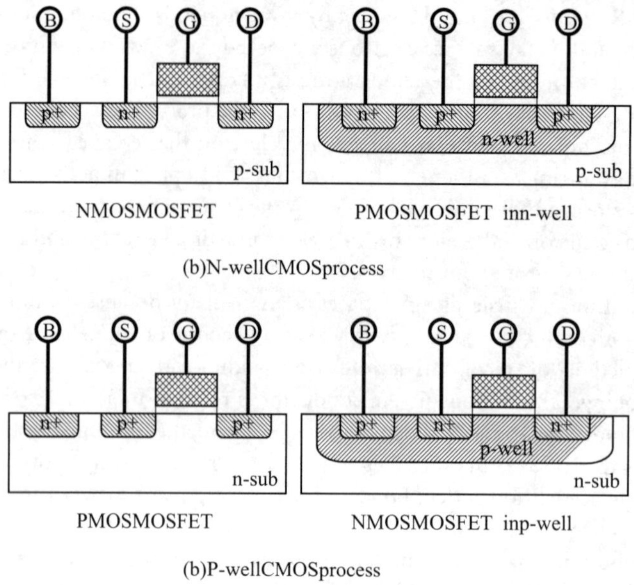

FIGURE 6.2 MOSFET terminals and active regions. (a) MOSFETs constructed on p-type substrate, (b) MOSFETs constructed on n-type substrate.

the channel by an insulating layer of SiO_2. The controlling capacitance in a MOSFET device is therefore due to the insulating oxide layer between the gate and the semiconductor surface of the channel. The conductivity of the channel region is controlled by the voltage applied across the gate oxide and channel region to the bulk material under the channel. The resulting electric field causes the redistribution of holes and electrons within the channel. For example, when a positive voltage is applied to the gate, it is possible that enough electrons are attracted to the region under the gate oxide, that this region experiences a local inversion of the majority carrier type. Although the bulk material is p-type (the majority carriers are holes and minority carriers are electrons), if enough electrons are attracted to the same region within the semiconductor material, the region becomes effectively n-type. Then the electrons are the majority carriers and the holes are the minority carriers. Under this condition electrons can flow from the n+ type drain to the n+ type source if a sufficient potential difference exists between the drain and source.

When the gate-to-source voltage exceeds a certain threshold, called V_T, the conductivity of the channel increases to the point where current may easily flow between the drain and the source. The value of V_T required for this to happen is determined largely by the dopant concentrations in the channel, but it also depends in part upon the voltage present on the bulk. This dependence of the threshold voltage upon the bulk voltage is known as the "body effect."

In MOSFETs, both holes and electrons can be used for conduction. As shown in Fig. 6.2, both n-type and p-type MOSFETs are possible. If, for an n-type MOSFET, all the n-type regions are replaced with p-type regions and all the p-type regions are replaced with n-type regions, the result is a p-type MOSFET. Since both the n-type and p-type MOSFETs require substrate material of the opposite type of doping, two distinct CMOS technologies exist, defined by whether the bulk is n-type or p-type. If the bulk material is p-type substrate, then n-type MOSFETs can be built directly on the substrate while p-type MOSFETs must be placed in an n-well. This type of process is illustrated in Fig. 6.2(a). Another possibility is that the bulk is composed of n-type substrate material, and in this case the p-type MOSFETs can be constructed directly on the bulk, while the n-type MOSFETs must be placed in an p-well, as in Fig. 6.2(b). A third type of process known as twin-well or twin-tub CMOS requires that both the p-type and n-type MOSFETs be placed in wells of the opposite type of doping. Other combinations of substrate doping and well types are in common use. For example, some processes offer a "deep-well" capability, which is useful for threshold adjustments and circuitry isolation.

Modern MOSFETs differ in an important respect from their counterparts developed in the 1960s. While the gate material used in the field effect transistors produced thirty years ago was made of metal, the use of this material was problematic for several reasons. At that time, the gate material was typically aluminium and was normally deposited by evaporating an aluminium wire by placing it in contact with a heated tungsten filament. Unfortunately, this method led to sodium ion contamination in the gate oxide, which caused the MOSFET's threshold voltage to be both high and unstable. A second problem with the earlier methods of gate deposition was that the gate was not necessarily correctly aligned with the source and drain regions. Matching between transistors was then problematic because of variability in the gate location with respect to source and drain for the various devices. Parasitics also varied greatly between devices because of this variability in gate location with respect to the source and drain regions. In the worst case, a nonfunctional device was produced because of the errors associated with the gate placement.

Devices manufactured today employ a different gate material, namely "polysilicon," and the processing stages used to produce a field-effect transistor with a poly gate are different than the processing stages required to produce a field-effect transistor with a metal gate. In particular, the drain and source wells are patterned using the gate and field oxide as a mask during the doping stage. Since the drain and source regions are defined in terms of the gate region, the source and drain are automatically aligned with the gate. CMOS manufacturing processes are referred to as self-aligning processes when this technique is used. Certain parasitics, such as the overlap parasitics considered later in this chapter, are minimized using this method. The use of a polysilicon gate tends to simplify the manufacturing process, reduces the variability in the threshold voltage, and has the additional benefit of automatically aligning the gate material with the edges of the source and drain regions.

The use of polysilicon for the gate material has one important drawback: the sheet resistance of polysilicon is much larger than that of aluminium and so the resistance of the gate is larger when polysilicon is used as the gate material. High-speed digital processes require fast switching time from the MOSFETs used in the digital circuitry, yet a large gate resistance hampers the switching speed of a MOSFET. One method commonly used to lower the gate resistance is to add a layer of silicide on top of the gate material. A silicide is a compound formed using silicon and a refractory metal, for example $TiSi_2$. Later in this chapter the importance of the gate resistance upon the radio frequency performance of the MOSFET is examined. In general, the use of silicided gates is required for reasonable radio frequency MOSFET performance.

Although the metal-oxide-semiconductor sandwich is no longer regularly used at the gate, the devices are still called MOSFETs. The term **I**nsulated-**G**ate **F**ield-**E**ffect **T**ransistor (**IGFET**) is also in common usage.

6.2.2 MOSFET Large Signal Current-Voltage Characteristics

When a bias voltage in excess of the threshold voltage is applied to the gate material, a sufficient number of charge carriers are concentrated under the gate oxide such that conduction between the source and drain is possible. Recall that the majority carrier component of the channel current is composed of charge carriers of the type opposite that of the substrate. If the substrate is p-type silicon then the majority carriers are electrons. For n-type silicon substrates, holes are the majority carriers.

The threshold voltage of a MOSFET depends on several transistor properties such as the gate material, the oxide thickness, and the silicon doping levels. The threshold voltage is also dependent upon any fixed charge present between the gate material and the gate oxides. MOSFETs used in most commodity products are normally the "enhancement mode" type. Enhancement mode n-type MOSFETs have a positive threshold voltage and do not conduct appreciable charge between the source and the drain unless the threshold voltage is exceeded. In contrast, "depletion mode" MOSFETs exhibit a negative threshold voltage and are normally conductive. Similarly, there exists enhancement mode and depletion mode p-type MOSFETs. For p-type MOSFETs the sign of the threshold voltage is reversed.

Equations that describe how MOSFET threshold voltage is affected by substrate doping, source and substrate biasing, oxide thickness, gate material, and surface charge density have been derived in the literature.[1–4] Of particular importance is the increase in the threshold voltage associated with a nonzero source-to-bulk voltage. This is known as the "body effect" and is quantified by the equation,

$$V_T = V_{T0} + \gamma\left(\sqrt{|2\phi_F| + V_{SB}} - \sqrt{|2\phi_F|}\right) \tag{6.1}$$

where V_{T0} is the zero-bias threshold voltage, γ is the body effect coefficient or body factor, ϕ_F is the bulk surface potential, and V_{SB} is the bulk-to-source voltage. Details on the calculation of each of the terms in Eq. (6.1) is discussed extensively in the literature.[5]

6.2.3 Operating Regions

MOSFETs exhibit fairly distinct regions of operation depending upon their biasing conditions. In the simplest treatments of MOSFETs, the three operating regions considered are subthreshold, triode, and saturation.

Subthreshold — When the applied gate-to-source voltage is below the device's threshold voltage, the MOSFET is said to be operating in the subthreshold region. For gate voltages below the threshold voltage, the current decreases exponentially toward zero according to the equation

$$i_{DS} = I_{DS0}\frac{W}{L}\exp\left(\frac{v_{GS}}{nkT/q}\right) \tag{6.2}$$

where n is given by

$$n = 1 + \frac{\gamma}{2\sqrt{\phi_j - v_{BS}}} \tag{6.3}$$

in which γ is the body factor, ϕ_j is the channel junction built-in voltage, and v_{BS} is the source-to-bulk voltage.

For radio frequency applications, the MOSFET is normally operated in its saturation region. The reasons for this will become clear later in this section. It should be noted, however, that some researchers feel that it may be advantageous to operate deep-submicron radio frequency MOSFETs in the moderate inversion (subthreshold) region.[6]

Triode — A MOSFET operates in its triode, also called "linear" region, when bias conditions cause the induced channel to extend from the source to the drain. When $V_{GS} > V_T$ the surface under the oxide is inverted and if $V_{DS} > 0$ a drift current will flow from the drain to the source. The drain-to-source voltage is assumed small so that the depletion layer is approximately constant along the length of the channel. Under these conditions, the drain source current for an NMOS device is given by the relation,

$$I_D = \mu_n C'_{ox}\frac{W}{L}\left(\left(V_{GS} - V_T\right)V_{DS} - \frac{V_{DS}^2}{2}\right)\Bigg|_{\substack{V_{GS}\geq V_T \\ V_{DS}\leq V_{GS}-V_T}} \tag{6.4}$$

where μ_n is the electron mobility in the channel and C'_{ox} is the per unit area capacitance over the gate area. Similarly, for PMOS transistors the current relationship is given as,

$$I_D = \mu_p C'_{ox} \frac{W}{L}\left(\left(V_{SG} - V_T\right)V_{SD} - \frac{V_{SD}^2}{2}\right)\Bigg|_{\substack{V_{SG} \geq V_T \\ V_{SD} \leq V_{SG} - V_T}} \tag{6.5}$$

in which the threshold voltage of the p-type MOSFET is assumed to be positive and μ_p is the hole mobility.

Saturation — The conditions $V_{DS} \leq V_{GS} - V_T$ in Eq. (6.4) and $V_{SD} \leq V_{SG} - V_T$ in Eq. (6.5) ensure that the inversion charge is never zero for any point along the channel's length. However, when $V_{DS} = V_{GS} - V_T$ (or $V_{SD} = V_{SG} - V_T$ in PMOS devices) the inversion charge under the gate at the channel-drain junction is zero. The required drain-to-source voltage is called $V_{DS,sat}$ for NMOS and $V_{SD,sat}$ for PMOS.

For $V_{DS} > V_{DS,sat}$ ($V_{SD} > V_{SD,sat}$ for PMOS), the channel charge becomes "pinched off," and any increase in V_{DS} increases the drain current only slightly. The reason that the drain currents will increase for increasing V_{DS} is because the depletion layer width increases for increasing V_{DS}. This effect is called channel length modulation and is accounted for by λ, the channel length modulation parameter. The channel length modulation parameter ranges from approximately 0.1 for short channel devices to 0.01 for long channel devices. Since MOSFETs designed for radio frequency operation normally use minimum channel lengths, channel length modulation is an important concern for radio frequency implementations in CMOS.

When a MOSFET is operated with its channel pinched off, in other words $V_{DS} > V_{GS} - V_T$ and $V_{GS} \geq V_T$ for NMOS (or $V_{SD} > V_{SG} - V_T$ and $V_{SG} \geq V_T$ for PMOS), the device is said to be operating in the saturation region. The corresponding equations for the drain current are given by,

$$I_D = \frac{1}{2}\mu_n C'_{ox} \frac{W}{L}\left(V_{GS} - V_T\right)^2\left(1 + \lambda\left(V_{DS} - V_{DS,sat}\right)\right) \tag{6.6}$$

for long-channel NMOS devices and by,

$$I_D = \frac{1}{2}\mu_p C'_{ox} \frac{W}{L}\left(V_{SG} - V_T\right)^2\left(1 + \lambda\left(V_{SD} - V_{SD,sat}\right)\right) \tag{6.7}$$

for long-channel PMOS devices.

Figure 6.3 illustrates a family of curves typically used to visualize a MOSFET's drain current as a function of its terminal voltages. The drain-to-source voltage spans the operating region while the gate-to-source voltage is fixed at several values. The bulk-to-source voltage has been taken as zero. As shown in Fig. 6.3, when the MOSFET enters the saturation region the drain current is essentially independent of the drain-to-source voltage and so the curve is flat. The slope is not identically zero however, as the drain-to-source voltage does have some effect upon the channel current due to channel modulation effects.

Three simplifying assumptions are advantageous when one is trying to gain an intuitive understanding of the low-frequency operation of a circuit containing MOSFETs. When operated in the linear region, the MOSFET can be treated much like a resistor with terminal voltage V_{DS}. When operated in the saturation region, the MOSFET may be considered a voltage-controlled current source where the controlling voltage is present at the gate. Finally, when operated below the voltage threshold, the MOSFET can be considered an open circuit from the drain to the source.

6.2.4 Nonideal and Short Channel Effects

The equations presented for the subthreshold, triode, and saturation regions of the MOSFET operating characteristic curves do not include the many nonidealities exhibited by MOSFETs. Most of these nonideal behaviors are more pronounced in deep submicron devices such as those employed in radio frequency designs, and so it is important for a designer to be aware of them.

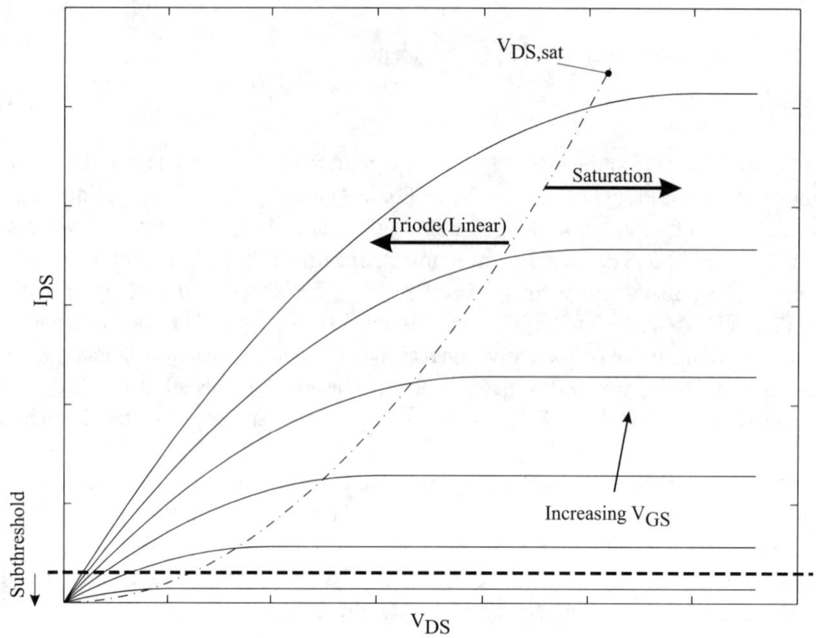

FIGURE 6.3 Characteristic drain-current versus terminal-voltage curves for a MOSFET.

Velocity saturation — Electron and hole mobility are not constants; they are a function of the applied electric field. Above a certain critical electric field strength the mobility starts to decrease, and the drift velocity of carriers does not increase in proportion to the applied electric field. Under these conditions the device is said to be velocity saturated. Velocity saturation has important practical consequences in terms of the current-voltage characteristics of a MOSFET acting in the saturation region. In particular, the drain current of a velocity saturated MOSFET operating in the saturation region is a linear function of V_{GS}. This is in contrast to the results given in Eqs. (6.6) and (6.7). The drain current for a short channel device operating under velocity saturation conditions is given by

$$I_D = \mu_{crit} C'_{ox} W \left(V_{GS} - V_T \right) \tag{6.8}$$

where μ_{crit} is the carrier mobility at the critical electric field strength.

Drain-induced barrier lowering — A positive voltage applied to the drain terminal helps to attract electrons under the gate oxide region. This increases the surface potential and causes a threshold voltage reduction. Since the threshold decreases with increasing V_{DS}, the result is an increase in drain current and therefore an effective decrease in the MOSFET's output resistance. The effects of drain-induced barrier lowering are reduced in modern CMOS processes by using lightly doped drain (LDD) structures.

Hot carriers — Velocity saturated charge carriers are often called hot carriers. Hot carriers can potentially tunnel through the gate oxide and cause a gate current, or they may become trapped in the gate oxide. Hot carriers that become trapped in the gate oxide change the device threshold voltage. Over time, if enough hot carriers accumulate in the gate oxide, the threshold voltage is adjusted to the point that analog circuitry performance is severely degraded. Therefore, depending upon the application, it may be unwise to operate a device so that the carriers are velocity saturated since the reliability and lifespan of the circuit is degraded.

6.2.5 Small Signal Models

Small signal equivalent circuits are useful when the voltage and current waveforms in a circuit can be decomposed into a constant level plus a small time-varying offset. Under these conditions, a circuit can be linearized about its DC operating point. Nonlinear components are replaced with linear components that reflect the bias conditions.

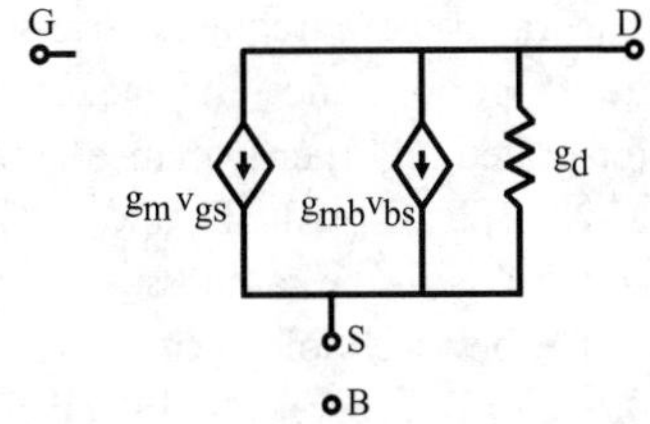

FIGURE 6.4 Low-frequency small signal MOSFET model.

The low-frequency small signal model for a MOSFET is shown in Fig. 6.4. Only the intrinsic portion of the transistor is considered for simplicity. As shown, the small signal model consists of three components: the small signal gate transconductance, g_m; the small signal substrate transconductance, g_{mb}; and the small signal drain, g_d. Mathematically, these three components are defined by,

$$g_m = \left.\frac{\partial I_D}{\partial V_{GS}}\right|_{V_{BS}, V_{DS} \text{ constant}} \tag{6.9}$$

$$g_{mb} = \left.\frac{\partial I_D}{\partial V_{BS}}\right|_{V_{GS}, V_{DS} \text{ constant}} \tag{6.10}$$

and

$$g_d = \left.\frac{\partial I_D}{\partial V_{DS}}\right|_{V_{GS}, V_{BS} \text{ constant}} \tag{6.11}$$

Equations (6.9), (6.10), and (6.11) can be evaluated using the relationships given earlier. For the saturation region, the small signal transconductances and the drain conductance are given by,

$$g_m = \mu C'_{ox} \frac{W}{L}\left(V_{GS} - V_T\right) \tag{6.12}$$

$$g_{mb} = g_m \eta \tag{6.13}$$

and

$$g_d = \frac{1}{2}\mu C'_{ox} \frac{W}{L}\left(V_{GS} - V_T\right)^2 \lambda \tag{6.14}$$

where η in Eq. (6.13) is a factor that describes how the threshold voltages changes with reverse body bias. For small V_{BS}, $\eta \approx 0$.

The small signal model shown in Fig. 6.4 is only valid at very low frequencies. At higher frequencies capacitances present in the MOSFET must be included in the small signal model, and at radio frequencies distributed effects must be taken into account. In the next section these two factors are explored and the small signal model is revised.

6.3 CMOS at Radio Frequencies

Integrated radio frequency transceiver design is an evolving field, particularly in terms of understanding device performance and maximizing integration level. Typical commercial implementations of highly-integrated high-performance wireless transceivers use a mixture of technologies, including CMOS, BiCMOS, BJTs, GaAs FETs, and HBTs. Regardless of the technology, all radio frequency integrated circuits contend with the same issues of noise, linearity, gain, and efficiency.

The best technology choice for an integrated radio frequency application must weigh the consequences of wafer cost, level of integration, performance, economics, and time to market. These requirements often lead designers into using several technologies within one transceiver system. Partitioning of transceiver functionality according to technology implies that the signal must go on-chip and off-chip at several locations. Bringing the signal off-chip and then on-chip again complicates the transceiver design because proper matching at the output and input terminals is required. Also, bringing the signal off-chip implies that power requirements are increased because it takes more power to drive an off-chip load than to keep the signal completely on the same integrated circuit. Generally, taking the signal off and then on-chip results in signal power loss accompanied by an undesirable increase in noise figure.

Recent trends apply CMOS to virtually the entire transceiver design, since CMOS excels in its level of integration. The level of integration offered by a particular technology determines the required die size, which in turn affects both the cost and the physical size of the final packaged circuitry. CMOS technology currently has the performance levels necessary to operate in the 900 MHz to 2.4 GHz frequency range, which is important for existing cellular and wireless network applications. Upcoming technologies should be able to operate in the 5 GHz ISM band, which is seen as the next important commodity frequency. CMOS devices manufactured with gate lengths of 0.18 μm will function at these frequencies, albeit with generous biasing currents. Future generations of CMOS scaled below the 100 nm gate length range are anticipated to provide the performance required to operate beyond the 5 GHz frequency range for receiver applications.

6.3.1 High Frequency Modeling

The majority of existing analog MOSFET models predate the use of CMOS in radio frequency designs and generally are unable to predict the performance of MOSFETs operating at microwave frequencies with the accuracy required and expected of modern simulators. These modeling shortcomings occur on two fronts. In the first case, existing models do not properly account for the distributed nature of the MOSFET, meaning that at high frequencies, geometry-related effects are ignored. In the second case, existing models do not properly model MOSFET noise at high frequencies. Typical problems with noise modeling include:

- not accounting for velocity-saturated carriers within the channel and the associated noise;
- discounting the significant thermal noise generated by the distributed gate resistance;
- ignoring the correlation between the induced gate noise and the MOSFET drain noise.

Accurate noise modeling is extremely important when low noise operation is essential, such as in a front-end low noise amplifier.

6.3.2 High Frequency Operation

MOSFET dimensions and physical layout are important determining factors for high frequency performance. As MOSFET operating frequencies approach several hundred MHz, the MOSFET can no longer be considered a lumped device. The intrinsic and extrinsic capacitance, conductance, and resistance are all distributed according to the geometry and physical layout of the MOSFET. The distributed nature of the MOSFET operating at high frequencies is particularly important for the front-end circuitry in a receiver, such as in the low noise amplifier and first stage mixer input MOSFETs. The devices used in

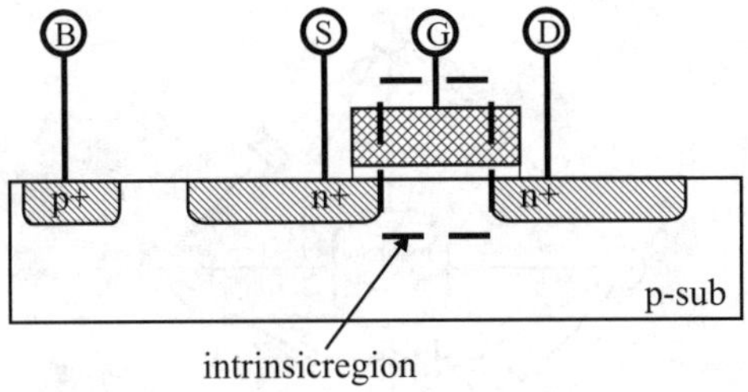

FIGURE 6.5 MOSFET intrinsic and extrinsic capacitance boundary.

these portions of the input circuitry are normally large, with high W/L ratios. Large W/L ratios are required because of the inherently low transconductance offered by CMOS, and in order to realize reasonable gain, the devices are therefore relatively wide compared to more conventional analog circuit layouts. Additionally, minimum gate lengths are preferred because the maximum operating frequency of the MOSFET scales as $1/L^2$. Shorter channels imply higher frequency because the time it takes the carriers to move from drain to source is inversely proportional to the length of the channel. Also, the mobility of the carriers is proportional to the electric field strength. Since the electric field strength along the length of the channel is inversely proportional to the distance between the source and the drain, the carrier mobility is inversely proportional to the length of the channel. Combined, these two effects have traditionally allowed the maximum operating frequency of the MOSFET to scale as $1/L^2$. It must be noted that in modern deep submicrometer MOSFETs experiencing velocity saturation, the maximum operating frequency no longer scales as $1/L^2$, but more closely to $1/L$. In any event, for maximum operating frequency, the device channel length should be the minimum allowable.

Since the gate width in RF front-end MOSFET devices is typically on the order of several hundred microns, the gate acts as a transmission line along its width. The gate acting as a transmission line is modeled similarly to a microstrip transmission line and can be analyzed by utilizing a distributed circuit model for transmission lines. Normally a transmission line is viewed as a two-port network in which the transmission line receives power from the source at the input port (source end) and delivers the power to the load of the output port (load end). In order to apply transmission line analysis to the gate of a MOSFET along its width, the width of the MOSFET gate is divided into many identical sections of incremental width Δx. Each portion of the transmission line with the width Δx is modeled by a resistance "R" per unit width, an inductance "L" per unit width, a capacitance "C" per unit width, and a conductance "G" per unit width. Normally the transmission line is assumed to be uniform, so that these parameters are constants along the transmission line's width. When analyzing signal propagation along the MOSFET gate width, it is important to note that there is no single output node. The transmission line cannot be treated as a two-port, since the gate couples to the channel in a distributed fashion.

6.3.3 Important Parasitics and Distributed Effects

Parasitic capacitances — At high operating frequencies the effects of parasitic capacitances on the operation of the MOSFET cannot be ignored. Transistor parasitic capacitances are subdivided into two general categories; extrinsic capacitances and intrinsic capacitances. The extrinsic capacitances are associated with regions of the transistor outside the dashed line shown in Fig. 6.5, while the intrinsic capacitances are all those capacitances located within the region illustrated in Fig. 6.5.

Extrinsic capacitances — Extrinsic capacitances are modeled by using small-signal lumped capacitances, each of which is associated with a region of the transistor's geometry. Seven small-signal capacitances are used, one capacitor between each pair of transistor terminals, plus an additional capacitor between the well and the bulk if the transistor is fabricated in a well. Figure 6.6(a) illustrates the seven extrinsic transistor capacitances added to an intrinsic small signal model, and Fig. 6.6(b) assigns a location to each capacitance within the transistor structure. In order of importance to high frequency performance, the extrinsic capacitances are as follows:

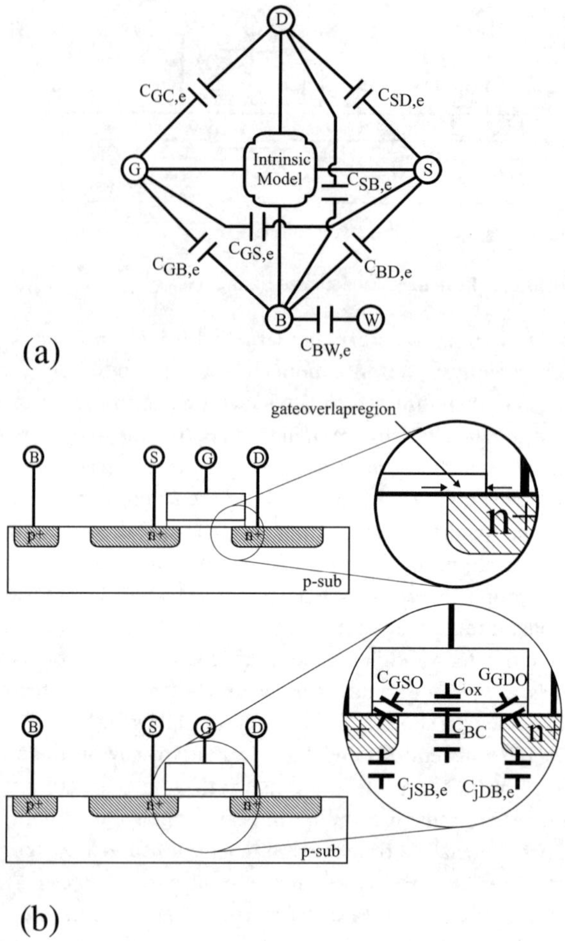

FIGURE 6.6 Extrinsic capacitances. (a) Lumped terminal capacitances added to the intrinsic model, (b) physical location of extrinsic capacitances.

Gate overlap capacitances — Although MOSFETs are manufactured using a self-aligned process, there is still some overlap between the gate and the source and the gate and the drain. This overlapped area gives rise to the gate overlap capacitances denoted by C_{GSO} and C_{GDO} for the gate-to-source overlap capacitance and the gate-to-drain overlap capacitance, respectively. Both capacitances C_{GSO} and C_{GDO} are proportional to the width, W, of the device and the amount that the gate overlaps the source and the drain, typically denoted as "LD" in SPICE parameter files. The overlap capacitances of the source and the drain are often modeled as linear parallel plate capacitors, since the high dopant concentration in the source and drain regions and the gate material implies that the resulting capacitance is largely bias independent. However, for MOSFETs constructed with a lightly doped drain (LDD-MOSFET), the overlap capacitances can be highly bias dependent and therefore nonlinear. For a treatment of overlap capacitances in LDD-MOSFETs, refer to Klein.[7] For non-lightly doped drain MOSFETs, the gate-drain and gate-source overlap capacitances are given by the expression $C_{GSO} = C_{GDO} = W\,LD\,C_{ox}$, where C_{ox} is the thin oxide field capacitance per unit area under the gate region.

When the overlap distances are small, fringing field lines add significantly to the total capacitance. Since the exact calculation of the fringing capacitance requires an accurate knowledge of the drain and source region geometry, estimates of the fringing field capacitances based on measurements are normally used.

TABLE 6.1 MOSFET Extrinsic Junction Capacitances

Extrinsic Source Capacitance	$C_{jBS,e} = C'_{jBS}A_S + C''_{jswBS}P_S$	$C'_{jBS} = \dfrac{C'_j}{\left(1-\dfrac{V_{BS}}{\phi_j}\right)^{m_j}}$	$C''_{jswBS} = \dfrac{C''_{jsw}}{\left(1-\dfrac{V_{BS}}{\phi_{jsw}}\right)^{m_{jsw}}}$
Extrinsic Drain Capacitance	$C_{jBD,e} = C'_{jBD}A_D + C''_{jswBD}P_D$	$C'_{jBD} = \dfrac{C'_j}{\left(1-\dfrac{V_{BD}}{\phi_j}\right)^{m_j}}$	$C''_{jswBD} = \dfrac{C''_{jsw}}{\left(1-\dfrac{V_{BD}}{\phi_{jsw}}\right)^{m_{jsw}}}$
Extrinsic Well Capacitance	$C_{jBW,e} = C'_{jBW}A_W + C''_{jswBW}P_W$	$C'_{jBW} = \dfrac{C'_j}{\left(1-\dfrac{V_{BW}}{\phi_j}\right)^{m_j}}$	$C''_{jswBW} = \dfrac{C''_{jsw}}{\left(1-\dfrac{V_{BW}}{\phi_{jsw}}\right)^{m_{jsw}}}$

Notes: m_j and m_{jsw} are process dependent, typically 1/3 … 1/2.

$$C'_j = \sqrt{\frac{\varepsilon_{si} q N_B}{2\phi_j}}$$

Where: ϕ_j and ϕ_{jsw} are the built-in junction potential and side wall junction potentials, respectively. ε_{si} is the dielectric constant of silicon, q is the electronic charge constant, and N_B is the bulk dopant concentration.

$$C''_{jsw} = \sqrt{\frac{\varepsilon_{si} q N_B}{2\phi_{jsw}}}$$

A_S, A_D, and A_W are the source, drain, and well areas, respectively.
P_s, P_d, and P_w are the source, drain, and well perimeters, respectively. The source and drain perimeters do not include the channel boundary.

Extrinsic junction capacitances — The bias-dependent junction capacitances that must be considered when evaluating the extrinsic lumped-capacitance values are illustrated in Fig. 6.6(a) and summarized in Table 6.1. At the source region there is a source-to-bulk junction capacitance, $C_{jBS,e}$, and at the drain region there is a drain-to-bulk junction capacitance, $C_{jBD,e}$. These capacitances can be calculated by splitting the drain and source regions into a "side wall" portion and a "bottom wall" portion. The capacitance associated with the side wall portion is found by multiplying the length of the side wall perimeter (excluding the side contacting the channel) by the effective side wall capacitance per unit length. Similarly, the capacitance for the bottom wall portion is found by multiplying the area of the bottom wall by the bottom wall capacitance per unit area. Additionally, if the MOSFET is in a well, a well-to-bulk junction capacitance, $C_{jBW,e}$, must be added. The well-bulk junction capacitance is calculated similar to the source and drain junction capacitances, by dividing the total well-bulk junction capacitance into side wall and bottom wall components. If more than one transistor is placed in a well, the well-bulk junction capacitance should only be included once in the total model.

Both the effective side wall capacitance and the effective bottom wall capacitance are bias dependent. Normally the per unit length zero-bias, side wall capacitance and the per unit area zero-bias, bottom wall capacitance are estimated from measured data. The values of these parameters for nonzero reverse-bias conditions are then calculated using the formulas given in Table 6.1.

Extrinsic source-drain capacitance — Accurate models of short channel devices may include the capacitance that exists between the source and drain region of the MOSFET. As shown in Fig. 6.6(a), the source-drain capacitance is denoted as $C_{sd,e}$. Although the source-drain capacitance originates in the region within the dashed line in Fig. 6.5, it is still referred to as an extrinsic capacitance.[1] The value of this capacitance is difficult to calculate because its value is highly dependent upon the source and drain geometries. For longer channel devices, $C_{sd,e}$ is very small in comparison to the other extrinsic capacitances, and is therefore normally ignored.

Extrinsic gate-bulk capacitance — As with the gate-to-source and gate-to-drain overlap capacitances, there is a gate-to-bulk overlap capacitance caused by imperfect processing of the MOSFET. The parasitic gate-bulk capacitance, $C_{GB,e}$, is located in the overlap region between the gate and the substrate (or well) material outside the channel region. The parasitic extrinsic gate-bulk capacitance is extremely small in comparison to the other parasitic capacitances. In particular, it is negligible in comparison to the intrinsic gate-bulk capacitance. The parasitic extrinsic gate-bulk capacitance has little effect on the gate input impedance and is therefore generally ignored in most models.

Intrinsic capacitances — Intrinsic MOSFET capacitances are significantly more complicated than extrinsic capacitances because they are a strong function of the voltages at the terminals and the field distributions within the device. Although intrinsic MOSFET capacitances are distributed throughout the device, for the purposes of simpler modeling and simulation, the distributed capacitances are normally represented by lumped terminal capacitances. The terminal capacitances are derived by considering the change in charge associated with each terminal with respect to a change in voltage at another terminal, under the condition that the voltage at all other terminals is constant. The five intrinsic small signal capacitances are therefore expressed as,

$$C_{gd,i} = \left.\frac{\partial Q_G}{\partial V_D}\right|_{V_G,V_S,V_B} \tag{6.15}$$

$$C_{gs,i} = \left.\frac{\partial Q_G}{\partial V_S}\right|_{V_G,V_D,V_B} \tag{6.16}$$

$$C_{bd,i} = \left.\frac{\partial Q_B}{\partial V_D}\right|_{V_G,V_S,V_B} \tag{6.17}$$

$$C_{bs,i} = \left.\frac{\partial Q_B}{\partial V_D}\right|_{V_G,V_D,V_B} \tag{6.18}$$

and

$$C_{gb,i} = \left.\frac{\partial Q_G}{\partial V_B}\right|_{V_G,V_S,V_D} \tag{6.19}$$

These capacitances are evaluated in terms of the region of operation of the MOSFET, which is a function of the terminal voltages. Detailed models for each region of operation were investigated by Cobbold.[8] Simplified expressions are given in Table 6.2, for the triode and saturation operating regions.

TABLE 6.2 Intrinsic MOSFET Capacitances

Operating Region	$C_{gs,i}$	$C_{gd,i}$	$C_{gb,i}$	$C_{bs,i}$	$C_{bd,i}$
Triode	$\approx \frac{1}{2} C_{ox}$	$\approx \frac{1}{2} C_{ox}$	≈ 0	$k_0 C_{ox}$	$k_0 C_{ox}$
Saturation	$\approx \frac{2}{3} C_{ox}$	≈ 0	$K_1 C_{ox}$	$K_2 C_{ox}$	≈ 0

Notes: Triode region approximations are for $V_{DS} = 0$.
k_0, k_1, and k_2 are bias dependent. See [1].

The total terminal capacitances are then given by combining the extrinsic capacitances and intrinsic capacitances according to,

$$
\begin{aligned}
C_{gs} &= C_{gs,i} + C_{gs,e} = C_{gs,i} + C_{gso} \\
C_{gd} &= C_{gd,i} + C_{gd,e} = C_{gd,i} + C_{gdo} \\
C_{gb} &= C_{gb,i} + C_{gb,e} = C_{gb,i} + C_{gbo} \\
C_{sb} &= C_{bs,i} + C_{sb,e} = C_{bs,i} + C_{jsb} \\
C_{db} &= C_{bd,i} + C_{db,e} = C_{bd,i} + C_{jdb}
\end{aligned}
\tag{6.20}
$$

in which the small-signal form of each capacitance has been used.

The contribution of the total gate-to-channel capacitance, C_{GC}, to the gate-to-drain and gate-to-source capacitances is dependent upon the operating region of the MOSFET. The total value of the gate-to-channel capacitance is determined by the per unit area capacitance C_{ox} and the effective area over which the capacitance is taken. Since the extrinsic overlap capacitances include some of the region under the gate, this region must be removed when calculating the gate-to-channel capacitance. The effective channel length, L_{eff}, is given by $L_{eff} = L - 2LD$ so that the gate-to-channel capacitance can be calculated by the formula $C_{GC} = C_{ox} W L_{eff}$. The total value of the gate-to-channel capacitance is apportioned to both the drain and source terminals according to the operating region of the device. When the device is in the triode region, the capacitance exists solely between the gate and the channel and extends from the drain to the source. Its value is therefore evenly split between the terminal capacitances C_{gs} and C_{gd} as shown in Table 6.2. When the device operates in the saturation region, the channel does not extend all the way from the source to the drain. No portion of C_{GC} is added to the drain terminal capacitance under these circumstances. Again, as shown in Table 6.2, analytical calculations demonstrated that an appropriate amount of C_{GS} to include in the source terminal capacitance is 2/3 of the total.[1]

Finally, the channel to bulk junction capacitance, C_{BC}, should be considered. This particular capacitance is calculated in the same manner as the gate-to-channel capacitance. Also similar to the gate-to-channel capacitance proportioning between the drain in the source when calculating the terminal capacitances, the channel-to-bulk junction capacitance is also proportioned between the source-to-bulk and drain-to-bulk terminal capacitances, depending on the region of operation of the MOSFET.

Wiring capacitances — Referring to Fig. 6.1, one can see that the drain contact interconnect overlapping the field oxide and substrate body forms a capacitor. The value of this overlap capacitance is determined by the overlapping area, the fringing field, and the oxide thickness. Reduction of the overlapping area will decrease the capacitance to a point, but with an undesirable increase in the parasitic resistance at the interconnect to MOSFET drain juncture. The parasitic capacitance occurring at the drain is particularly troublesome due to the Miller effect, which effectively magnifies the parasitic capacitance value by the gain of the device. The interconnects between MOSFET devices also add parasitic capacitive loads to the each device. These interconnects may extend across the width of the IC in the worst case, and must be considered when determining the overall circuit performance.

Modern CMOS processes employ thick field-oxides that reduce the parasitic capacitance that exists at the drain and source contacts, and between interconnect wiring and the substrate. The thick field-oxide also aids in reducing the possibility of unintentional MOSFET operation in the field region.

Distributed gate resistance — Low-frequency MOSFET models treat the gate as purely capacitive. This assumption is invalid for frequencies beyond approximately 1 GHz, because the distributed gate resistance is typically larger than the capacitive reactance present at the gate input for frequencies beyond 1 GHz.

The impact of the distributed gate resistance upon the high frequency performance of MOSFETs has been investigated both experimentally and analytically by several researchers.[1,6,9–15] The distributed gate resistance affects the radio frequency performance of the MOSFET in three primary ways. In the first

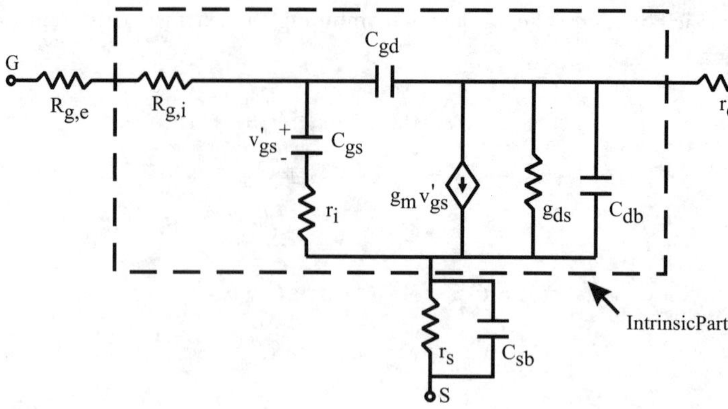

FIGURE 6.7 MOSFET small-signal high-frequency model.

case, discounting the gate resistance causes nonoptimal power matching to off-chip source impedances. In the second case, discounting the distributed gate resistance in noise figure calculations causes an underestimation of the noise figure of the transistor. Finally, in the third case, since the power gain of the MOS transistor is strongly governed by the gate resistance, discounting the gate resistance causes an overestimation of the MOSFET's available power gain and maximum oscillation frequency. The gate resistance of MOSFET transistors therefore deserves important consideration during the design phase of integrated RF CMOS receivers.

Nonzero gate resistances have been factored into recent successful designs. Rofougaran et al.[16] noted that matching, input noise, and voltage gain are all ultimately limited by transistor imperfections such as gate resistance. The effects were most recently quantified by Enz.[6]

Channel charging resistance — Charge carriers located in the channel cannot instantaneously respond to changes in the MOSFET gate-to-source voltage. The channel charging resistance is used to account for this non-quasi-static behavior along the channel length. In Bagheri et al.,[15] the channel charging resistance was shown to be inversely proportional to the MOSFET transconductance, $r_i \approx (kg_m)^{-1}$. For long channel devices, with the distributed nature of the channel resistance between the source and drain taken into account, the constant of proportionality, k, was shown to equal five. Measurements of short channel devices indicate that the proportionality constant can go as low as one.

The channel charging resistance of a MOSFET is important because it strongly influences the input conductance and the forward transconductance parameters of the device. Both the input conductance and the forward transconductance are monotonically decreasing functions of the channel charging resistance. Since the transconductance of even a large MOSFET is small, on the order of 10 mS, the charging resistance of typical front-end transistors is large, potentially on the order of hundreds of ohms.

Transconductance delay — MOSFET transconductance does not respond instantaneously to changes in gate voltage. The time it takes for the charge in the channel to be redistributed after an excitation of the gate voltage is dictated by the time constant due to the gate-to-source capacitance and the channel charging resistance. This time constant is denoted as τ, and is given by the expression $\tau \approx r_i C_{gs}$. The transconductance delay is generally ignored for frequencies less than $2\pi/\tau$.

6.3.4 Small Signal Models

Several high frequency small signal models incorporating the effects described in the previous sections have been proposed in the literature. The small signal model presented in Fig. 6.7 is useful for MOSFETs operating in saturation in a common-source configuration. Evident in the figure are the various lumped terminal capacitances, the drain conductance, and the output transconductance. Note that the transconductance g_{mb} is taken as zero because $V_{SB} = 0$ is assumed. Also evident in the model are the high-frequency related elements, namely the charging resistance, the transconductance delay, and the extrinsic and intrinsic gate resistances.

TABLE 6.3 MOSFET Small-Signal Y-Parameters

Parameter	Y-Parameters for the Intrinsic MOSFET Model of Fig. 6.7	Distributed MOSFET Y-Parameters*
y_{11}	$\dfrac{s(\kappa+C_{gd})}{1+s(\kappa+C_{gd})R_g}$	$\dfrac{s(C_{gs}+C_{gd})+s^2C_{gs}C_{gd}r_i}{1+sC_{gs}r_i}\dfrac{\tanh(\gamma W)}{\gamma W}$
y_{12}	$\dfrac{-sC_{gd}}{1+s(\kappa+C_{gd})R_{g,i}}$	$-sC_{gd}\dfrac{\tanh(\gamma W)}{\gamma W}$
y_{21}	$\dfrac{g_m-sC_{gd}}{1+s(\kappa+C_{gd})R_{g,i}}$	$\left(\dfrac{g_m}{1+sC_{gs}r_i}-sC_{gd}\right)\dfrac{\tanh(\gamma W)}{\gamma W}$
y_{22}	$g_d+sC_{gd}+\dfrac{(g_m-s(\kappa-C_{gd}))sC_{gd}R_{g,i}}{1+s(\kappa+C_{gd})R_{g,i}}$	$g_d+sC_{gd}+\dfrac{\left(\dfrac{g_m}{1+sC_{gs}r_i}-sC_{gd}\right)C_{gd}}{C_{gd}+\dfrac{C_{gs}}{1+sC_{gs}r_i}}\left(1-\dfrac{\tanh(\gamma W)}{\gamma W}\right)$
Notes:	$\kappa=\dfrac{C_{gs}}{1+sC_{gs}r_i}$	$\gamma=\dfrac{\sqrt{sC_{gd}r_i\dfrac{sC_{gs}r_i}{1+sC_{gs}r_i}}}{W}$

**Source:* Abou-Allam, E. and Manku, T., An Improved Transmission-Line Model for MOS Transistors, *IEEE Transactions on Circuits and Systems II: Analog and Digital Signal Processing,* 46, 1380–1387, Nov. 1999.

6.3.5 MOSFET Small Signal Y-parameters

Small-signal y-parameters are useful in radio frequency design work involving MOSFETs. As discussed later in this section, radio frequency MOSFETs are typically laid out in a "fingered" style, and if the y-parameters are found for a single finger, they are easily combined for the complete device.

The y-parameters corresponding to the small signal equivalent circuit shown in Fig. 6.7 have been evaluated and appear in Table 6.3. These y-parameters cannot accurately portray the distributed nature of the MOSFET at high frequencies because the model presented in Fig. 6.7 is composed of lumped elements. Recall that the gate resistance in MOS transistors operating at GHz frequencies in conjunction with intrinsic and extrinsic device capacitances acts as a distributed RC network. This is shown schematically in Fig. 6.8, where a MOSFET is represented as a network of smaller MOSFETs, interconnected by gate material.

Several models developed over the last three decades incorporate high frequency and distributed effects. Noise arising from the distributed gate was modeled by Jindal,[12] but high frequency effects were not incorporated in this model. Distributed geometry effects were recognized as important,[14,15] but the distributed nature of the gate was not fully explored analytically.[10] As the viability of CMOS in RF transceiver applications improved, significant progress in modeling wide devices, such as those required for RF applications, was made by Kim et al.[11] and Razavi et al.,[9] in which wide transistors were treated as arrays of smaller transistors interconnected by resistors representing the gate. Recently in a paper by Abou-Allam,[17,18] a closed-form, small signal model incorporating the distributed gate for wide transistors was derived, taking into account the distributed nature of the gate resistance and intrinsic capacitances. The y-parameters developed in Abou-Allam's paper[17] appear here in Table 6.3.

A parallel between the results presented by Razavi et al.[9] and Abou-Allam[18] was drawn in a paper by Tin et al.,[19] in which a useful small signal lumped circuit model was presented and leads to the model here in Fig. 6.9. The lumped model shown in Fig. 6.9 incorporates the distributed effects represented by the $\tanh(\gamma W)/\gamma W$ factor within the expressions for the y-parameters presented in Table 6.3.

The distributed gate resistance appears as a lumped resistor of value $R_g/3$ and the distributed intrinsic capacitances appear as a lumped capacitor with value $C_g/5$. It is important to note that these expressions

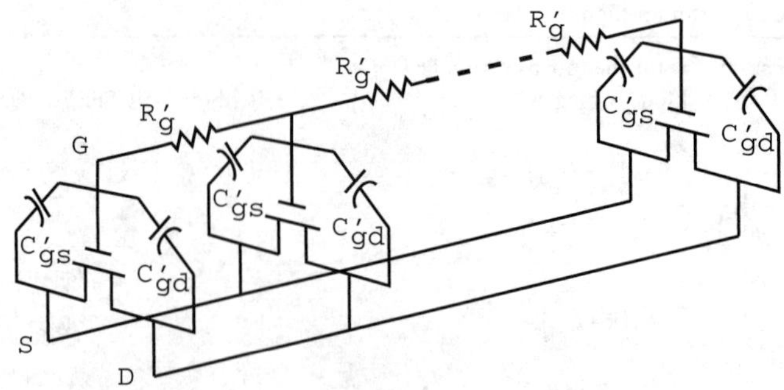

FIGURE 6.8 Illustration of distributed MOSFET network model.

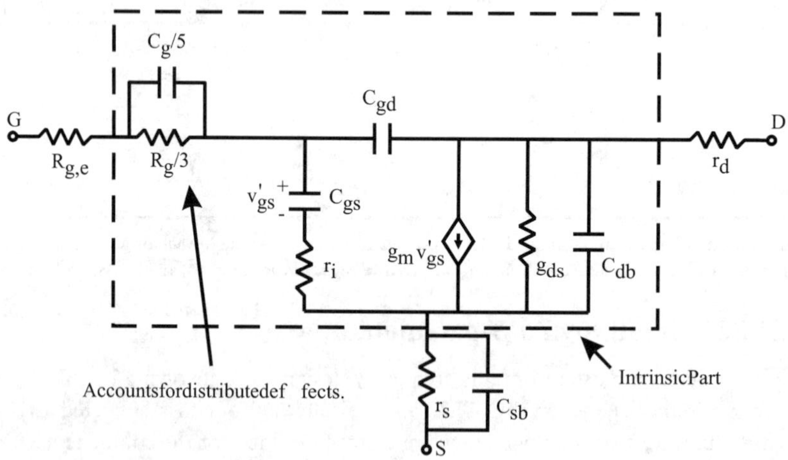

FIGURE 6.9 A lumped small-signal MOSFET model that accounts for distributed gate resistance effects.

were derived for a gate connected from one end only. For example, when the gate is connected from both ends, the equivalent resistor changes to $R_g/12$.

The performance limitations imposed by distributed effects at radio frequencies was summarized by Manku.[20] Analysis of a two-port constructed from the y-parameters as given in Table 6.3 for a traditional MOSFET yields several important device performance metrics, as now discussed.

6.3.6 Unity Current Gain Frequency: f_t

The unity current gain frequency is defined as the signal input frequency at which the extrapolated small-signal current gain of the MOSFET equals one. The small-signal current gain is defined as the amplitude of the small-signal drain current to the small-signal gate current. The symbol used in the literature to denote the unity current gain frequency is f_t and is read as the "transit frequency."

The unity current gain frequency, or transit frequency, is used as a benchmark to describe the speed of the intrinsic device. The performance of the complete device, which includes the additional effects of the extrinsic parasitics, is always lower. For linear amplifier configurations, the small-signal unity current gain is of primary concern since it determines the maximum achievable gain-bandwidth product of the amplifier. Small signal linear two-port models are useful for estimating the unity current gain frequency. The value of f_t is most easily found from the y-parameters using the relation,[13]

$$\left|\frac{y_{21}}{y_{11}}\right| = 1 \tag{6.21}$$

which holds when the current gain is unity. From y-parameters given in Table 6.3, the value of f_t is found as,

$$f_t = \frac{g_m}{2\pi\sqrt{C_g^2 - \left(g_m R_{gi} C_{gs} - C_{gs}\right)^2}} \approx \left.\frac{g_m}{2\pi C_g}\right|_{C_g = C_{gs} + C_{gd}} \tag{6.22}$$

where $C_g^2 \gg (g_m R_{gi} C_{gs} - C_{gs})^2$ is assumed.

Note that the unity current gain frequency is independent of the distributed gate resistance. The most important determining factors of the unity current gain frequency are the device transconductance and the gate-to-source and gate-to-drain capacitances. Since the device f_t is directly proportional to g_m, the analog circuit designer can trade power for speed by increasing the device bias current and therefore g_m. Note that f_t cannot be arbitrarily increased by an increase in drain-source bias current; eventually g_m becomes independent of I_{DS}, and therefore f_t becomes independent of the bias current. MOSFETs used at radio frequencies are normally operated in saturation, because g_m is maximum for a given device in the saturation region. Recall that g_m can also be increased by increasing the width of the transistor, but this will not increase f_t because the parasitic capacitance C_g is proportional to the width of the device.

The transit frequency must not be confused with another often used performance metric, the "intrinsic cutoff frequency," f_τ. While both the "t" in f_t and the "τ" in f_τ refer to the carrier transit time along the length of the device channel, the two symbols have decidedly different meanings. The intrinsic cutoff frequency is given by $f_{\tau = (2\pi\tau)^{-1}}$ where τ is the mean transit time of the carriers along the channel length. Typically, f_τ is five or six times larger than f_t.[6]

6.3.7 Maximum Available Power Gain: G_{max}

Maximum port-to-port power gain within a device occurs when both the input and the output ports are matched to the impedance of the source and the load, respectively. The maximum available power gain of a device, G_{max}, provides a fundamental limit on how much power gain can be achieved with the device. The maximum available power gain achievable within a linear two-port network can be described in terms of the two-port y-parameters as,[13]

$$G_{max} = \frac{1}{4}\frac{\left|y_{21} - y_{12}\right|^2}{\mathrm{Re}\left(y_{11}\right)\mathrm{Re}\left(y_{22}\right) - \mathrm{Re}\left(y_{21}\right)\mathrm{Re}\left(y_{12}\right)} \tag{6.23}$$

where the two-port system is assumed to be unconditionally stable. Treating a common-source MOSFET as a two-port and using the y-parameters shown in Table 6.3 and the relation for f_t given in Eq. (6.22), the MOSFET G_{max} is derived in terms of the intrinsic device parameters as,

$$\begin{aligned} G_{max} &\approx \frac{\left(f_t/f\right)^2}{4R_{g,i}\left(g_{ds} + g_m C_{gd}/C_g\right) + 4r_i g_{ds}} \\ &\approx \frac{f_t}{8\pi R_{g,i} C_{gd} f^2} \end{aligned} \tag{6.24}$$

in which the simplifying assumption $(R_{g,i} + r_i)g_{ds} \ll g_m R_{g,i} C_{gd}/C_g$ is made. To first order, the maximum achievable power gain of a MOSFET is proportional to the device's f_t, and the maximum achievable power gain is inversely proportional to the device's intrinsic gate resistance, $R_{g,i}$. Note that linear amplification is assumed in Eqs. (6.23) and (6.24). Equation (6.24) therefore applies to small-signal analyses such as for a receiver front-end, and would possibly be inadequate for describing the maximum power gain achievable using a nonlinear power amplifier in the transmitter portion of a transceiver.

6.3.8 Unity Power Gain Frequency: f_{max}

The third important figure of merit for MOSFET transistors operating at radio frequencies is the maximum frequency of oscillation, f_{max}. This is the frequency at which the maximum available power gain of the transistor is equal to one. An estimate of f_{max} for MOSFET transistors operating in saturation is found from Eq. (6.24) by setting $G_{max} = 1$ and is given by,

$$f_{max} = \sqrt{\frac{f_t}{8\pi R_{g,i} C_{gd}}} \tag{6.25}$$

From a designer's perspective, f_{max} can be optimized by a combination of proper device layout and careful choice of the operating bias point. From Eq. (6.22) $f_t \propto g_m$, therefore $f_{max} \propto \sqrt{g_m}$, and hence g_m should be maximized. Maximizing g_m requires some combination of increasing the gate-source overdrive, increasing device bias current I_{DS}, or increasing the device width. Note that increasing the device width will increase $R_{g,i}$ and reduce f_{max} unless a fingered layout is used.

6.4 MOSFET Noise Sources

The intrinsic and extrinsic noise sources in a MOSFET operating at microwave frequencies are predominantly thermal in origin. The extrinsic noise arises from the parasitic resistances found at the MOSFET terminal connections, and the metal to semiconductor junctions found at the contacts at these terminals. The intrinsic noise arises from three important sources:

- Drain channel noise, $\overline{i_d^2}$, which is the noise generated by the carriers in the channel region and appears as a noise current;
- Gate resistance noise, $\overline{v_{rg}^2}$, which is the thermal noise due to the resistance of the distributed gate material;
- Induced gate noise, $\overline{i_g^2}$, which is a gate noise current that is capacitively coupled onto the gate from the distributed noise generated by the carriers in the channel and the distributed channel charging resistance. Induced gate noise is one of the main components of noise within the intrinsic portion of a MOSFET transistor.

6.4.1 MOSFET Noise Models

The noise sources discussed previously can be added to an appropriate lumped network model of a MOSFET transistor. The small-signal equivalent circuit shown in Fig. 6.10(a) incorporates the noise sources that are important at radio frequencies. The model is essentially the same model as presented in Fig. 6.7 except for the inclusion of the thermally generated noise. It is possible to refer all of the internal MOSFET noise sources to the gate input as shown in Fig. 6.10(b). This procedure is discussed in a later section for optimum noise matching. The various noise sources are now described.

6.4.2 Gate Resistance Noise

The resistance of the gate material, given by $R_g = (W \times R_\square)/(N \times L)$, where $R_\square$ is the sheet resistance of the gate material, contributes to the thermal noise present in the device. The lumped resistor shown at

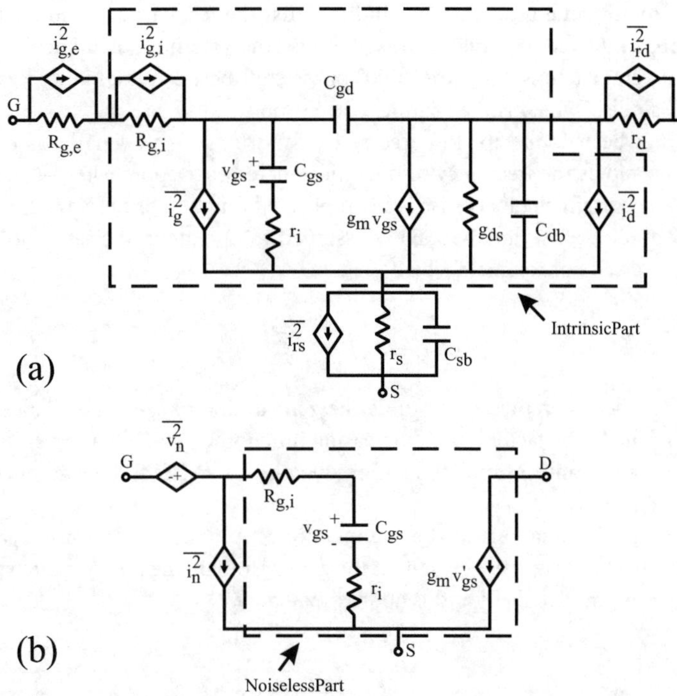

FIGURE 6.10 MOSFET small-signal noise model. (a) High-frequency small-signal model augmented with noise sources, (b) input-referred noise model.

the gate of the MOSFET in Figs. 6.7 and 6.10(a) is intended to represent the thermal noise resistance of the distributed gate material. The value of this resistor is dependent upon whether the gate is connected from one end or both ends, and can be approximated analytically by treating the gate as a transmission line along its width, since for practical radio frequency MOSFET dimensions $W \gg L$. In the case of the gate connected from one end only, $R_{g,i} = R_g/3$. When the gate is connected from both ends, $R_{g,i} = R_g/12$. In terms of the equivalent gate resistance and the transistor's dimensions, the gate resistance noise is given by,

$$\overline{v_{g,i}^2} = 4kTR_{g,i}\Delta f \tag{6.26}$$

where k is Boltzmann's constant and T is the temperature in Kelvin. Note that the gate resistance noise is proportional to the width, inversely proportional to the length of the device, and scales in inverse proportion to the number of fingers used in the transistor layout.

6.4.3 Thermal Channel Noise

Thermal noise within the channel produces both drain channel noise and induced gate noise. Since both the channel drain noise and the induced gate noise are generated by the same physical noise sources, they exhibit a degree of correlation. The normalized correlation coefficient between the drain current noise and the gate current noise is in general a complex quantity given by the expression,[21]

$$c = \frac{\overline{i_g i_d^*}}{\sqrt{\overline{i_g i_g^*} \cdot \overline{i_d i_d^*}}} \tag{6.27}$$

Simulations and experimental measurements indicate that the real part of "c" in Eq. (6.27) is approximately equal to zero. Intuitively this makes sense because the gate noise is induced from the channel current noise capacitively. There is, therefore, a 90° phase shift between the induced gate current noise and the source of this noise, which is the channel current noise.

The value of "c" has been found for low frequencies and longer gate lengths as c = –j0.395.[21] For submicron MOSFETs, and as the frequency of operation increases, c approaches –j0.3.

Channel noise — The drain channel noise is a complicated function of the transistor bias conditions. For radio frequency applications, the MOSFET is assumed to operate in the saturation region and the drain channel noise is given approximately by

$$\overline{i_d^2} = 4kT\gamma g_{do}\Delta f \tag{6.28}$$

where γ is a bias dependent parameter and g_{do} is the zero drain voltage conductance of the channel. Usually $g_{do} = g_m$ is assumed. The factor γ is an increasing function of V_{DS}, but a value of 2/3 is often used for hand calculations and simple simulations. For quasi-static MOSFET operation, $\overline{i_d^2}$ is essentially independent of frequency.

Induced gate noise — Fluctuations in the channel are coupled to the transistor gate via the oxide capacitance. This produces a weak noise current at the gate terminal. The mean square value of this noise current was evaluated by van der Ziel[21] and is approximated by,

$$\overline{i_g^2} = 4kT\beta g_{do}\frac{\left(\omega C_{gs}\right)^2}{k_{gs}g_{do}}\Delta f \tag{6.29}$$

where β is a bias-dependent parameter typically greater than or equal to 4/3. The factor $1/k_{gs}$ arises from a first-order expansion that gives $k_{gs} = 5$ for long channel devices. Interestingly, the induced gate noise is proportional to the square of the frequency. Clearly this expression cannot hold as the frequency becomes extremely large. The expression given in Eq. (6.29) is valid up to approximately $\frac{2}{3}f_t$.

6.4.4 1/f-Noise

Experimental measurements of the noise spectral density in MOSFETs demonstrate that the noise increases with decreasing frequency. The noise spectral density at very low frequencies exceeds the noise levels predicted for purely thermally generated noise. This excess noise is evident up to a corner frequency of approximately 100 kHz to 1 MHz for MOSFETs. The low-frequency excess noise is generally known as a flicker noise, and its spectral density is inversely proportional to the frequency raised to some power. Due to this inverse relationship to frequency, flicker noise is also called 1/f-noise or "pink" noise.

There are two dominant theories on the origins of 1/f-noise in MOSFETs. First, there is the carrier density fluctuation theory in which flicker noise is taken as a direct result of the random trapping and release of charges by the oxide traps near the Si-SiO_2 interface beneath the gate. The channel surface potential fluctuates because of this charge fluctuation, and the channel carrier density is in turn modulated by the channel surface potential fluctuation. The carrier density fluctuation theory predicts that the input-referred flicker noise is independent of the gate bias voltage and the noise power is proportional to the interface trap density. The carrier density fluctuation model is supported by experimental measurements that demonstrate the correlation between the flicker noise power and the interface trap density.

The second major theory on the origins of flicker noise is the mobility fluctuation theory. This theory treats flicker noise as arising from the fluctuation in bulk mobility based on Hooge's empirical relation for the spectral density of flicker noise in a homogenous medium. In contrast with the charge density fluctuation theory, the mobility fluctuation theory does predict that the power spectral density of 1/f-noise is dependent upon the gate bias voltage.

Neither of these two main theories satisfactorily accounts for the observed 1/f-noise power spectral density in MOSFETs under all conditions. Current thinking applies both models for an overall understanding of 1/f-noise in MOSFETs. Expressions for the MOSFET 1/f-noise have been derived by various researchers and normally some amount of "fitting" is required to agree with theory. Common expressions for the 1/f-noise of a MOSFET operating in saturation include

$$\overline{di_f^2} = K_f \frac{g_m^2}{C'_{ox} W L} \frac{df}{f^a} \tag{6.30}$$

for the flicker-noise current, and

$$\overline{dv_{in,f}^2} = K_f \frac{1}{C'_{ox} W L} \frac{df}{f^a} \tag{6.31}$$

for the equivalent input noise voltage. The value of α is typically close to unity, and K_f is in general a bias-dependent parameter, on the order of 10^{-14} C/m^2 for PMOS devices and 10^{-15} C/m^2 for NMOS devices.

Although 1/f-noise is negligible at radio frequencies, it is still an important consideration for transceiver design. For example, 1/f-noise can be a limiting factor in direct conversion receivers. Since the direct conversion receiver directly translates the signal channel of interest to base band, the 1/f-noise corrupts the desired information content. Modern modulation formats place important information signal content near the center of a channel. When such a channel is directly downconverted to DC, this important central region experiences the worst of the 1/f-noise contamination. If the 1/f-noise degrades the signal-to-noise ratio sufficiently, reception becomes impossible.

A second area of transceiver design in which 1/f-noise can create problems and cannot be ignored involves oscillator design. The close-in phase noise of an oscillator is normally dominated by 1/f-noise. At frequencies close to the oscillator center frequency, the phase noise fall off is typically –30 dB per decade, while further from the center frequency the phase noise fall off is normally –20 dB per decade. The additional phase noise close to the center frequency can be attributed to flicker noise present in the active devices comprising the oscillator circuitry.

Designers seeking to minimize 1/f-noise have four options. First, the choice of technology and process is important since the flicker noise exhibited by devices of equal sizes across different processes can vary significantly. Second, a designer can opt to use PMOS devices instead of NMOS devices since PMOS devices typically exhibit one tenth the amount of flicker noise a comparable NMOS device produces. Thirdly, devices operating in weak inversion exhibit markedly lower 1/f-noise than devices operated in saturation. However, operating in weak inversion is normally not an option for devices working at radio frequencies, since the device f_t is too low in the weak inversion region. Finally, as shown in Eqs. (6.30) and (6.31), the area of the device plays an important role in determining the flicker noise produced by the device, and so the area should be maximized. Unfortunately, for radio frequency designs there are constraints on the maximum practical width and length of the MOSFETs. Both the width and length determine the highest usable operating frequency of the device and so they are normally kept small. Small width and length implies that the flicker noise of devices designed for operation at radio frequencies is comparatively large.

6.4.5 Extrinsic Noise

The primary extrinsic noise sources of concern are due to the parasitic resistances at the MOSFETs terminals. The corresponding mean-square noise currents due to these parasitic resistances are given by,

$$\overline{i_{D,e}^2} = \frac{4kT}{R_{D,e}}\Delta f$$

$$\overline{i_{S,e}^2} = \frac{4kT}{R_{S,e}}\Delta f \tag{6.32}$$

$$\overline{i_{G,e}^2} = \frac{4kT}{R_{G,e}}\Delta f$$

$$\overline{i_{B,e}^2} = \frac{4kT}{R_{B,e}}\Delta f$$

where $R_{D,e}$, $R_{S,e}$, $R_{G,e}$ and $R_{B,e}$, are the extrinsic parasitic resistances at the drain, source, gate, and bulk, respectively.

6.5 MOSFET Design for RF Operation

Radio frequency integrated circuit design with CMOS requires numerous choices regarding device dimensions and bias voltages and currents. A combination of microwave theory and analog design fundamentals must be pooled with a firm knowledge of the overall system-level functional requirements. System-level calculations will normally dictate the allowable gain, linearity, and noise of the basic circuitry. Two important aspects of radio frequency design in CMOS involve MOSFET input impedance matching for power gain, and impedance matching for minimum noise figure. These two important topics are discussed next.

6.5.1 MOSFET Impedance Matching

Circuitry that accepts a signal from off-chip, such as the input to a receiver's low noise amplifier, must be conjugately matched to the source impedance. The concept of conjugate matching is often unfamiliar to digital designers of CMOS, but is a common theme in microwave design. When a source and load are conjugately matched, maximum power transfer from the source to the load occurs. Input impedance matching is also important because pre-select filters preceding the low noise amplifier may be sensitive to the quality of their terminating impedances.

A load is conjugately matched to its driving impedance when the input impedance of the load equals the complex conjugate of the driving impedance. Since the input impedance of any two-port network is given by,

$$Z_{in} = \frac{y_{22}}{y_{11}y_{22} - y_{21}y_{12}} \tag{6.33}$$

the y-parameters given in Table 6.3 can be used in conjunction with circuit analysis to give the total circuit input impedance. Note that a linear system has been tacitly assumed in Eq. (6.33). For small-signal applications, such as in the receiver portion of a transceiver, linearity is a reasonable assumption. However, for some applications, such as a nonlinear power amplifier, conjugate matching is not possible and so a resistive load line is used instead.

While matching can be augmented by a high-Q matching network constructed off-chip, this method adds expense to mass-market products, in which commodity radio frequency ICs may terminate an unknown length of transmission line. It is therefore desirable to provide as close to a stable 50 Ω input impedance as possible within the integrated circuit. The quality of the external matching network can then be relaxed.

As an example of input impedance matching, consider the common source low noise amplifier configuration shown in Fig. 6.11. The matching components include L_1, which is typically implemented as a bond wire having a value of 1 to 2 nH. Inductor L_1 is used for producing the real part of the input impedance for matching. For simplicity, the bonding pad capacitance is ignored. The LNA is assumed to be operating in a cascode configuration and so the Miller capacitance of the MOSFET is ignored. Conjugate matching to the source resistance R_S requires that the approximate matching condition $R_g + \frac{g_m}{C_{gs}} L_1 = R_S$ is satisfied. Proper choice of the device width, biasing, and number of fingers used for construction allow this relation to be met. Note that $2\pi f_t \approx g_m/C_{gs}$, and as discussed in the section on Maximum Available Power Gain, the maximum achievable power gain of a MOSFET is proportional to the device's f_t. Also, as shown in the next section, the device noise figure is inversely proportional to f_t. Hence the MOSFET biasing should be designed such that the resulting f_t is adequate for the application.

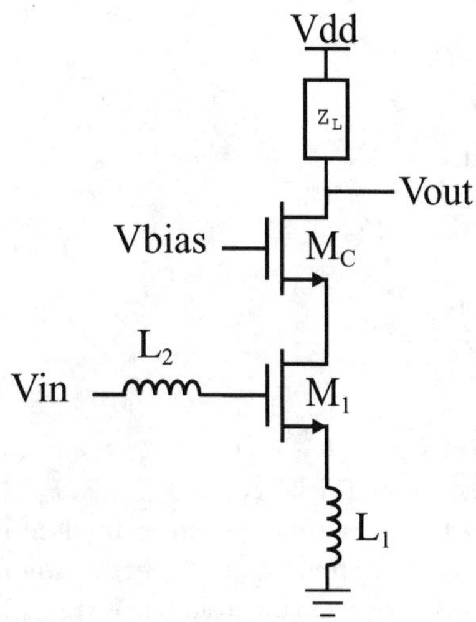

FIGURE 6.11 Common-source low-noise amplifier configuration.

6.5.2 Matching for Noise

Classical noise matching analysis is applicable to the MOSFET shown in Fig. 6.10(b). Classical methods are directly applicable when all MOSFET internal noise sources are referred to the input. For the common source MOSFET configuration shown in Fig. 6.10(b), the input referred noise sources v_n and i_n are given by,

$$v_n = v_{g,i} - \frac{i_d}{g_m} + \left(R_{g,i} + r_i\right) i_g - j\frac{f}{f_t}\left(R_{g,i} + r_i\right) i_d \tag{6.34}$$

and

$$i_n = i_g - j\frac{f}{f_t} i_d \tag{6.35}$$

Recall from the section on Thermal Channel Noise that the two noise sources in Eqs. (6.34) and (6.35) are correlated since i_d and i_g arise from the same physical process.

The noise performance of a transistor with driving impedance $Z_{SS} = R_{SS} + jX_{SS}$ is fully described in terms of four noise parameters: the minimum noise figure (F_{min}), the noise conductance (G_n), and the optimum driving impedance (Z_{opt}), which consists of a resistance (R_{opt}) and a reactance (X_{opt}). Following the procedure of Gonzalez[13] the noise parameters are found as,

$$G_n \approx \left(\frac{f}{f_t}\right)^2 \gamma g_{do} F_1 \tag{6.36}$$

$$R_{opt} \approx \sqrt{\left(R_{g,i} + r_i\right)^2 + \left(\frac{f_t}{f}\right)^2 \frac{R_{g,i} F_4}{\gamma g_{do}} + \left(\frac{f_t}{f}\right)^2 \frac{F_2^2 F_5}{\gamma g_{do}^2 F_1}} \tag{6.37}$$

$$X_{opt} \approx \frac{F_2}{2\pi f C_{gs}} \approx -F_2 X_{in} \tag{6.38}$$

and

$$\begin{aligned} F_{\min} &\approx 1 + 2G_n\left(R_{opt} + R_{g,i} + r_i\right) \\ &\approx 1 + 2F_3 \frac{f}{f_t}\sqrt{\gamma g_{do} R_{g,i}} \end{aligned} \tag{6.39}$$

The factors $\{F_1, F_2, F_3, F_4, F_5\}$ result from algebraic manipulations of Eqs. (6.34) and (6.35). If induced gate noise is ignored, then $\{F_1, F_2, F_3, F_4\}$ are all equal to one, otherwise they are less than one. The fifth factor, F_5, is equal to zero if induced gate noise is ignored, and equal to one if induced gate noise is included. From Eq. (6.39) the minimum noise figure is proportional to $\sqrt{R_{g,i}}$. Therefore to realize a low noise figure the intrinsic gate resistance must be made as small as practical. Noting that g_{do} in Eq. (6.39) is approximately equal to g_m, and using the results for f_t from Eq. (6.22), the minimum noise figure is shown to be inversely proportional to $\sqrt{g_m}$. Hence, increasing the g_m of the device decreases the minimum noise figure. Since both g_m and γ are functions of the transistor bias conditions, an optimum biasing point can be found that gives the lowest minimum noise figure.[20]

For optimum noise matching, the optimum noise resistance should equal the driving source resistance. For MOSFETs, near-optimum noise matching is possible by correctly choosing the number of gate fingers used to construct the MOSFET. In Fig. 6.10(b), the equivalent input noise current i_{eq} and noise voltage v_{eq}, yield a transistor noise figure given by

$$F = F_{\min} + \frac{G_n}{R_S}\left|R_S - Z_{opt}\right|^2 \tag{6.40}$$

where R_S is the driving source resistance, G_n is the noise conductance of a single finger, and Z_{opt} is the optimum noise impedance. Neglecting interconnect parasitics, if N MOSFETs are placed in parallel, the equivalent noise current is $i_{eq}^{(N)} = i_{eq}\sqrt{N}$ and the equivalent noise voltage is $v_{eq}^{(N)} = v_{eq} = v_{eq}/\sqrt{N}$. The noise figure for the complete MOSFET consisting of N fingers is then given by

$$F = F_{\min} + \frac{NG_n}{R_S}\left|R_S - \frac{Z_{opt}}{N}\right|^2 \tag{6.41}$$

in which the minimum noise finger is independent of the number of fingers, but the noise conductance scales as N, and the optimum noise impedance scales as 1/N. Therefore, by using fingered MOSFETs, the transistor can be matched for noise using the number of fingers as a design parameter.

6.6 MOSFET Layout

Radio frequency MOSFET layout is a nontrivial task. A poor transistor layout adversely affects every aspect of the transistor's radio frequency performance. Recall that a MOSFET's $f_{\max}$, $NF_{\min}$, and input impedance-matching characteristics are all determined by the transistor's W/L ratio and the number of fingers used to construct the device. Additionally, both matching and flicker noise are affected by the area of the gate. Devices with a large gate area will match better and have lower flicker noise.

The various scaling rules for MOSFETs operating in saturation appear in Table 6.4. The MOSFET length is assumed to be the minimum allowed. Also shown in Table 6.4 is the effect of biasing upon the

TABLE 6.4 MOSFET Scaling Rules for the Saturation Region

	Parameter	Number of Fingers N	Finger Width W	Bias Current $I_{DS} < I_{DS,sat}$	Bias Current $I_{DS} > I_{DS,sat}$
Performance	g_m	N	W	Increases with I_{DS}	~constant
	f_t	Independent	Independent	Increases with I_{DS}	~constant
	f_{max}	Independent	~1/W	Increases with I_{DS}	~constant
Parasitics	C_{gs}, C_{gd}	N	W	~constant	~constant
	r_d, r_s, r_i	1/N	1/W	Independent	Independent
	$R_{g,i}$	1/N	W	Independent	Independent
Noise	F_{min}	Independent	~1/W	Decreases with I_{DS}	Increases with I_{DS}
	G_n	N	W	Increases with I_{DS}	~constant
	R_{opt}	1/N	~1/W	Decreases with I_{DS}	~constant
	X_{opt}	1/N	1/W	Independent	Independent

MOSFET. A typical design trade-off is between maximum gain and maximum f_{max}, assuming power requirements are a secondary concern. Using multiple MOSFET fingers is a convenient way of improving the device performance.

6.6.1 Fingered Layout

MOSFETs designed for radio frequency applications normally have large W/L ratios. This is primarily so that the MOSFET has a large transconductance and therefore has a large f_t so that its noise figure is minimal. As shown in Table 6.4, increasing the MOSFET width causes a decrease in f_{max}. To counteract this effect, the MOSFET is laid out as a parallel connection of narrower MOSFETs. The resulting fingered structure is shown in Fig. 6.12. Neglecting the interconnect resistance between fingers, the intrinsic gate resistance scales as $1/N$, where N is the number of fingers.

6.6.2 Substrate Connections

Parasitic resistance to the substrate affects both the impedance matching and the noise figure of MOSFETs. The substrate resistance is in general bias dependent. Local variations in threshold voltage are therefore possible, and matching between devices is reduced. This may affect the performance of differential pairs and differential circuitry. Voltage fluctuations induced in the substrate, especially by digital clocks and switching circuitry, will raise the noise figure of radio frequency analog MOSFETs. Adequate substrate connections are therefore important for low-noise, precision MOSFET design. A substrate connection method is shown in Fig. 6.12.

6.7 The Future of CMOS

CMOS has definite potential for applications in commodity radio frequency electronics, and is expected to find increased use in radio frequency integrated circuits. Although this section has considered CMOS operating at radio frequencies mostly in terms of analog small-signal performance, it is important to note that existing implementations of digital logic are also operating at radio frequencies. Microprocessors operating above 1 GHz are now commonplace. The modeling of radio frequency logic (RFL) and the study of radio frequency digital design will be important contributing factors for future high-speed microprocessors, digital signal processors, and ASICs. Additionally, as the demand for extremely high-speed digital microprocessors increases, the base-level digital CMOS processes will improve to accommodate the performance expectations of designers. Exotic process features that were once considered purely for analog design use are now appearing in digital designs.

CMOS is not the ideal radio frequency technology, but it does compare favorably with bipolar and BiCMOS. For example, although not discussed in this section, communications circuits constructed with

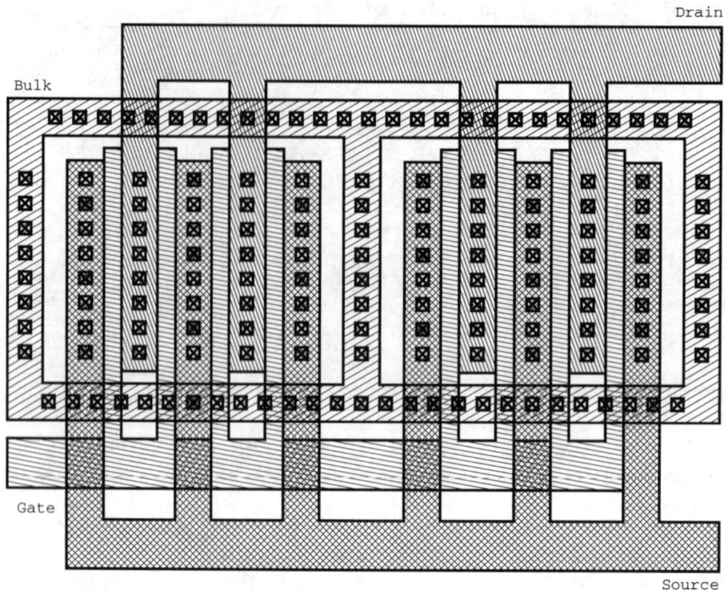

FIGURE 6.12 MOSFET finger layout style for high frequency applications.

CMOS devices often exhibit higher linearity than their counterparts do when constructed with other technologies such as bipolar. Furthermore, in applications where the MOSFET's square-law characteristic applies, differential circuitry can effectively reduce second-order distortion. For an LNA constructed using deep submicron MOSFETs operating in velocity saturation, the device transconductance is essentially constant with v_{gs}, which implies an increased level of linearity.

CMOS may become the technology of choice in the near term for consumer radio frequency applications in the sub-10 GHz bands. The extent to which this becomes possible is dependent on the ability of the device performance to improve as the technology scales and improves. With this in mind, this section concludes with a brief examination of scaling issues in radio frequency CMOS, and recent advances in MOSFET manufacture.

6.7.1 The Impact of Technology Scaling and Process Improvements

A common measure for the gain efficiency of a MOSFET is its transconductance per drain current g_m/I_{DS}. MOSFETs require a large W/L ratio and a high DC biasing current to offer a transconductance comparable to that of other RF device technologies, such as bipolar. In bipolar transistors, the transconductance-to-bias current ratio equals kT/q, which was proved by Johnson[22] to be the maximum transconductance for both MOSFET and bipolar devices. As MOSFET processing improves and the minimum gate length decreases, the g_m/I_{DS} ratio has been found to tend toward the maximum value of kT/q.

Continual reduction in MOSFET minimum gate length has enabled commensurate increases in device f_t and f_{max}. Existing 0.18 μm devices have demonstrated f_t's in excess of 60 GHz, and experimental processes with gate lengths less than 0.1 μm have demonstrated f_t values in the order of 150 GHz. These increases in device f_t also mean that the overall device noise performance improves as the gate length scales down.

Although the radio frequency performance of MOSFETs has traditionally scaled in accordance with a decreasing minimum gate length, this trend will probably not continue along its present path. For example, the gate oxide thickness scales approximately linearly with the minimum gate length. The gate source capacitance therefore remains roughly constant per unit width of the gate for minimum channel length devices. As shown in Eq. (6.22), the MOSFET f_t is inversely proportional to the gate source capacitance, and if the gate source capacitance does not scale in inverse proportion to the gate length, it becomes a limiting factor. Other limitations for increased radio frequency performance of MOSFETs are found in the extrinsic parasitics, such as the gate overlap capacitances. For example, the gate-source overlap capacitance does not scale as 1/L for submicron devices, and C_{gso} becomes an increasingly large proportion of the total gate-to-source terminal capacitance, therefore limiting the high frequency gain of the device.

Certain process enhancements have a profound effect upon the performance of integrated MOSFET circuitry. The use of copper interconnects between MOSFETs is expected to increase the operating frequency of both digital and analog circuitry when used in place of aluminum. Copper has two primary advantages over aluminum when used as an interconnect material. First, the resistivity of copper is approximately 40% lower than that of aluminum. Second, electromigration effects are lower in copper interconnects implying that copper interconnects will exhibit higher reliability, longer life spans, and greater current-handling capability.

Since copper has a lower resistivity than aluminum, interconnect lines can be thinner and yet still allow for the same circuit performance. Alternatively, maintaining the same interconnect thickness gives a lower series resistance. Decreasing the series resistance of metal lines improves the quality factor of integrated inductors and capacitors.

Recent CMOS processes have given designers six or more interconnect metal layers. The increased number of metal layers has important consequences for radio frequency design. For example, several metal layers can be linked in parallel in order to decrease the series resistance of metal lines used to create inductors. Additionally, the top-level metal layers may be used so that the metal-to-substrate capacitance is reduced and therefore the quality factor of the integrated circuit is improved.

While not yet offered on any commodity CMOS processes, substrate etching has been used to increase the quality factor of integrated inductors by removing a section of substrate below the metal lines. Isolation between lower-level metal lines and substrate has also been achieved using patterned ground shields and spun-on thick dielectrics. Future process enhancements will increase the applicability of CMOS to radio frequency applications. Sub-micron mixed-signal processes featuring six metal layers, metal-insulator-metal capacitors, deep n-wells, and threshold voltage-adjust capability are available now, and hint at what will become available in the near future.

To conclude, CMOS is a mature, easily sourced, and reliable technology. If the performance of CMOS can be improved via scaling to within that of bipolar, BiCMOS, and GaAs, there is no reason why it will not supplant those technologies in commodity radio frequency integrated circuit applications.

References

1. Tsividis, Y. P., *Operation and Modeling of the MOS Transistor*, New York: McGraw-Hill, 1987.
2. Sze, S. *Physics of Semiconductor Devices*, New York: Wiley and Sons, 1981.
3. Grove, A. *Physics and Technology of Semiconductor Devices*, New York: Wiley and Sons, 1967.
4. Muller, R. and Kamins, T. *Device Electronics for Integrated Circuits*, New York: Wiley and Sons, 1986.
5. Laker, K. R. and Sansen, W. M. C. *Design of Analog Integrated Circuits and Systems*, New York: McGraw-Hill, 1994.
6. Enz, C. C. and Cheng, Y., MOS Transistor Modeling for RF IC Design, *IEEE Journal of Solid-State Circuits*, 35, 186–201, Feb. 2000.
7. Klein, P., A Compact-Charge LDD-MOSFET Model, *IEEE Transactions on Electron Devices*, 44, 1483–1490, Sept. 1997.
8. Cobbold, R. S. C. *Theory and Applications of Field-Effect Transistors*, New York: Wiley-Interscience, 1970.
9. Razavi, B., Yan, R.-H., and Lee, K. F., Impact of Distributed Gate Resistance on the Performance of MOS Devices, *IEEE Trans. Circuits and Systems I*, 41, 750–754, 1994.
10. Park, H. J., Ko, P. K., and Hu, C., A Non-Quasi-static MOSFET Model for SPICE-AC Analysis, *IEEE Trans. Computer Aided Design*, 11, 1247–1257, 1992.

11. Kim, L.-S. and Dutton, R. W., Modeling of the Distributed Gate RC Effect in MOSFETs, *IEEE Trans. Computer Aided Design*, 8, 1365–1367, 1989.
12. Jindal, R. P., Noise Associated with Distributed Resistance of MOSFET Gate Structure in Integrated Circuits, *IEEE Trans. Electron Devices*, ED-31, 1505–1509, 1984.
13. Gonzalez, G. *Microwave Transistor Amplifiers Analysis and Design*, Prentice-Hall, Englewood Cliffs, NJ, 1997.
14. Das, M. B., High Frequency Network Properties of MOS Transistors Including the Substrate Resistivity Effects, *IEEE Trans. Electron Devices*, ED-16, 1049–1069, 1969.
15. Bagheri, M. and Tsividis, Y., A Small-Signal DC-to-High-Frequency Non-Quasistatic Model for Four-Terminal MOSFETs Valid in All Regions of Operation, *IEEE Trans. Electron Devices*, ED-32, 2383–2391, 1985.
16. Rofougaran, A., Chang, J. Y. C., Rofougaran, M., and Abidi, A. A., A 1GHz CMOS RF Front-End IC for a Direct-Conversion Wireless Receiver, *IEEE Journal of Solid-State Circuits*, 31, 880–889, July 1996.
17. Abou-Allam, E. and Manku, T., An Improved Transmission-Line Model for MOS Transistors, *IEEE Transactions on Circuits and Systems II: Analog and Digital Signal Processing,* 46, 1380–1387, Nov. 1999.
18. Abou-Allam, E. and Manku, T., A Small-Signal MOSFET Model for Radio Frequency IC Applications, *IEEE Transactions on Computer-Aided Design of Integrated Circuits and Systems*, 16, 437–447, May 1997.
19. Tin, S. F., Osman, A. A., and Mayaram, K., Comments on A Small-Signal MOSFET Model for Radio Frequency IC Applications, *IEEE Trans. Computer-Aided Design*, 17, 373–374, 1998.
20. Manku, T., Microwave CMOS-Device Physics and Design, *IEEE Journal of Solid-State Circuits,* 34, 277–285, Mar. 1999.
21. Aldert van der Ziel. *Noise in Solid State Devices and Circuits*, John Wiley & Sons, New York, 1986.
22. Johnson, E. O., The Insulating-Gate Field Effect Transistor — a Bipolar Transistor in Disguise, *RCA Review,* 34, 80–94, 1973.

7

Metal Semiconductor Field Effect Transistors

Michael S.Shur
Rensselaer Polytechnic Institute

7.1 Introduction

Silicon Metal Oxide Semiconductor Field Effect Transistors (MOSFETs) dominate modern microelectronics. Gallium Arsenide Metal Semiconductor Field Effect Transistors (GaAs MESFETs) are "runners-up," and they find many important niche applications in high-speed or high frequency circuits. After the first successful fabrication of GaAs MESFETs by Mead in 1966[1] and after the demonstration of their performance at microwave frequencies in 1967 by Hooper and Lehrer,[2] these devices emerged as contenders with silicon MOSFETs and bipolar transistors. In the late 1970s and early 1980s, high quality semi-insulating substrates and ion-implantation processing techniques made it possible to fabricate GaAs MESFET VLSI circuits, such as 16 × 16 multipliers with a multiplication time of 10.5 ns and less than 1 W power dissipation.[3]

Today GaAs MESFETs play an important role in both analog and digital applications, such as for satellite and fiber-optic communication systems, in cellular phones, in automatic IC test equipment, and for other civilian and military uses. The microwave performance of GaAs MESFETs approaches that of Heterostructure Field Effect Transistors (HFETs).[4] As discussed below, the record maximum frequency of oscillations and the record cutoff frequency, f_T, for GaAs MESFETs reached 190 and 168 GHz, respectively. Even though the record numbers of f_{max} and f_T for GaAs- and InP-based HFETs reach 400 GHz and 275 GHz, respectively, their more typical f_{max} and f_T are well within the reach of GaAs MESFET technology. The integration scale of GaAs MESFET integrated circuits approaches 1,000,000 transistors.

Emerging materials for MESFET applications are SiC and GaN wide bandgap semiconductors that have a much higher breakdown voltage, a higher thermal conductivity, and a higher electron velocity than GaAs. SiC MESFETs are predicted to reach breakdown voltages up to nearly 100 kV.[5]

However, SiGe and even advanced, deep submicron Si technologies emerge as a serious competitor to GaAs MESFET technology at relatively low frequencies (below 40 GHz or so). This trend toward SiGe

0-8493-1562-X/03/$0.00+$1.50

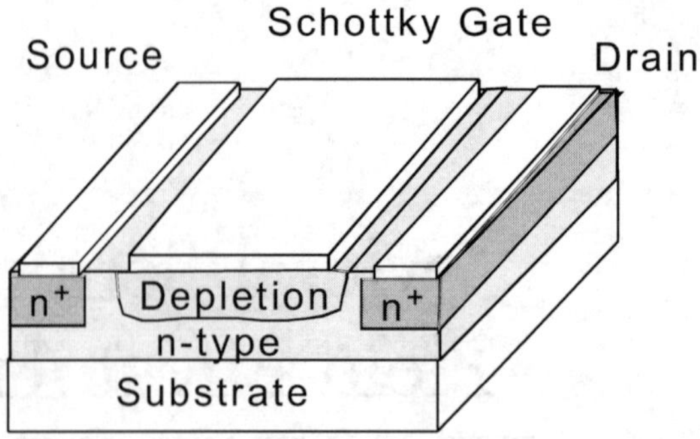

FIGURE 7.1 Schematic MESFET structure.

and Si might be alleviated by a shift toward 150 mm GaAs substrates, which are now used by the leading GaAs IC manufacturers, such as Vitesse, Anadigics, Infinion, Motorola, Tektronix, and RFMD.

In this section, we first discuss the MESFET principles of operation. Then we review the material properties of semiconductors competing for applications in MESFETs and the properties of Schottky barrier contacts followed by a brief review of MESFET fabrication, and MESFET modeling. We also consider wide bandgap semiconductor MESFETs, new emerging hetero-dimensional MESFETs, and discuss applications of the MESFET technology.

7.2 Principle of Operation

Figure 7.1 shows a schematic MESFET structure. In n-channel MESFETs, an n-type channel connects n^+ drain and source regions. The depletion layer under the Schottky barrier gate contact constricts the current flow across the channel between the source and drain. The gate bias changes the depletion region thickness, and, hence, modulates the channel conductivity.

This device is very different from a silicon Metal Oxide Semiconductor Field Effect Transistor (MOSFET), where a silicon dioxide layer separates the gate from the channel. MOSFETs are mainstream devices in silicon technology, and silicon MESFETs are not common. Compound semiconductors, such as GaAs, do not have a stable oxide, and a Schottky gate allows one to avoid problems related to traps in the gate insulator, such as hot electron trapping in the gate insulator, threshold voltage shift due to charge trapped in the gate insulator, and so on.

In **normally-off** (**enhancement mode**) MESFETs, the channel is totally depleted by the gate built-in potential even at zero gate voltage (see Fig. 7.2). The threshold voltage of normally-off devices is positive. In **normally-on** (**depletion mode**) MESFETs, the conducting channel has a finite cross-section at zero gate voltage. The drawback of normally-off MESFET technology is a limited gate voltage swing due to the low turn-on voltage of the Schottky gate. This limitation is much less important in depletion mode FETs with a negative threshold voltage. Also, this limitation is less important in low power digital circuits operating with a low supply voltage.

Usually, the source is grounded, and the drain is biased positively. A schematic diagram of the depletion region under the gate of a MESFET for a finite drain-to-source voltage is shown in Fig. 7.3.

The depletion region is wider closer to the drain because the positive drain voltage provides an additional reverse bias across the channel-to-gate junction. With an increase in the drain-to-source bias, the channel at the drain side of the gate becomes more and more constricted. Finally, the velocity of electrons saturates leading to the current saturation (see Fig. 7.4 that shows typical MESFET current-voltage characteristics).

MESFETs have been fabricated using many different semiconductor materials. However, GaAs MESFETs are mainstream MESFET devices. In many cases, GaAs MESFETs are fabricated by direct ion implantation

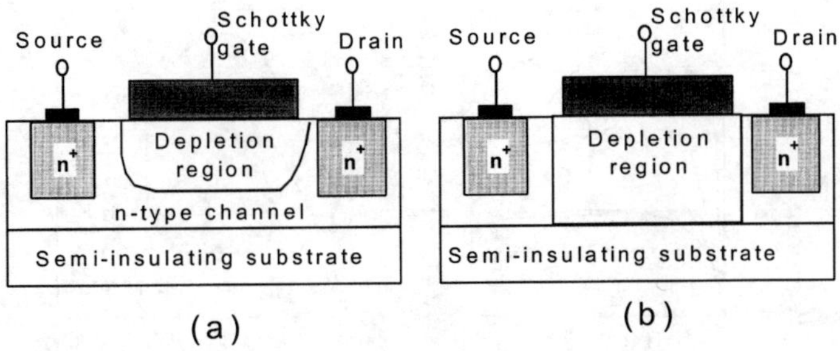

FIGURE 7.2 Normally-on and normally-off MESFETs at zero gate bias.

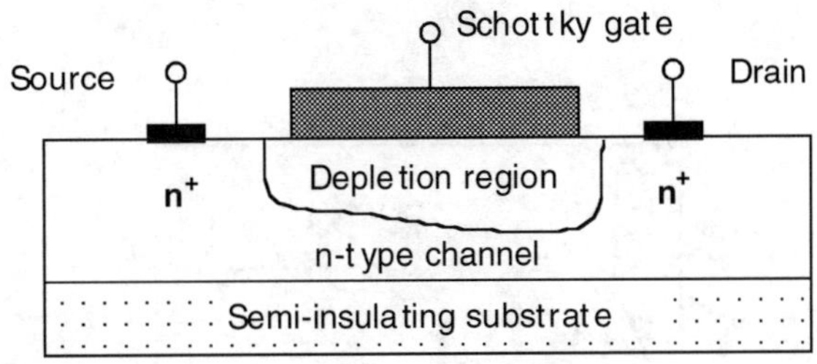

FIGURE 7.3 Depletion region in MESFET with positive drain bias.

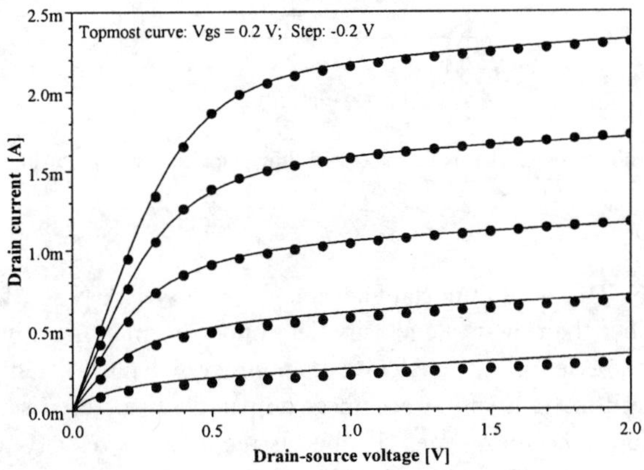

FIGURE 7.4 Measured (symbols) and simulated (using AIM-Spice, lines) drain current characteristics of MESFET operating at room temperature. (After Ytterdal et al.[6])

into a GaAs semi-insulating substrate, making GaAs IC fabrication less complicated than silicon CMOS fabrication.

7.3 Properties of Semiconductor Materials Used in MESFET Technology

The effective mass of electrons in GaAs is very small (0.067 m_e in GaAs compared to 0.98 m_e longitudinal effective mass and 0.19 m_e transverse effective mass in Si, where m_e is the free electron mass). This leads to a much higher electron mobility in GaAs — approximately 8500 cm²/Vs in pure GaAs at room temperature compared to 1500 cm²/Vs in Si. As shown in Fig. 7.5, the electron velocity in GaAs exceeds

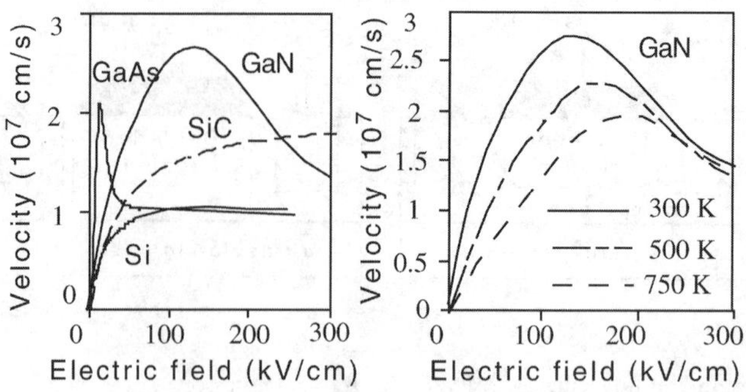

FIGURE 7.5 (a) Electron drift velocity at 300 K in GaN, SiC, and GaAs. (b) Electron drift velocity in GaN at 300 K, 500 K, and 750 K.[12]

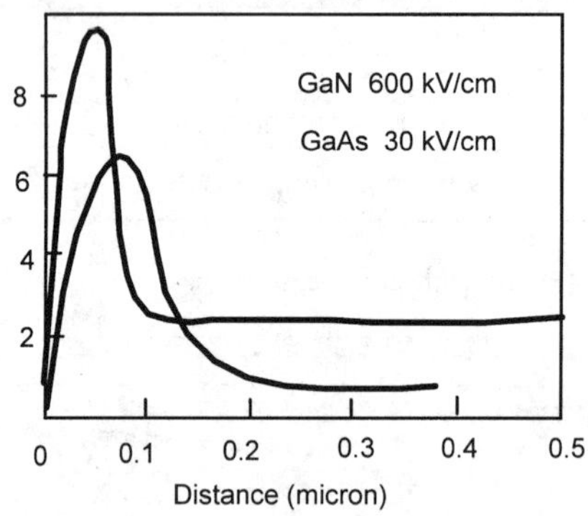

FIGURE 7.6 Computed velocity of electrons injected with low velocities into a constant electric field region into GaN and GaAs (after Foutz et al.[13]).

that for the electrons in Si. This is an important advantage for modern day short channel devices, where the electric fields are higher than the peak velocity field under normal operating conditions. The light electrons in GaAs also experience so-called overshoot or even ballistic transport in short channel devices,[7–11] where the electron transit time becomes comparable to or even smaller than the electron energy or even momentum relaxation time. This boosts the electron velocity well above the expected steady-state values (see Fig. 7.6). GaN also has a high electron velocity and pronounced overshoot effects (see Figs. 7.5 and 7.6).

Another important advantage of GaAs and related compound semiconductors is the availability of semi-insulating material that could serve as a substrate for device and circuit fabrication. A typical resistivity of semi-insulating GaAs is 10^7 Ω-cm or larger, compared to 2.5×10^5 Ω-cm for intrinsic silicon at room temperature. The semi-insulating GaAs is used as a substrate for fabricating GaAs MESFETs and other devices. Passive elements can also be fabricated on the same substrate, which is a big advantage for fabricating Monolithic Microwave Integrated Circuits (MMICs). As mentioned above, an important advantage of the GaAs MESFET is the possibility of fabricating these devices and integrated circuits using a direct implantation into the semi-insulating GaAs substrate.

Since GaAs, InP, and related semiconducting materials are direct gap materials, they are widely used in optoelectronic applications. Hence, electronic and photonic devices can be integrated on the same chip for use in optical interconnects or in optoelectronic circuits.

The direct band gap leads to a high recombination rate, which improves radiation hardness. GaAs-based devices can survive over 100 megarads of ionizing radiation.[14]

TABLE 7.1 Material Properties of Si, GaAs, α-SiC, and GaN

Property	Si	GaAs	α-SiC(6H)	GaN
Energy gap (eV)	1.12	1.42	2.9	3.4
Lattice constant(a) Å	5.43107	5.6533	3.081	3.189
Lattice constant (c), Å	—	—	15.117	5.185
Density (g/cm³)	2.329	5.3176	3.211	6.1
Dielectric constant	11.7	12.9	9.66(⊥) 10.03(‖)	9.5 (8.9)
Electron mobility (cm²/Vs)	1450	8500	330	1200
Hole mobility (cm²/Vs)	500	400	60	<30
Saturation velocity (m/s)	10^5	1.2×10^5	$2–2.5 \times 10^5$	$2–2.5 \times 10^5$
Electron effective mass ratio	0.92/0.19	0.067	0.25/1.5	0.22
Light hole mass ratio	0.16	0.076	0.33	0.7
Optical phonon energy (eV)	0.063	0.035	0.104	0.092
Thermal conductivity (W/cm°C)	1.31	0.46	4.9	1.5

GaAs, and especially SiC and GaN MESFETs, are suitable for use in high-temperature electronics and as power devices (because of a small intrinsic carrier concentration and a high breakdown field).

Table 7.1 summarizes important material properties of semiconductors used in MESFET technology.

7.4 Schottky Barrier Contacts

Figure 7.7 shows the energy band diagram of a GaAs Schottky barrier metal-semiconductor contact at zero bias. The Fermi level will be constant throughout the entire metal-semiconductor system, and the energy band diagram in the semiconductor is similar to that for an n-type semiconductor in a p^+-n junction.

Energies Φ_m and Φ_s shown in Figure 7.7 are called the metal and the semiconductor **work functions.** The **work function** is equal to the difference between the **vacuum level** (which is defined as a free electron energy in vacuum) and the Fermi level. The **electron affinity** of the semiconductor, X_s (also shown in

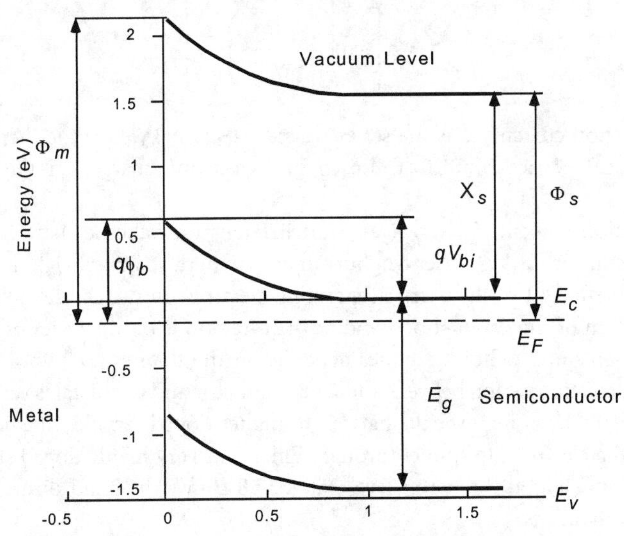

FIGURE 7.7 Simplified energy diagram of GaAs metal-semiconductor barrier $q\phi_b$ is the barrier height (0.75 eV), X_s is the electron affinity in the semiconductor, Φ_s and Φ_m are the semiconductor and the metal work functions, and V_{bi} (0.59 V) is the built-in voltage. Donor concentration in GaAs is 10^{15} cm^{-3} (From Shur[15]).

Fig. 7.7), corresponds to the energy separation between the vacuum level and the conduction band edge of the semiconductor.

In the idealized picture of the Schottky junction shown in Fig. 7.7, the energy barrier between the semiconductor and the metal is

$$q\phi_b = \Phi_m - X_s \tag{7.1}$$

Here X_s is the electron affinity in the semiconductor, Φ_s and Φ_m are the metal work functions, and ϕ_b is the Schottky barrier height, q is the electronic charge [in Eq. (7.1) and in Fig. 7.7, ϕ_b is measured in eV, and Φ_s and Φ_m are measured in Joules). Since $\Phi_m > \Phi_s$ the metal is charged negatively. The positive net space charge in the semiconductor leads to a band bending

$$qV_{bi} = \Phi_m - \Phi_s, \tag{7.2}$$

where V_{bi} is called the **built-in voltage**, in analogy with the corresponding quantity in a *p-n* junction.

Equation (7.1) and Fig. 7.7 are not quite correct. In reality, a change in the metal work function, Φ_m, is not equal to the corresponding change in the barrier height, ϕ_b, as predicted by Eq. (7.1). In actual Schottky diodes, ϕ_b increases with an increase in Φ_m, but only by 0.1 to 0.3 eV when Φ_m increases by 1 to 2 eV. This difference is caused by interface states and is determined by the properties of a thin interfacial layer. However, even though a detailed and accurate understanding of Schottky barrier formation remains a challenge, many properties of Schottky barriers may be understood independently of the exact mechanism determining the barrier height. In other words, we can simply determine the effective barrier height from experimental data.

A forward bias decreases the potential barrier for electrons moving from the semiconductor into the metal and leads to an exponential rise in current. At high forward biases (approaching the built-in voltage), the voltage drop across the series resistance (comprised of the contact resistance and the resistance of the neutral region between the ohmic contact and the depletion region) becomes important, and the overall current-voltage characteristic of a Schottky diode can be described by the following **diode equation**

$$I = I_S\left[\exp\left(\frac{V - IR_S}{\eta V_{th}}\right) - 1\right], \tag{7.3}$$

where I_s is the saturation current, R_s is the series resistance, $V_{th} = k_B T/q$ is the thermal voltage, η is the ideality factor (η typically varies from 1.02 to 1.6), q is the electronic charge, k_B is the Boltzmann constant, and T is temperature.

The diode saturation current, I_s, is typically much larger for Schottky barrier diodes than in *p-n* junction diodes since the Schottky barrier height is smaller than the barrier height in *p-n* junction diodes. For a *p-n* junction, the height of the barrier separating electrons in the conduction band of the *n*-type region from the bottom of the conduction band in the *p*-region is on the order of the energy gap.

The current mechanism in Schottky diodes depends on the doping level. In a relatively low-doped semiconductor, the depletion region between the semiconductor and the metal is very wide, and electrons can only reach the metal by going over the barrier. In higher doped samples, the barrier near the top is narrow enough for the electrons to tunnel through. Finally, in very highly doped structures, the barrier is thin enough for tunneling at the Fermi level. Figure 7.8 shows the band diagrams illustrating these three conduction mechanisms.

For low-doped devices, the saturation current density, j_{ss}, in a Schottky diode is given by

$$j_{ss} = A * T^2 \exp\left(-\frac{q\phi_b}{k_B T}\right), \tag{7.4}$$

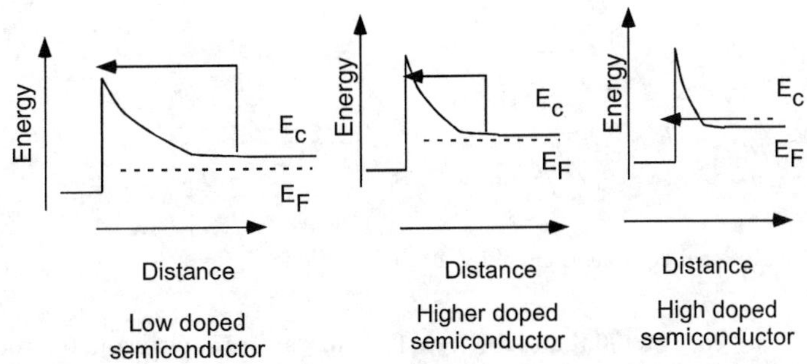

FIGURE 7.8 Current paths in low-doped, higher doped, and high-doped Schottky diodes.

where A^* is called the Richardson constant, T is temperature (in degrees Kelvin), and k_B is the Boltzmann constant. For a conduction band minimum with the spherical surface of equal energy (such as the Γ minimum in GaAs),

$$A^* = \alpha \frac{m_n q k_B^2}{2\pi^2 \hbar^3} \approx 120\, \alpha \frac{m_n}{m_e}\left(\frac{\text{A}}{\text{cm}^2\text{K}^2}\right), \tag{7.5}$$

where m_n is the effective mass, m_e is the free electron mass, $\hbar$ is the Planck constant, and α is an empirical factor on the order of unity. The Schottky diode model described by Eqs. (7.4) and (7.5) is called the **thermionic emission model**. For Schottky barrier diodes fabricated on the {111} surfaces of Si, $A^* = 96$ A/(cm^{2}K^2). For GaAs, $A^* = 4.4$ A/(cm^{2}K^2).

As stated above, in higher doped semiconductors, the depletion region becomes so narrow that electrons can tunnel through the barrier near the top. This conduction mechanism is called thermionic-field emission.

The current-voltage characteristic of a Schottky diode in the case of thermionic-field emission (i.e., for higher doped semiconductors) under forward bias is given by:

$$j = j_{stf} \exp\left(\frac{qV}{E_o}\right) \tag{7.6}$$

where

$$E_o = E_{oo} \coth\left(\frac{E_{oo}}{k_B T}\right) \tag{7.7}$$

$$E_{oo} = \frac{qh}{4\pi}\sqrt{\frac{N_d}{m_n \varepsilon_s}} = 1.85 \times 10^{-11}\left[\frac{N_d\left(\text{cm}^{-3}\right)}{\left(m_n/m_e\right)\left(\varepsilon_s/\varepsilon_o\right)}\right]^{1/2}\left(\text{eV}\right) \tag{7.8}$$

$$j_{stf} = \frac{A^* T\sqrt{\pi E_{oo}\left(\phi_b - qV - E_c + E_{Fn}\right)}}{k_B \cosh\left(E_{oo}/k_B T\right)} \exp\left[-\frac{E_c - E_{Fn}}{k_B T} - \frac{\left(\phi_b - E_c + E_{Fn}\right)}{E_o}\right] \tag{7.9}$$

Here E_c is the bottom of the conduction band in a semiconductor (outside of the depletion region and E_{Fn} is the electron quasi-Fermi level. In GaAs Schottky diodes, the thermionic-field emission becomes

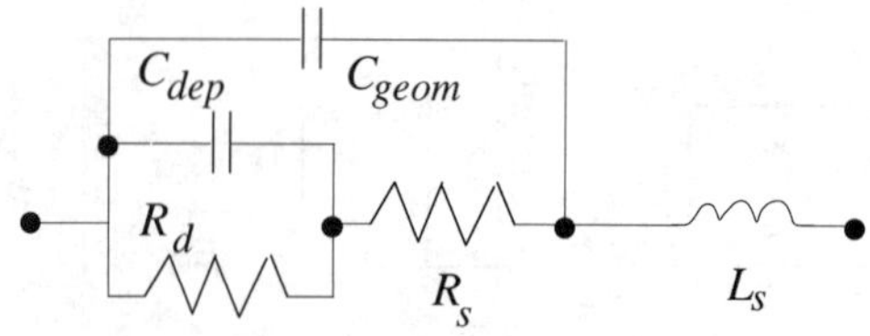

FIGURE 7.9 The small signal equivalent circuit of a Schottky diode.

important for $N_d > 10^{17}$ cm^{-3} at 300 K and for $N_d > 10^{16}$ cm^{-3} at 77 K. In silicon, the corresponding values of N_d are several times larger.

In degenerate semiconductors, especially in semiconductors with a small electron-effective mass, such as GaAs, electrons can tunnel through the barrier near or at the Fermi level, and the tunneling current is dominant. This mechanism is called **field emission.** The resistance of the Schottky barrier in the field emission regime is quite low. Metal-n^+ contacts operated in this regime are used as ohmic contacts.

Figure 7.9 shows a small signal equivalent circuit of a Schottky diode, which includes a parallel combination of the differential resistance of the Schottky barrier

$$R_d = \frac{dV}{dI} \tag{7.10}$$

and the differential capacitance of the space charge region:

$$C_{dep} = S\sqrt{\frac{qN_d\varepsilon_s}{2(V_{bi} - V)}} \tag{7.11}$$

Here V and I are the voltage drop across the Schottky diode and the current flowing through the Schottky diode, respectively, V_{bi} is the built-in voltage of the Schottky barrier, and N_d is the ionized donor concentration in the semiconductor. The equivalent circuit also includes the series resistance, R_s, which accounts for the contact resistance and the resistance of the neutral semiconductor region between the ohmic contact and the depletion region, the equivalent series inductance, L_s, and the device geometric capacitance:

$$C_{geom} = \varepsilon_s S/L \tag{7.12}$$

where L is the device length and S is the device cross-section.

7.5 MESFET Technology

The most popular MESFETs (GaAs MESFETs) found in applications in both analog microwave circuits (including applications in Microwave Monolithic Integrated Circuits) and in digital integrated circuits. Ion implanted GaAs MESFETs represent the dominant technology for applications in digital integrated circuits. They also have microwave applications. Figure 7.10 shows a typical process sequence for ion implanted GaAs MESFETs (developed in late 1970s[16,17]).

In a typical fabrication process, a GaAs semi-insulating substrate is coated with a thin silicon nitride (Si_3N_4) film. Implantation steps shown in Fig. 7.10 are carried out through this layer. As shown in Fig. 7.10, the first implant defines the active layer including the MESFET channel. A deeper and a higher dose implant is used for ohmic contacts. This implant is often done as a self-aligned implant. In this case, a temperature-stable refractory metal-silicide gate (typically tungsten silicide) is used as a mask for implanting the n^+ source and drain contacts. This technique reduces parasitic resistances. Also, this

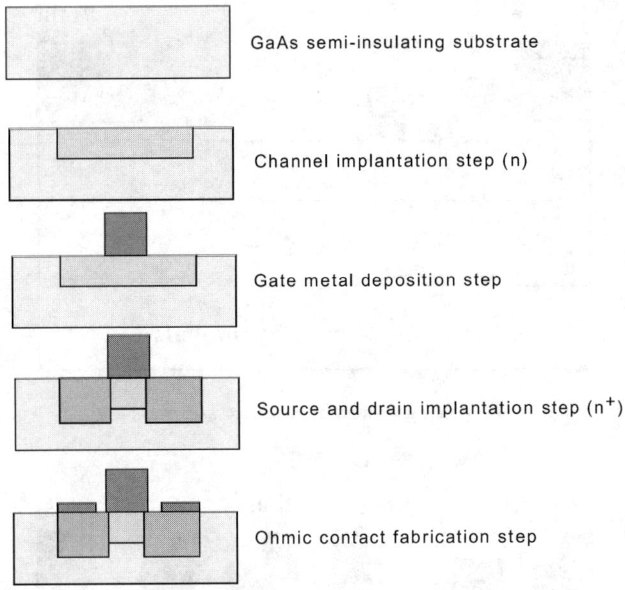

FIGURE 7.10 Fabrication steps for self-aligned GaAs MESFET.

fabrication process is planar. However, the n^+ implant straggle under the gate might increase gate leakage current and also cause carrier injection into the substrate.[18]

After the implants, an additional insulator is deposited in order to cap the GaAs surface for the subsequent annealing step. This annealing (at 800°C or above) activates the implants. For microwave applications, the devices are often grown by molecular beam epitaxy. In this design, a top of n^+ layer doping extending from the source and drain contacts helps minimize the series resistances.

Figure 7.11 shows the recessed gate MESFET structure, where the thickness of the active layer under the gate is reduced. A thick n-doped layer between the gate and the source and drain contacts leads to a relatively low series source and drain resistances. The position and the shape of the recess are very important, since they strongly affect the electric field distribution and the device breakdown voltage.

In power devices, the gate contact in the recess is usually closer to the source than to the drain (see Fig. 7.12). Such placement reduces the source parasitic resistance and enhances the drain-source breakdown voltage by allowing additional expansion space for the high-field region at the drain side of the gate.

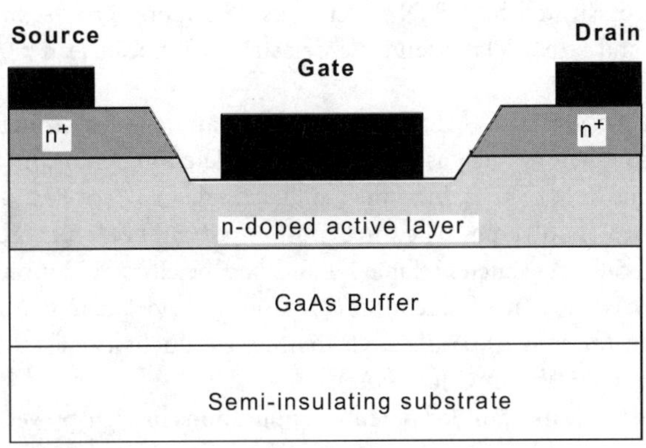

FIGURE 7.11 Recessed gate structure.

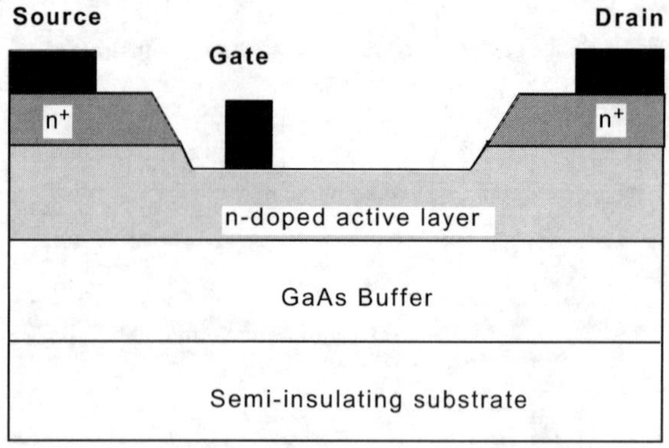

FIGURE 7.12 Recessed structure with offset gate for power devices.

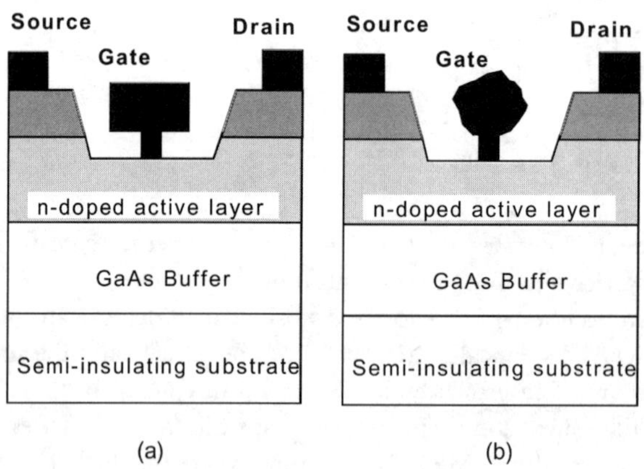

FIGURE 7.13 T-gate (a) and mushroom gate (b) for gate series resistance reduction.

Another important issue is the reduction of the gate series resistance. This can be achieved by using a T-shape gate or a so-called mushroom gate (which might be obtained by side etching the gate), see Fig. 7.13.

In this design, the gate series resistance is reduced without an increase in the gate length, which determines the device cutoff frequency.

MESFETs are usually passivated by a Si_3N_4 layer. This passivation affects the surface states and the surface depletion layer, and stress-related and piezoelectric effects can lead to shifts in the threshold voltage.[19]

A more detailed discussion of GaAs MESFET fabrication can be found in References 20 and 21.

Wide band gap semiconductors, such as SiC, GaN, and related materials, might potentially compete with GaAs for applications in MESFETs and other solid-state devices (see Fig. 7.14).

SiC exists in more than 170 polytypes. The three most important polytypes are hexagonal 6H (α-SiC) and 4H, and cubic 3C (β-SiC). As stated in Table 7.1, SiC has the electron saturation drift velocity of 2×10^7 cm/s (approximately twice that of silicon), a breakdown field larger than 2,500 to 5,000 kV/cm (compared to 300 kV/cm for silicon), and a high thermal conductivity of 4.9 W/cm°C (compared to 1.3 W/cm°C for silicon and 0.5 W/cm°C for GaAs).

These properties make SiC important for potential applications in high-power, high frequency devices as well as in devices operating at high temperatures and/or in harsh environments.

Palmour et al.[23] reported operation of α-SiC MESFETs at a temperature of 773°K. In a 6H-SiC MESFET fabricated by CREE (gate length 24 μm, channel depth 600 nm, doping 6.5×10^{16} cm^{-3}), the room

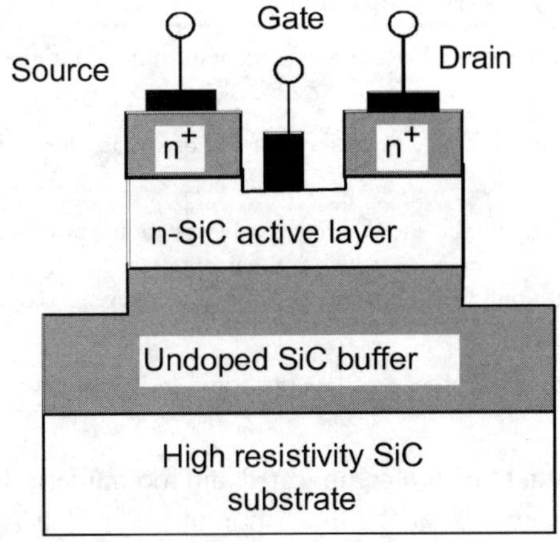

FIGURE 7.14 Cross-section of SiC MESFET (From Shur[22]).

temperature transconductance was approximately 4 mS/mm. At elevated temperatures, the device transconductance decreases owing to the decrease in mobility. MESFETs did not exhibit breakdown even with drain voltages up to 100 V. Using the square law MESFET model (described below), Kelner and Shur[24] estimated that the field effect mobility in these MESFETs was approximately 300 cm^2/Vs. β-SiC MESFETs have also been fabricated[25] but α-SiC MESFETs exhibit better performance because of better material quality.

α-SiC MESFETs have achieved microwave operation.[26] A cutoff frequency of 5 GHz, 12 dB gain at 2 GHz, and a breakdown voltage of 200 V was demonstrated in an α-SiC MESFET with a 0.4 μm gate length.[27]

GaN is another material that is potentially important for MESFET applications. For GaN at room temperature and with an n-type doping density of 10^{17} cm^{-3}, Monte Carlo simulations predict a high peak velocity (2.7×10^5 m/s), a high saturation velocity (1.5×10^5 m/s), and a high electron mobility (1000 cm^2/Vs).[28–31] Khan et al.[32] and Binari et al.[33] reported on microwave performance of GaN MESFETs. However, most of the research on GaN-based FETs has concentrated on GaN-based Heterostructure Field Effect Transistors.[34,35]

7.6 MESFET Modeling

MESFET modeling has been done at several different levels. Most advanced numerical simulation techniques rely on self-consistent simulation based on the Monte Carlo approach. In this approach, random number generators are used to simulate random electron scattering processes. The motion of these electrons is simulated in the electric field that is calculated self-consistently by solving the Poisson equation iteratively. The particle movements between scattering events are described by the laws of classical mechanics, while the probabilities of the various scattering processes and the associated transition rates are derived from quantum mechanical calculations. Some of the results obtained by using this approach were reviewed in Reference 36. Table 7.2 from Reference 36 describes the Monte Carlo algorithm in more detail.

Self-consistent Monte Carlo simulations are very useful for revealing the device physics and verifying novel device concepts and ideas.[22,37–41]

A less rigorous, but also less numerically demanding approach relies on solving the balance equations. These partial differential equations describe conservation laws derived from the Boltzmann Transport Equation.[42,43] Two-dimensional device simulators based on the balance equations and on the drift-diffusion model can be used to optimize device design and link the device characteristics to the device fabrication process.[44–46]

A more simplistic and easier approach is to use conventional drift-diffusion equations implemented in commercial two-dimensional and three-dimensional device simulators, such as ATLAS or MEDICI.

TABLE 7.2 Monte Carlo Algorithm[36]

Generate random number r and determine the duration of the free flight.
↓
Record the time the particle spends in each cell of k-space during the free flight.
↓
Generate random numbers to determine which scattering process has occurred, and which is the final state.
Repeat until the desired number of scattering events is reached.
↓
Calculate the distribution function, the drift velocity, the mean energy, etc.

However, even this approach might be too complicated and too numerically involved for the simulation of MESFET-based digital VLSI and/or for the simulation of MESFET-based analog circuits.

The simplest model that relates the MESFET current-voltage characteristics to the electron mobility, the electron saturation velocity, the device dimensions, and applied voltages is called the square-law model. This model predicts the following equation for the drain saturation current:

$$I_{sat} = \beta\left(V_{GS} - V_T\right)^2, \tag{7.13}$$

where[47,48]

$$\beta = \frac{2\varepsilon_s \mu_n v_s W}{A\left(\mu_n V_{po} + 3v_s L\right)} \tag{7.14}$$

is the transconductance parameter,

$$V_T = V_{bi} - V_{po} \tag{7.15}$$

is the threshold voltage, V_{GS} is the intrinsic gate-to-source voltage, and

$$V_{po} = \frac{qN_d A^2}{2\varepsilon_s} \tag{7.16}$$

is the pinch-off voltage. Here A is the channel thickness, μ_n is a low field mobility, and v_s is the electron saturation velocity.

This "square law" model is fairly accurate for devices with relatively low pinch-off voltages ($V_{po} = V_{bi} - V_T \le 1.5 \sim 2$ V). For devices with higher pinch-off voltages, the model called the **Raytheon model** (which is implemented in many versions of SPICE) yields a better agreement with experimental data:

$$I_{sat} = \frac{\beta\left(V_{GS} - V_T\right)^2}{1 + t_c\left(V_{GS} - V_T\right)} \tag{7.17}$$

Here t_c is an empirical parameter that depends on the doping profile in the MESFET channel. Another empirical model (called the **Sakurai-Newton model**) is also quite useful for MESFET modeling:

$$I_{sat} = \beta_{sn}\left(V_{GS} - V_T\right)^{m_{sn}} \tag{7.18}$$

The advantage of this model is simplicity. The disadvantage is that the empirical parameters β_{sn} and m_{sn} cannot be directly related to the device and material parameters. (The Sakurai-Newton model is implemented in several versions of SPICE. In AIM-Spice,[49,50] this model is implemented as Level 6 MOSFET model.)

The source and drain series resistances, R_S and R_d, may play an important role in determining the current-voltage characteristics of GaAs MESFETs. The intrinsic gate-to-source voltage, V_{GS}, is given by

$$V_{GS} = V_{gs} - I_{ds} R_S \tag{7.19}$$

where V_{gs} is the applied (extrinsic) gate-to-source voltage. Substituting Eq. (7.19) into Eq. (7.16) and solving for I_{sat} we obtain

$$I_{sat} = \frac{2\beta V_{gt}^2}{1 + 2\beta V_{gt} R_S + \sqrt{1 + 4\beta V_{gt} R_S}} \tag{7.20}$$

In device modeling suitable for computer-aided design, one has to model the current-voltage characteristics in the entire range of drain-to-source voltages, not only in the saturation regime. In 1980, Curtice proposed the use of a hyperbolic tangent function for the interpolation of MESFET current-voltage characteristics

$$I_d = I_{sat}\left(1 + \lambda V_{ds}\right) \tanh\left(\frac{g_{ch}}{I_{sat}}\right), \tag{7.21}$$

where

$$g_{ch} = \frac{g_i}{1 + g_i\left(R_S + R_d\right)} \tag{7.22}$$

is the MESFET conductance at low drain-to-source voltages, and

$$g_i = g_{cho}\left(1 - \sqrt{\frac{V_{bi} - V_{GS}}{V_{po}}}\right) \tag{7.23}$$

is the intrinsic channel conductance at low drain-to-source voltages predicted by the Shockley model.

The constant λ in Eq. (7.21) is an empirical constant that accounts for the output conductance in the saturation regime. This output conductance may be related to short channel effects and also to parasitic leakage currents in the substrate. Hence, output conductance may be reduced by using a heterojunction buffer layer between the device channel and the substrate or by using a *p*-type buffer layer. Such a layer creates an additional barrier, which prevents carrier injection into the substrate.[18]

The **Curtice model** is implemented in PSpice™. The Curtice model and the Raytheon model [see Eq. (7.23)] have become the most popular models used for MESFET circuit modeling. A more sophisticated model, which describes both subthreshold and above-threshold regimes of MESFET operation, is implemented in AIM-Spice. This model accurately reproduces current-voltage characteristics over several decades of currents and is suitable for both analog and digital circuit simulations.[51] One of the simulation results obtained using this model is depicted in Fig. 7.15.

In order to simulate MESFET circuits, one also needs to have a model describing the MESFET capacitances. Meyer[53] proposed a simple charge-control model, in which capacitances ($C_{ij} = C_{ji}$) were obtained as derivatives of the gate charge with respect to the various terminal voltages. Fjeldly et al.[52,54]

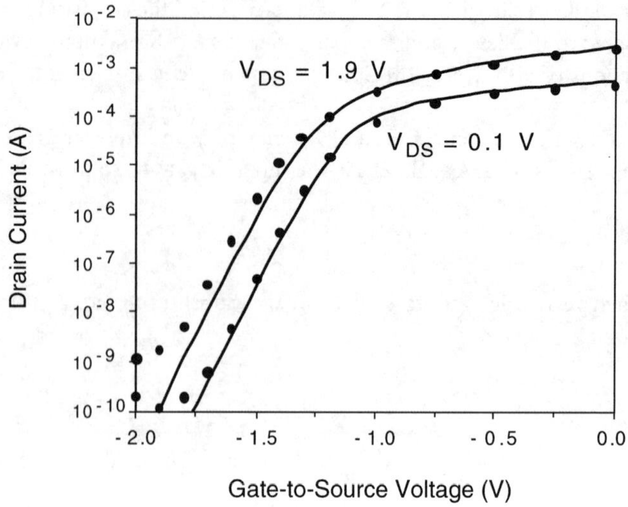

FIGURE 7.15 Subthreshold experimental (symbols) and calculated (solid lines) *I-V* characteristics for ion-implanted MESFET with nominal gate length $L = 1\ \mu m$.[52]

approximated a unified gate-channel capacitance C_{gc} of a MESFET at zero drain-source bias by the following combination of the above-threshold capacitance C_a and the below-threshold capacitance C_b:

$$C_{gc} = \frac{C_a C_b}{C_a + C_b}. \tag{7.24}$$

This approach in conjunction with Meyer's model, leads to the following expressions for the gate-to-source (C_{gs}) and gate-to-drain capacitance (C_{gd}) valid for the sub-threshold and the above-threshold regimes:

$$C_{gs} = \frac{2}{3} C_{gc} \left[1 - \left(\frac{V_{sat} - V_{dse}}{2V_{sat} - V_{dse}} \right)^2 \right], \tag{7.25}$$

$$C_{gs} = \frac{2}{3} C_{gc} \left[1 - \left(\frac{V_{sat}}{2V_{sat} - V_{dse}} \right)^2 \right]. \tag{7.26}$$

Here, V_{sat} is the extrinsic saturation voltage and V_{dse} is an effective extrinsic drain-source voltage that equals V_{ds} for $V_{ds} < V_{sat}$ and V_{sat} for $V_{ds} > V_{sat}$.

A more accurate model of the intrinsic capacitances requires an analysis of the variation of the charge distribution in the channel versus terminal bias voltages. For the MESFET, the depletion charge under the gate has to be partitioned between the source and drain terminals.[54]

Finally, the gate leakage current has to be modeled in order to accurately reproduce MESFET current voltage characteristics in the entire range of bias voltages including positive gate biases. To a first order approximation, the gate leakage current can be described in terms of simple diode equations assuming that each "diode" represents half of the gate area:

$$I_g = J_{ss} \frac{LW}{2} \left[\exp\left(\frac{V_{gs}}{m_{gs} V_{th}} \right) + \exp\left(\frac{V_{gd}}{m_{gd} V_{th}} \right) - 2 \right]. \tag{7.27}$$

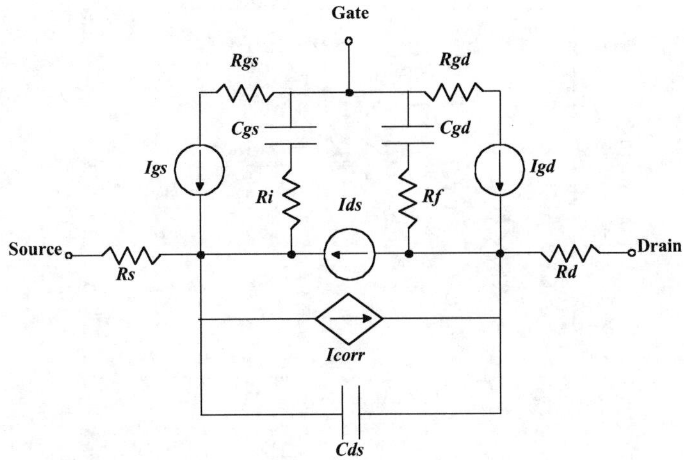

FIGURE 7.16 MESFET equivalent circuit. (After Shur et al.[52])

Here L and W are the gate length and width, respectively, J_{ss} is the saturation current, V_{gs} and V_{gd} are gate-to-source and gate-to-drain voltages, V_{th} is the thermal voltage, and m_{gs} and m_{gd} are the ideality factors. Figure 7.16 shows a more accurate equivalent circuit, which accounts for the effect of the leakage current on the drain current. In the equivalent circuit shown in Fig. 7.16, this effect is accounted for by the current controlled current source, I_{corr}.

Here, J_{ss} is the reverse saturation current density, and m_{gs} and m_{gd} are the gate-source and gate-drain ideality factors, respectively.

A more accurate description proposed by Berroth et al.[55] introduced effective electron temperatures at the source side and the drain side of the channel. The electron temperature at the source side of the channel T_s is taken to be close to the lattice temperature, and the drain side electron temperature T_d is assumed to increase with the drain-source voltage to reflect the heating of the electrons in this part of the channel. The resulting gate leakage current can be written as

$$I_g = J_{gs}\frac{LW}{2}\left[\exp\left(\frac{V_{gs}}{m_{gs}V_{ths}}\right)-1\right]+\frac{LW}{2}\left[J_{gd}\exp\left(\frac{V_{gd}}{m_{gd}V_{thd}}\right)-J_{gs}\right], \tag{7.28}$$

where J_{gs} and J_{gd} are the reverse saturation current densities for the gate-source and the gate-drain diodes, respectively, and $V_{ths} = k_BT_s/q$ and $V_{thd} = k_BT_d/q$. The second term in Eq. (7.28) accounts for the gate-drain leakage current and for the fact that the effective temperature of the electrons in the metal is maintained at the ambient temperature.

In GaAs MESFETs, the reverse gate saturation current is usually also dependent on the reverse bias.[56] The following expression accounts for this dependence:[52]

$$\begin{aligned} I_g = J_{gs}\frac{LW}{2}\left[\exp\left(\frac{V_{gs}}{m_{gs}V_{ths}}\right)-1\right]+\frac{LW}{2}g_{gs}V_{gs}\exp\left(-\frac{qV_{gs}\delta_g}{k_BT_s}\right)+ \\ \frac{LW}{2}\left[J_{gd}\exp\left(\frac{V_{gd}}{m_{gd}V_{thd}}\right)-J_{gs}\right]+\frac{LW}{2}g_{gd}V_{gd}\exp\left(-\frac{qV_{gd}\delta_g}{k_BT_s}\right), \end{aligned} \tag{7.29}$$

where g_{gs} and g_{gd} are the reverse diode conductances and δ_g is the reverse bias conductance parameter. However, using the above expressions directly will cause a kink in the gate current and a discontinuity

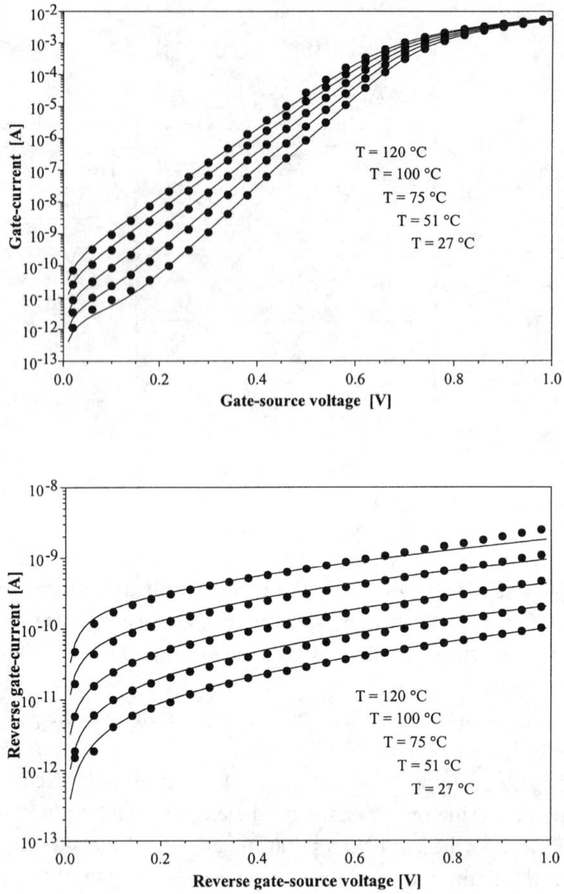

FIGURE 7.17 Measured (symbols) and simulated (lines) gate current versus gate bias for (a) positive and (b) negative gate-source voltages at different temperatures. Temperature parameters: Φ_{b1} = 0.96 meV/K, ξ = 0.033 K^{-1}. (After Ytterdal et al.[6])

in its derivatives at zero applied voltage. Equation (7.29) is valid for both negative and positive values of V_{gs} and V_{gd}.

Figure 7.17 compares the measured gate leakage current with the model implemented in AIM-Spice and described above.[6]

Figure 7.18 shows that the GaAs MESFET model implemented in AIM-Spice accurately reproduces the differential characteristics of the devices. Therefore, this model is suitable for the simulations of analog, microwave, and mixed-mode circuits.

7.7 Hetero-Dimensional (2D MESFETs)

Hetero-dimensional MESFET technology utilizes the Schottky contact between a 3D metal and a 2D electron-gas in a semiconductor. This technology holds promise for the fabrication of high-speed devices with low power consumption.[57–62] However, this is still a very immature technology that has not found its way into production.

Figure 7.19 shows the 3D-2D Schottky barrier junction.

The depletion width d_{dep} of the semiconductor 2D electron gas for a reverse biased 3D-2D Schottky barrier shows a linear instead of a square root dependence of voltage:[58]

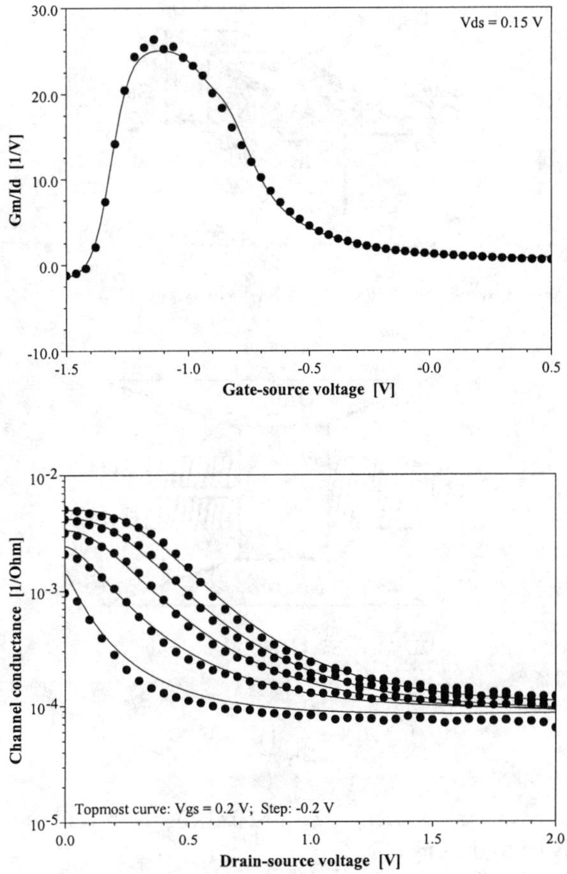

FIGURE 7.18 Measured (symbols) and simulated using AIM-Spice (lines) drain current characteristics of device A operating at room temperature; (a) ratio of transconductance and drain current, (b) channel conductance. (After Ytterdal et al.[6])

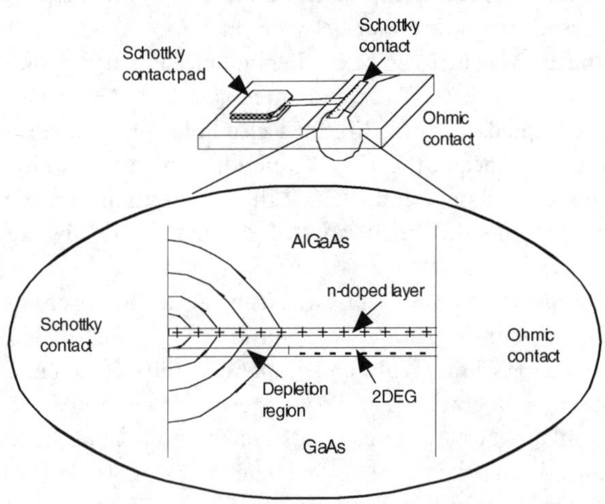

FIGURE 7.19 Schematic structure of a 3D-2D Schottky diode. (After Peatman et al.[63])

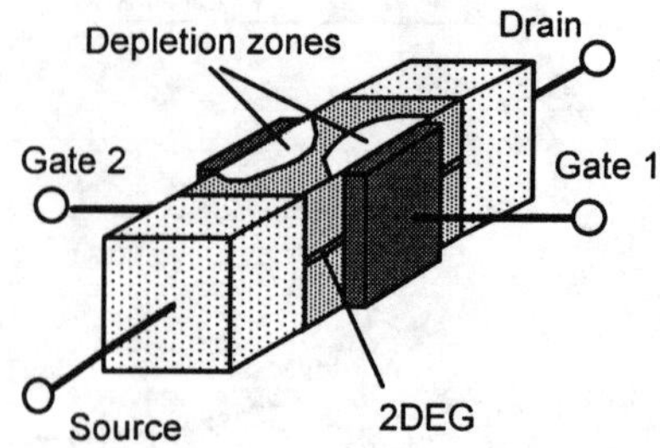

FIGURE 7.20 Schematic structure of a 2D MESFET. (After Peatman et al.[67])

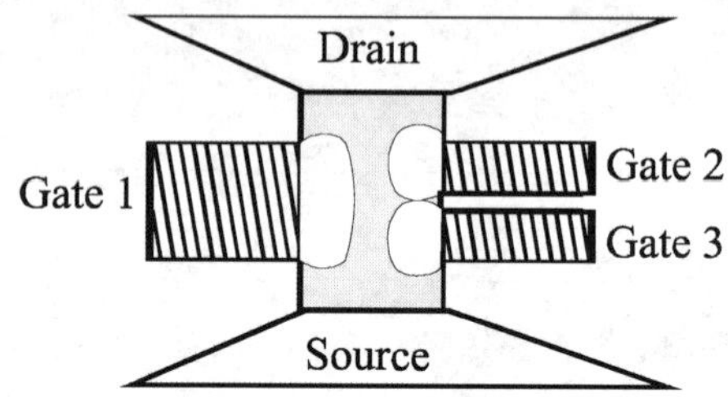

FIGURE 7.21 Structure of a three-gate 2D MESFET. (After Iñiguez et al.[69])

$$d_{dep} = \frac{\varepsilon}{qn_s}\left(V_{bi} - V\right) \tag{7.30}$$

Here n_s is the sheet density in the 2D electron gas (2-DEG), V_{bi} is the built-in voltage of the junction, and V is the voltage applied to the junction.

Figure 7.20 shows the two-dimensional metal-semiconductor field-effect transistor (2D MESFET). This transistor utilizes Schottky gates on both sides of a degenerate 2-DEG channel to laterally modulate the current between the drain and source.[65–66]

The novel geometry of this 2D MESFET eliminates or reduces parasitic effects associated with top planar contacts of conventional FETs, such as narrow-channel and short-channel effects. The output conductance in the saturation regime is quite small, and the junction capacitance of the 3D-2D Schottky diode is also small. This results in a low power-delay product.

The functionality of the 2D MESFET can be further enhanced by using multiple gates on both sides of the channel, as shown in Fig. 7.21. Two- and three-gate 2D MESFETs with excellent electrical performance have been fabricated.[68] The 2D MESFET also holds promise for microwave analog applications, where the small capacitance of the 3D-2D junction should lead to low channel and amplifier noise. Also, the high transconductance, and the fact that the transconductance and the output conductance do not vary much across the broad range of gate biases, are advantageous factors for linear amplification.

The 2D MESFET prototype devices were fabricated using a pseudomorphic $Al_{0.25}Ga_{0.75}As/In_{0.2}Ga_{0.8}As$ hetero-structure grown on a semi-insulating GaAs substrate.[68–70,72–74] Ni/Ge/Au ohmic contacts were formed using standard contact UV lithography and evaporation/lift-off techniques. The gate pattern was defined using electron beam lithography. The Pt/Au gates were deposited into the gate trench using capacitor discharge electroplating. Cr/Au contact pads were evaporated on the wafer and a wet etch was used to isolate the ohmic and Schottky pads (see Reference 64 for more details.)

A unified 2D MESFET is described in Reference 69. Figures 7.22 and 7.23 show the comparison between the measured and calculated 2D-MESFET I-V characteristics.

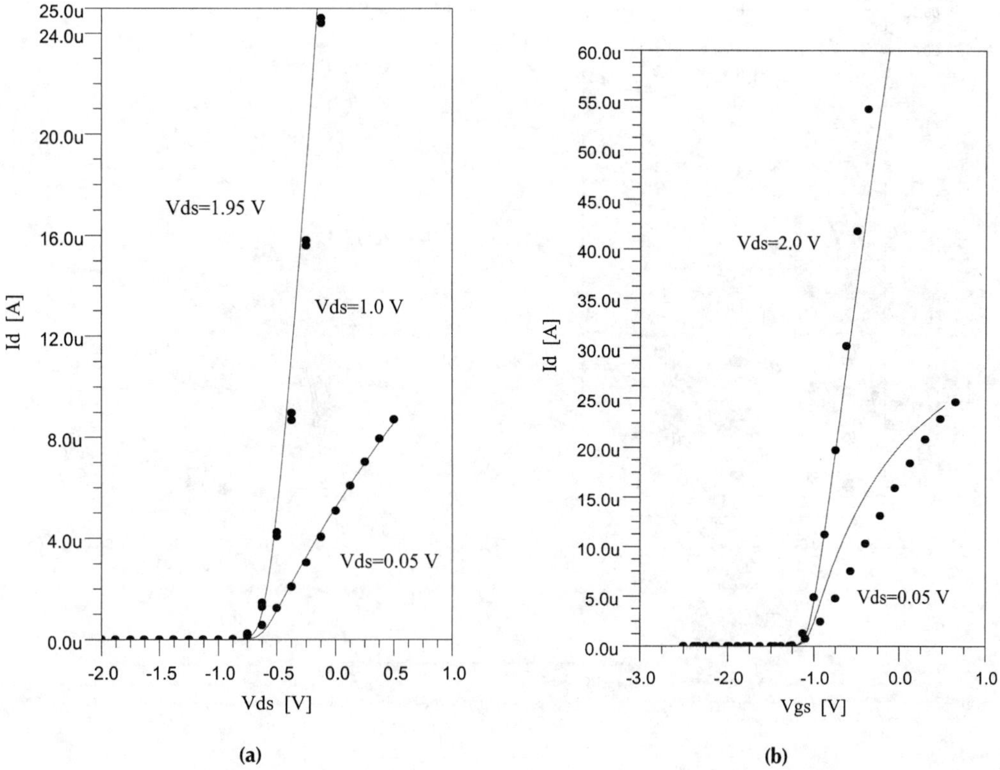

FIGURE 7.22 I-V characteristics for 2D MESFETs with the gates tied together for (a) L = 3 μm, (b) L = 0.5 μm. (From Iñiguez et al.[69])

7.8 Applications

GaAs MESFETs play an important role in both analog and digital applications, such as in satellite and fiber-optic communication systems, in cellular phones and other wireless equipment, in automatic IC test equipment, and for other diverse civilian and military uses. GaAs MESFETs have been used in highly efficient microwave power amplifiers, since they combine low on resistance and high cutoff frequency. GaAs semi-insulating substrates also present a major advantage for microwave applications, since they decrease parasitic capacitance and allow for fabrication of passive elements with low parasitics for microwave monolithic integration. GaAs MESFETs have also found applications in linear low-noise amplifiers.

Figure 7.24 compares the cutoff frequencies and maximum frequencies of oscillations for different GaAs technologies.

As can be seen from the figure, GaAs MESFETs exhibit quite respectable microwave performance and, given a lower cost of GaAs MESFETs compared to more advanced hetero-structure devices, they could capture a sizeable portion of the microwave market.

GaAs MESFET technology has also been used in efficient DC-to-DC converters that demonstrated a high switching speed.[71] These devices are capable of operating at a higher switching speeds and require a less complex circuitry.

Vitesse Semiconductor Corporation is one of the leaders in digital GaAs MESFET technology. VSC8141 and the VSC8144 SONET/SDH OC-48 multi-rate transceivers include multiplexer and demultiplexer with integrated clock generation capabilities for the physical layer.[72] Both ICs dissipate the lowest power available in the industry today — 1.2 W typically. These integrated circuits are suitable for transmission

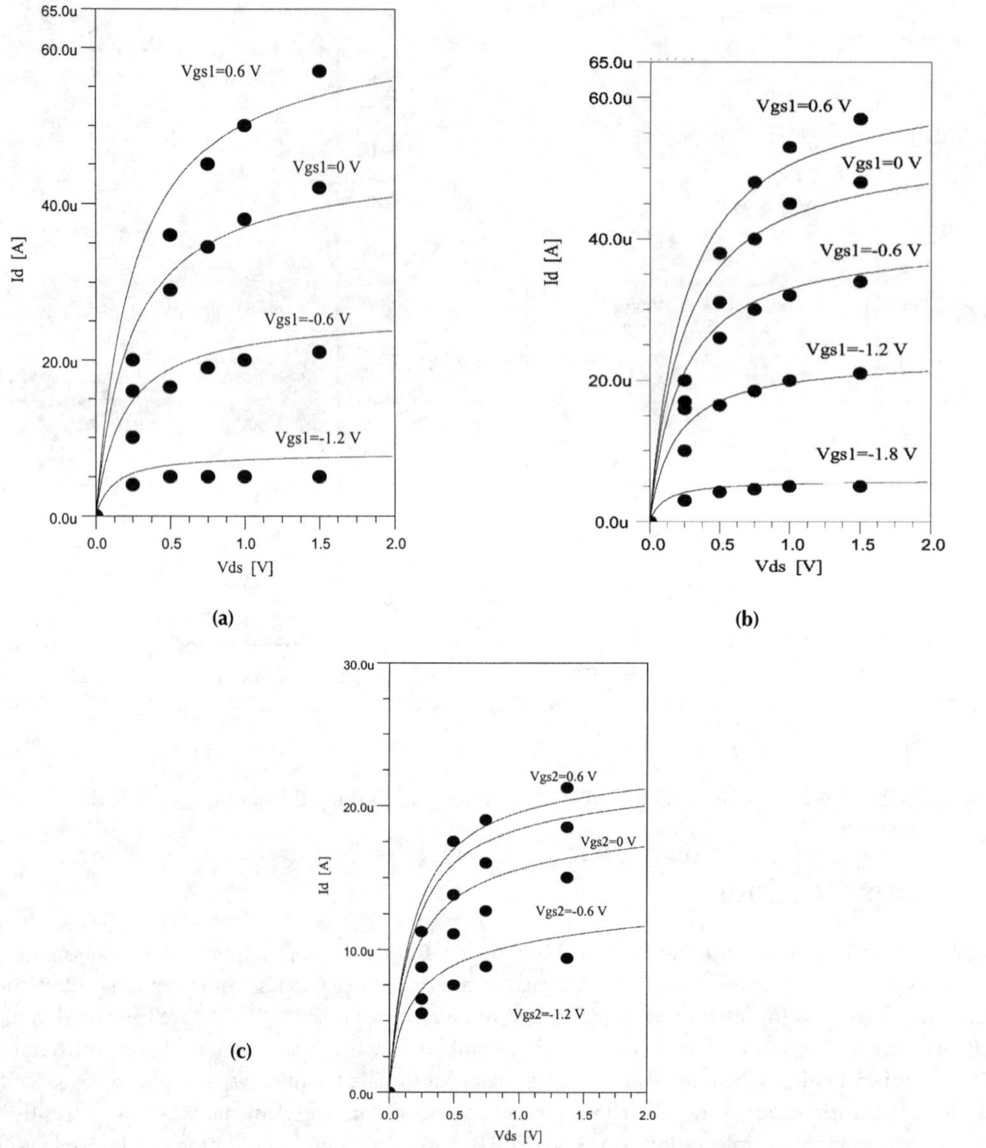

FIGURE 7.23 *I-V* characteristics for a three-gate 2D MESFET with $L = 2$ μm and $W = 1$ μm; (a) $V_{GS1} = V_{GS2} = V_{GS2} = V_{GS3}$, (b) $V_{GS2} = V_{GS3} = 0.6$ V. (c) $V_{GS1} = -1.2$ V, $V_{GS3} = 0.6$ V. Symbols: measurements. Solid lines: AIM-Spice simulations.

systems, optical networking equipment, networking and digital cross-connect systems, and they have 20% lower power dissipation than competing products.

Acknowledgments

The author is grateful to Mr. Tobias Werner for useful comments.

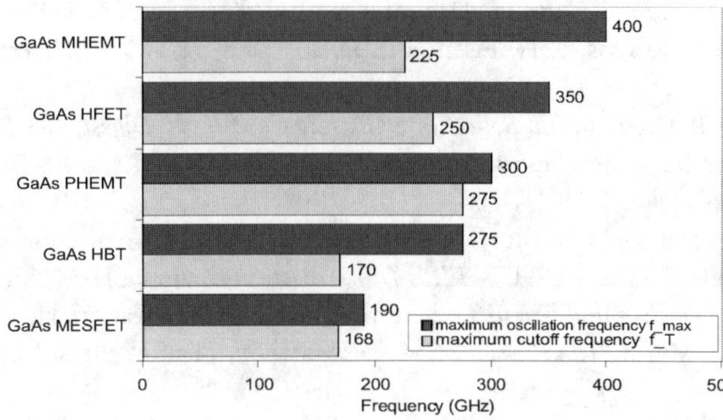

FIGURE 7.24 Maximum frequency of oscillation f_{max} and maximum cutoff frequency f_T of different GaAs technologies.[70]

References

1. C. A. Mead, Schottky barrier gate field effect transistor, *Proc. IEEE*, 54, 2, 307–308, Feb. 1966.
2. W. W. Hooper and W. I. Lehrer, An epitaxial GaAs field-effect transistor, *Proc. IEEE*, 55, 7, 1237–1238, July 1967.
3. Y. Nakayama, K. Suyama, H. Shimizu, N. Yokoyama, H. Ohnishi, A. Shibatomi, and H. Ishikawa, A 16 × 16 bit parallel multiplier, *IEEE J. Solid-State Circuits*, SC-18, 599–603, 1983.
4. M. Feng, C. L. Lau, V. Eu, and C. Ito, Does the two-dimensional electron gas effect contribute to high-frequency and high speed performance of field-effect transistors?, *Appl. Phys. Lett.*, 57, 1233, 1990.
5. A. Mills, The GaAs IC business never so healthy! *GaAs IC Report, III-Vs Review,* 13, 1, 35–39, Jan. 2000.
6. T. Ytterdal, B-J. Moon, T. A. Fjeldly, and M. S. Shur, Enhanced GaAs MESFET CAD model for a wide range of temperatures, *IEEE Trans. Electron Devices,* 42, 10, 1724–1734, 1995.
7. M. S. Shur and L. F. Eastman, Ballistic transport in semiconductors at low-temperatures for low power high speed logic, *IEEE Trans. Electron Devices*, 26, 11, 1677–1683, Nov. 1979.
8. M. Heiblum, M. I. Nathan, D. C. Thomas, and C. M. Knoedler, Direct observation of ballistic transport in GaAs, *Phys. Rev. Lett.*, 55, 2200, 1985.
9. A. F. J. Levi, J. R. Hayes, P. M. Platzman, and W. Wiegmann, Injected hot electron transport in GaAs, *Phys. Rev. Lett.*, 55, 2071–2073, 1985.
10. G. Ruch, Electronics dynamics in short channel field-effect transistors, *IEEE Trans. Electron Devices*, ED-19, 652–654, 1972.
11. A. Cappy, B. Carnes, R. Fauquembergue, G. Salmer, and E. Constant, *IEEE Trans. Electron Devices*, ED-27, 2158–2168, 1980.
12. M. S. Shur and M. Asif Khan, Electronic and Optoelectronic AlGaN/GaN Heterostructure Field Effect Transistors, in *Proceedings of the Symposium on Wide Band Gap Semiconductors and the Twenty-Third State-of-the-Art Program on Compound Semiconductors* (SOTAPOCS XXIII), F. Ren, D. N. Buckley, S. J. Pearton, P. Van Daele, G. C. Chi, T. Kamijoh, and F. Schuermeyer, eds., Proceedings Volume 95-21, 128–135, The Electrochemical Society, Inc., New Jersey, 1995.
13. Foutz, B. E., L. F. Eastman, U. V. Bhapkar, and M. S. Shur, Comparison of high electron transport in GaN and GaAs, *Appl. Phys. Lett.*, 70, 21, 2849–2851, 1997.
14. S. Roosild, in *Microprocessor Design for GaAs Technology*, V. Milutinovic, Editor, Prentice Hall, Englewood Cliffs, NJ, 1990.

15. M. S. Shur, *Introduction to Electronic Devices,* John Wiley and Sons, New York, 1996.
16. J. A. Higgins, R. L. Kuvaas, F. H. Eisen, and D. R. Chen, *IEEE Trans. Electron Devices*, ED-25, 587–596, 1978.
17. B. M. Welch and R. C. Eden, *Int. Solid State Circuits Conf. Tech. Digest*, 205–208, 1977.
18. L. F. Eastman and M. S. Shur, Substrate Current in GaAs MESFET's, *IEEE Trans. Electron Devices,* ED-26, 9, 1359–61, Sept. 1979.
19. C. H. Chen, A. Peczalski, M. S. Shur, and H. K. Chung, Orientation and ion-implanted transverse effects in self-aligned GaAs MESFETs, *IEEE Trans. Electron Devices,* ED-34, 7, 1470–1481, July 1987.
20. M. S. Shur, *GaAs Devices and Circuits,* Plenum Publishing Corporation, New York, 1987.
21. I. Brodie and J. J. Muray, *The Physics of Microfabrication,* Plenum Press, New York and London, 1982.
22. M. S. Shur, SiC Transistors, in *SiC Materials and Devices,* Vol. 52, Y. S. Park, ed., Academic Press, New York, 161–193, 1998.
23. J. W. Palmour, H. S. Kong, D. G. Waltz, J. A. Edmond, and C. H. Carter, Jr., *Proc. of First Intern. High Temperature Electronics Conference,* Albuquerque, NM, 511–518, 1991.
24. G. Kelner and M. Shur, SiC Devices, in *Properties of Silicon Carbide*, G. Harris, ed., M. Faraday House, IEE, England, 1995.
25. G. Kelner, S. Binari, K. Sleger, and H. Kong, *IEEE Electr. Dev. Lett.,* 8, 428, 1987.
26. J. W. Palmour and J. A. Edmond, *Proc. 14th IEEE Cornell Conf.*, Ithaca, NY, 1991.
27. S. Sriram, R. C. Clarke, M. H. Hanes, P. G. McMullin, C. D. Brandt, T. J. Smith, A. A. Burk, Jr., H. M. Hobgood, D. L. Barrett, and R. H. Hopkins, SiC Microwave Power MESFETS, *Inst. Phys. Conf. Ser.*, 137, 491–494, 1993.
28. R. F. Davis, G. Kelner, M. Shur, J. W. Palmour, and J. A. Edmond, Thin Film Deposition and Microelectronic and Optoelectronic Device Fabrication and Characterization in Monocrystalline Alpha and Beta Silicon Carbide, *Proc. of IEEE*, 79, 5, 677–701, May 1991 .
29. H. Morkoc, S. Strite, G. B. Gao, M. E. Lin, B. Sverdlov, and M. Burns, Large-band-gap SiC, III-V Nitride, and II-VI ZnSe-based semiconductor device technologies, *J. Appl. Phys.*, 76, 3, 1363–1398, Aug. 1994.
30. M. A. Littlejohn, J. R. Hauser, and T. H. Glisson, *Appl. Phys. Lett.*, 26, 625, 1975.
31. B. Gelmont, K. S. Kim, and M. Shur, *J. Appl. Phys.*, 74, 1818, 1993.
32. M. A. Khan, J. N. Kuznia, D. T. Olson, W. Schaff, G. Burm, M. S. Shur, and C. Eppers presented at Device Research Conference, Boulder, Colorado, 1994.
33. S. Binari, L. B. Rowland, W. Kruppa, G. Kelner, K. Doverspike, and D. K. Gatskill, *Electronics Letters,* 30, 15, 1248, July 1994.
34. R. Gaska, M. S. Shur, and A. Khan, GaN-based HEMTs, Gordon & Breach Science Publishers, O. Manasreh an Ed Yu, eds., in press.
35. M. S. Shur and M. A. Khan, *GaN and AlGaN Devices: Field Effect Transistors and Photodetectors,* S. Pearton, ed., Gordon and Breach Science Publishers, Amsterdam, 1999, 47–86.
36. G. U. Jensen, B. Lund, M. S. Shur, and T. A. Fjeldly, Monte Carlo simulation of semiconductor devices, *Computer Physics Communications,* 67, 1, 1–61, 1991.
37. K. Hess K and C. Kizilyalli, Scaling and transport properties of high electron mobility transistors, *IEDM Technical Digest*, Los Angeles, 556–558, 1986.
38. A. Afzalikushaa and G. Haddad, High-frequency characteristics of MESFETs, *Solid-State Electronics*, 38, 2, 401–406, Feb. 1995.
39. C. Jacoboni and P. Lugli, *The Monte Carlo Method for Semiconductor Simulation*, Springer Verlag, Vienna, 1989.
40. C. Moglestue, *Monte Carlo Simulation of Semiconductor Devices*, Chapman & Hall, London, 1993.
41. K. Tomizawa, *Numerical Simulation of Submicron Semiconductor Devices*, Artech House, Boston, 1993.
42. M. Lundstrom, *Fundamentals of Carrier Transport*, Addison-Wesley, Reading, MA, 1990.

43. K. Bløtekjær, Transport equations for two-valley semiconductors, *IEEE Trans. Electron Devices*, 17, 38–47, 1970.
44. J. Jyegal and T. A. Demassa, New nonstationary velocity overshoot phenomenon in submicron Gallium-Arsenide field-effect transistors, *J. of Appl. Physics*, 75, 6, 3169–3175, Mar. 1994.
45. S. H. Lo and C. P. Lee, Analysis of surface-state effect on gate lag phenomena in GaAs-MESFETs, *IEEE Trans. Electron Devices*, 41, 9, 1504–1512, Sept. 1994.
46. BLAZE, *Atlas II User's Manual*, Silvaco, 1993.
47. M. S. Shur, Analytical Models of GaAs FETs, *IEEE Trans. Electron Devices*, ED-32, 1, 70–72, Jan. 1985.
48. M. S. Shur, *Introduction to Electronic Devices*, John Wiley and Sons, New York, 1996.
49. K. Lee, M. S. Shur, T. A. Fjeldly, and T. Ytterdal, *Semiconductor Device Modeling for VLSI*, Prentice Hall, Englewood Cliffs, NJ, 1993.
50. T. Fjeldly, T. Ytterdal, and M. S. Shur, *Introduction to Device and Circuit Modeling for VLSI*, John Wiley and Sons, New York, 1998.
51. B. Iñiguez, T. A. Fjeldly, M. S. Shur, and T. Ytterdal, Spice modeling of compound semiconductor devices, in Special Issue Silicon and Beyond. Advanced Device Models and circuit simulators, *IJHSES*, M. S. Shur and T. A. Fjeldly, eds., in press.
52. M. S. Shur, T. Fjeldly, Y. Ytterdal, and K. Lee, Unified GaAs MESFET model for circuit simulations, *International Journal of High Speed Electronics*, 3, 2, 201–233, June 1992.
53. J. E. Meyer, MOS models and circuit simulation, *RCA Review*, 32, 42–63, 1971.
54. M. Nawaz and T. A. Fjeldly, A new charge conserving capacitance model for GaAs MESFETs, *IEEE Trans. Electron Devices*, 44, 11, 1813–1821, 1997.
55. M. Berroth, M. Shur, and W. Haydl, Experimental studies of hot electron effects in GaAs MESFETs, in *Extended Abstracts of the 20th International Conf. on Solid State Devices and Materials* (SSDM-88), Tokyo, 255–258, 1988.
56. C. Dunn, *Microwave Semiconductor Devices and Their Applications*, H. A. Watson, ed., McGraw Hill, New York, 1969.
57. B. Gelmont, M. Shur, and C. Moglestue, Theory of junction between two-dimensional electron gas and p-type semiconductor, *IEEE Trans. Electron Devices*, 39, 5, 1216–1222, 1992.
58. S. G. Petrosyan and Y. Shik, Contact phenomena in a two dimensional electron gas, *Sov. Phys.-Semicond.*, 23, 6, 696–697, 1989.
59. W. C. B. Peatman, T. W. Crowe, and M. S. Shur, A novel Schottky/2-DEG diode for millimeter and submillimeter wave multiplier applications, *IEEE Electron Device Lett.*, 13, 11–13, 1992.
60. W. C. B. Peatman, H. Park, B. Gelmont, M. S. Shur, P. Maki, E. R. Brown, and M. J. Rooks, Novel metal/2-DEG junction transistors, *Proc. 1993 IEEE/Cornell Conference*, Ithaca, NY, 314–319, 1993.
61. W. C. B. Peatman, H. Park, and M. Shur, Two-dimensional metal-semiconductor field effect transistor for ultra low power circuit applications, *IEEE Electron Device Lett.*, 15, 7, 245–247, 1994.
62. M. S. Shur, W. C. B. Peatman, H. Park, W. Grimm, and M. Hurt, Novel heterodimensional diodes and transistors, *Solid-State Electronics*, 38, 9, 1727–1730, 1995.
63. W. C. B. Peatman, T. W. Crowe, and M. S. Shur, A novel Schottky/2-DEG diode for millimeter and submillimeter wave multiplier applications, *IEEE Electron Device Lett.*, 13, 1, 1992.
64. W. C. B. Peatman, M. J. Hurt, H. Park, T. Ytterdal, R. Tsai and M. Shur, Narrow channel 2-D MESFET for low power electronics, *IEEE Trans. Electron Devices*, 42, 9, 1569–1573, 1995.
65. W. C. B. Peatman, R. Tsai, T. Ytterdal, M. Hurt, H. Park, J. Gonzales, and M. Shur, Sub-half-micrometer width 2-D MESFET, *IEEE Electron Device Lett.*, 17, 2, 40–42, 1996.
66. M. Hurt, W. C. B. Peatman, R. Tsai, T. Ytterdal, M. Shur, and B. J. Moon, An ion-implanted 0.4 μm wide 2-D MESFET for low-power electronics, *Electronics Lett.*, 32, 8, 772–773, 1996.
67. W. C. B. Peatman, H. Park, B. Gelmont, M. S. Shur, P. Maki, E. R. Brown, and M. J. Rooks, Novel metal/2-DEG junction transistors, in *Proc. 1993 IEEE/Cornell Conf.*, Ithaca, NY, 314, 1993.

68. J. Robertson, T. Ytterdal, W. C. B. Peatman, R. Tsai, E. Brown, and M. Shur, RTD/2-D MESFET/RTD logic elements for compact, ultra low-power electronics, *IEEE Trans. Electron Devices,* 44, 7, 1033–1039, 1997.
69. B. Iñiguez, J. –Q. Lü, M. Hurt, W. C. B. Peatman, and M. S. Shur, Modeling and simulation of single and multiple gate 2-D MESFETs, *IEEE Trans. Electron Devices,* 46, 8, 1999.
70. T. Werner and M. S. Shur, GaAs microwave transistors, unpublished.
71. http://www.anadigics.com/GaAsline/mesfets.html
72. http://www.vitesse.com/news/101199.htm

8

High Electron Mobility Transistors

Prashant Chavarkar
CREE Lighting

Umesh K. Mishra
University of California

8.1 Introduction

The concept of modulation doping was first introduced in 1978.[1] In this technique electrons from remote donors in a higher bandgap material transfer to an adjacent lower gap material. The electrostatics of the heterojunction results in the formation of a triangular well at the interface, which confines the electrons in a two-dimensional (2D) electron gas (2DEG). The separation of the 2DEG from the ionized donors significantly reduces ionized impurity scattering resulting in high electron mobility and saturation velocity.

Modulation-doped field effect transistors (MODFETs) or high electron mobility transistors (HEMTs), which use the 2DEG as the current conducting channel have proved to be excellent candidates for microwave and millimeter-wave analog applications and high-speed digital applications. This progress has been enabled by advances in crystal growth techniques such as molecular beam epitaxy (MBE) and metal-organic chemical vapor deposition (MOCVD) and advances in device processing techniques, most notably electron beam lithography, which has enabled the fabrication of HEMTs with gate lengths down to 0.05 μm.

0-8493-1562-X/03/$0.00+$1.50

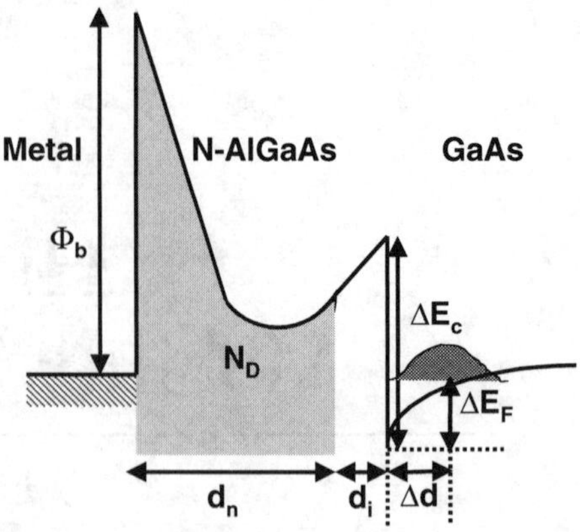

FIGURE 8.1 Schematic of conduction band diagram at the AlGaAs/GaAs interface.

However, using a high electron mobility channel alone does not guarantee superior high-frequency performance. It is crucial to understand the principles of device operation and to take into consideration the effect of scaling to design a microwave or millimeter-wave HEMT device. The advantages and limitations of the material system used to implement the device also need to be considered. This section therefore begins with a discussion on the device operation of a HEMT. This is followed by a discussion of scaling issues in HEMT, which are of prime importance, as the reduction of gate length is required to increase the operating frequency of the device.

The first HEMT was demonstrated in the AlGaAs/GaAs material system in 1981. It demonstrated significant performance improvements over the GaAs MESFET at microwave frequencies. However, the high-frequency performance was not sufficient for operation at millimeter-wave frequencies. In the past twenty years, the AlGaAs/InGaAs psuedomorphic HEMT on GaAs substrate (referred to as GaAs pHEMT) and the AlInAs/GaInAs HEMT on InP substrate (referred to as InP HEMT) have emerged as premier devices for microwave and millimeter-wave circuit applications. This highlights the importance of choosing the appropriate material system for device implementation. This will be discussed in the section on Material Systems for HEMT Devices.

The next two sections will discuss the major advances in the development of the GaAs pHEMT and InP HEMT. Traditionally these devices have been used in low-volume, high-performance and high-cost military and space-based electronic systems. Recently the phenomenal growth of commercial wireless and optical fiber-based communication systems has opened up new applications for these devices. This also means that new issues like manufacturability and operation at low bias voltage have to be addressed.

8.2 HEMT Device Operation and Design

8.2.1 Linear Charge Control Model

The current control mechanism in the HEMT is control of the 2DEG density at the heterojunction interface by the gate voltage. Figure 8.1 shows the band diagram along the direction perpendicular to the heterojunction interface using the AlGaAs/GaAs interface as an example.

The first HEMT charge control model was proposed by Delagebeaudeuf and Linh in 1982.[2] The potential well at the AlGaAs/GaAs interface is approximated by a triangular well. The energy levels in this triangular well and the maximum 2DEG density, n_{sm} can be calculated by solving the Schrödinger equation in the triangular well and Poisson equation in AlGaAs donor layers.[3] For $0 < n_s < n_{sm}$, the sheet charge density n_s as a function of gate voltage V_g can be expressed as

$$qn_s = C_s\left(V_g - V_{th}\right) \tag{8.1}$$

where C_s is the 2DEG capacitance per unit area and is given by the following expression:

$$C_s = \frac{\varepsilon}{d_n + d_i + \Delta d} \tag{8.2}$$

Here Δd is the distance of the centroid of the 2DEG distribution from the AlGaAs/GaAs interface and is typically or the order of 80 Å for $n_s \sim 10^{12}/\text{cm}^2$. Here V_{th} is the threshold voltage or pinch-off voltage and is given by,

$$V_{th} = \phi_b - \frac{qN_D}{2\varepsilon} d_n^2 - \Delta E_c + \Delta E_F \tag{8.3}$$

where ϕ_b, N_D, and d_n are the Schottky barrier height on the donor layer, doping density, and doped layer thickness as illustrated in Fig. 8.1. Here, ΔE_F is the Fermi potential of the 2DEG with respect to the bottom of the conduction band. It can be expressed as a function of 2DEG density as follows

$$\Delta E_F = \Delta E_{FO}(T) + an_s \tag{8.4}$$

where $\Delta E_{FO}(T) = 0$ at 300 K, $a = 0.125 \times 10^{-16}$ V/m^2.

This simplified version of charge control is accurate only at low temperature. At room temperature, apart from the 2DEG charge density n_s, the gate voltage also modulates the bound carrier density, n_{bound} in the donor layer and the free electrons, n_{free} in the donor layers. This results in premature saturation of the sheet charge and degradation of device performance. A more accurate model for charge control, which solves Poisson's and Schrödinger's equations in a self-consistent manner was proposed by Vinter.[4]

8.2.2 Modulation Efficiency

The parasitic modulation of charge in the higher bandgap donor layer reduces the efficiency of the gate voltage to modulate the drain current, as the carriers in the donor layers do not contribute the drain current. The modulation efficiency (η) of the FET is proportional to ratio between the change in drain current (δI_{ds}) and the change in total charge (δQ_{tot}) required to cause this change.[5] This ratio is defined as follows,

$$\eta \,\alpha\, \frac{\delta I_{ds}}{\delta Q_{tot}} = \frac{\delta(qv_{sat} n_s)}{\delta q(n_s + n_{bound} + n_{free})} \tag{8.5}$$

Dividing the numerator and denominator by the change in gate voltage, δV_g that is required to cause this change, the following expression is obtained,

$$\frac{\delta \mathrm{I_{ds}}}{\delta \mathrm{Q_{tot}}} = v_{sat} \frac{\delta(n_s)/\delta V_g}{\delta(n_s + n_{bound} + n_{free})/\delta V_g} \tag{8.6}$$

The modulation efficiency is defined as the ratio of the rate of change of the useful charge, i.e., the 2DEG over that of the total charge,

$$\eta = \frac{\delta(n_s)/\delta V_g}{\delta(n_s + n_{bound} + n_{free})/\delta V_g} = \frac{\delta(n_s)/\delta V_g}{C_{TOT}} = \frac{C_s}{C_{TOT}} \tag{8.7}$$

The relation between the modulation efficiency and high frequency performance of the FET is evident in the expressions for transconductance (g_m) and current gain cutoff frequency (f_T).

$$g_m = \frac{\delta I_{ds}}{\delta V_g} = \frac{\delta(qv_{sat}n_s)}{\delta V_g} = qv_{sat}(\delta n_s/\delta V_g);$$
$$C_{gs} = C_{TOT}L_g; \tag{8.8}$$
$$f_T = \frac{g_m}{2\pi C_{gs}} = \frac{qv_{sat}(\delta n_s/\delta V_g)}{2\pi L_g C_{TOT}} = \frac{v_{sat}}{2\pi L_g}\eta$$

Hence, to improve the high frequency performance it is essential to improve the modulation efficiency.

Equation (8.8) must be used with caution in case of short gate length HEMTs. The saturation velocity v_{sat} may be replaced by the effective velocity v_{eff}. Usually v_{eff} is higher than v_{sat} due to high field and velocity overshoot effects. Using v_{sat} in this case may lead to values of modulation efficiency that are greater than 100%.

8.2.3 Current-Voltage (I-V) Models for HEMTs

By assuming linear charge control, gradual channel approximation, and a 2-piece linear velocity-field model, the expression for the saturated drain current I_{DSS} in a HEMT is given by,[2]

$$I_{DSS} = C_s v_{sat}\left(\sqrt{(E_c L_g)^2 + (V_g - V_c(0) - V_{th})^2} - E_c L_g\right) \tag{8.9}$$

Here E_c is defined as the critical electric field at which the electrons reach their saturation velocity v_{sat} and $V_c(0)$ is the channel potential at the source end of the gate. For a long gate length HEMT, Eq. (8.9) is valid until the onset of donor charge modulation, that is, $0 < n_s < n_{sm}$. The intrinsic transconductance of the device obtained by differentiating this expression with respect to the gate voltage and is expressed as follows:

$$g_{mo} = \frac{\delta I_{ds}}{\delta V_g} = C_s v_{sat}\frac{V_g - V_c(0) - V_{th}}{\sqrt{(V_g - V_c(0) - V_{th})^2 + (E_c L_g)^2}} \tag{8.10}$$

For a short gate length HEMT, the electric field in the channel is much greater in magnitude than the critical electric field E_c. Assuming that the entire channel of the FET operates in saturated velocity mode, we can make the following assumption, that is, $V_g - V_c(0) - V_{th} \gg E_c L_g$. Then using Eqs. (8.1), (8.9), and (8.10) are reduced to the following:

$$I_{DSS} = qn_s v_{sat} \tag{8.11}$$

$$g_m = C_s v_{sat} \tag{8.12}$$

More insight can be obtained in terms of device parameters if the equation for charge control [Eq. (8.1)] is substituted in the expressions for I_{ds} and V_g as follows:[6]

$$I_{ds} = qv_{sat}n_s\left[\sqrt{1 + \left(\frac{n_c}{n_s}\right)^2} - \frac{n_c}{n_s}\right] \tag{8.13}$$

$$g_{mo} = C_s v_{sat} \frac{1}{\sqrt{1+\left(n_c/n_s\right)^2}} \tag{8.14}$$

where $n_c = E_c C_s L_g / q$ and $0 < n_s < n_{sm}$. Dividing both sides of Eq. (8.14) by $C_s v_{sat}$ the following expression for modulation efficiency is obtained:

$$\eta = \frac{1}{\sqrt{1+\left(n_c/n_s\right)^2}} \tag{8.15}$$

Hence it is necessary to maximize the 2DEG density n_s to maximize the current drive, transconductance, and modulation efficiency of the HEMT. Although this is in contrast with the saturated-velocity model, it agrees with the experimental results. The foregoing results can also be used to select the appropriate material system and layer structure for the fabrication of high-performance microwave and millimeter-wave HEMTs.

Although the analytical model of device operation as was described here provides great insight into the principles of device operation and performance optimization, it fails to predict some of the nonlinear phenomena such as reduction of g_m at high current levels (g_m compression) and soft pinch-off characteristics. A model has been developed to explain these phenomena.[5] The total charge in the HEMT is divided into three components. The first, Q_{SVM}, is the charge required to support a given I_{ds} under the saturated velocity model (SVM). This charge is uniformly distributed under the gate. In reality this is not the case as the electron velocity under the gate varies. To maintain the current continuity under the gradual channel approximation (GCA), extra charge under the channel has to be introduced. This is defined as Q_{GCA} and is maximum at the source end of the gate and minimum at the drain end.

The excess charge in the wide bandgap electron supply layer is denoted by Q_{SL}. Figure 8.2 shows the location and distribution of these charges in the HEMT. Only Q_{SVM} supports current density and thus contributes to the transconductance of the HEMT. The other two components contribute only to the total capacitance of the device. Hence the modulation efficiency (ME) of the HEMT in terms of these charges is expressed as

$$\eta = \frac{\delta Q_{SVM}}{\delta\left(Q_{SVM} + Q_{GCA} + Q_{SL}\right)} \tag{8.16}$$

and the transconductance can be expressed as $g_m = C_s v_{sat} \eta$.

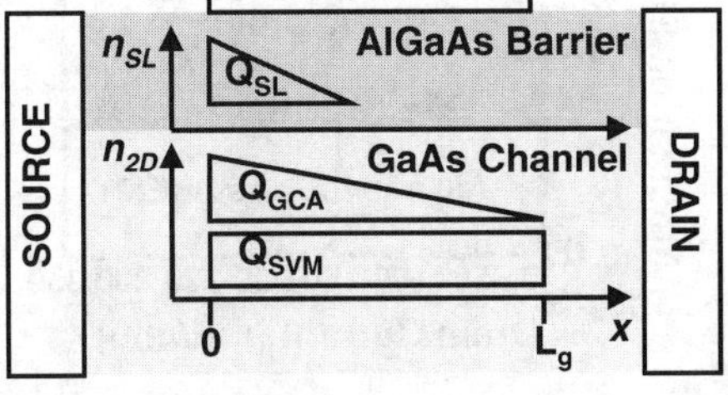

FIGURE 8.2 Schematic diagram showing the location and distribution of Q_{SVM}, Q_{GCA}, and Q_{SL} in a HEMT (Foisy et al. 1988).

Figure 8.3 shows the variation of ME as a function of drain current density for an AlGaAs/GaAs HEMT and an AlGaAs/InGaAs pHEMT. At low current density, ME is low as most of the charge in the 2DEG channel has to satisfy the gradual channel approximation. This low value of ME results in low transconductance and soft pinch-off characteristics at low drain current densities. In the high current regime, modulation of Q_{SL} reduces the ME, resulting in gain compression. In the intermediate current regime the ME is maximum. However, if there exists a bias condition where both Q_{GCA} and Q_{SL} are modulated (as in the low band offset AlGaAs/GaAs system), it severely affects the ME.

For optimal high-power and high-frequency performance, it is necessary to maximize the range of current densities in which ME is high. The drop off in ME due to parasitic charge modulation in the donor layers can be pushed to higher current density by increasing the maximum 2DEG density n_{sm}. The 2DEG density can be maximized by using planar doping in the donor layer and by increasing the conduction band discontinuity at the barrier/2DEG interface. The drop off in ME due to operation in gradual channel mode can be pushed to lower current densities by reducing the saturation voltage V_{Dsat}. This is achieved by increasing the mobility of the electrons in the 2DEG channel and by reducing the gate length. As seen from Fig. 8.3 higher modulation efficiency is achieved over a larger range of current density for the AlGaAs/InGaAs pHEMT, which has higher sheet charge density, mobility, and band discontinuity at the interface than the AlGaAs/GaAs HEMT.

8.2.4 Small Signal Equivalent Circuit Model of HEMT

The small signal equivalent circuit model of the HEMT is essential for designing HEMT-based amplifiers. The model can also provide insights into the role of various parameters in the high-frequency performance of the device. Figure 8.4 shows the small signal equivalent circuit for a HEMT. The grey box

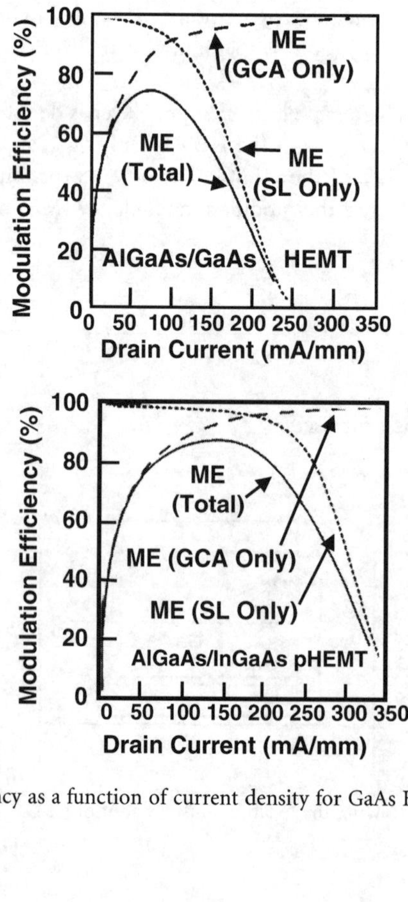

FIGURE 8.3 Modulation efficiency as a function of current density for GaAs HEMT and GaAs pHEMT.

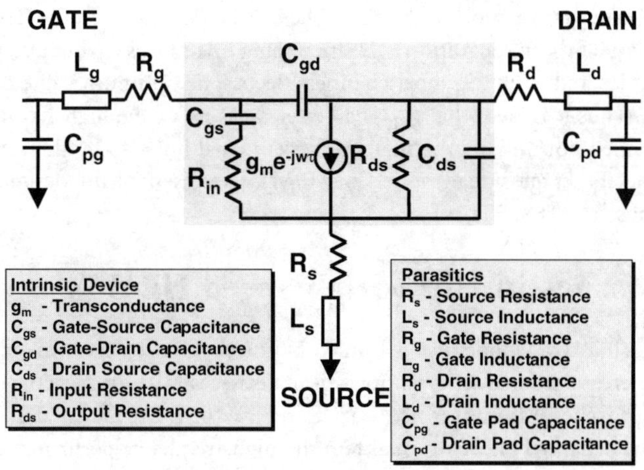

FIGURE 8.4 Small signal equivalent circuit of a HEMT.

highlights the intrinsic device. The circuit elements in the preceding model are determined using microwave S-parameter measurements.[7,8] The intrinsic circuit elements are a function of the DC bias, whereas the extrinsic circuit elements or parasitics are independent of it. The two measures of the high frequency performance of a FET can now be defined in terms of the small signal model of the device as follows. The current gain cutoff frequency, f_T can be defined as

$$f_T = \frac{g_m}{2\pi\left(C_{gs} + C_{gd}\right)} \tag{8.17}$$

Hence, to increase the current gain cutoff frequency it is essential to increase the g_m and reduce C_{gs} and C_{gd}. Referring to Eq. (8.18), it is clear that this can be achieved by increasing electron velocity in the channel and reducing gate length. The current gain cutoff frequency is mainly a physical measure of device performance. A more practical measure of high-frequency device performance is $f_{\max}$, the power gain cutoff frequency. This is the frequency at which the power gain of the FET is unity. It is defined as follows,[9]

$$f_{\max} = \frac{f_T}{\sqrt{4g_{ds}\left(R_{in} + \dfrac{R_s + R_g}{1 + g_m R_s}\right) + \dfrac{4}{5}\dfrac{C_{gd}}{C_{gs}}\left(1 + \dfrac{2.5C_{gd}}{C_{gs}}\right)\left(1 + g_m R_s\right)^2}} \tag{8.18}$$

A simple form of Eq. (8.18) is:

$$f_{\max} = f_T\sqrt{\frac{R_{ds}}{4R_{in}}} = \frac{f_T}{\sqrt{4g_{ds}R_{in}}} \tag{8.19}$$

To improve the $f_{\max}$ of the device it is necessary to minimize the quantities in the denominator of Eq. (8.18). The crucial parameters here are the output conductance of the device g_{ds}, the source and gate parasitic resistances R_s and R_g, and the gate-drain feedback capacitance C_{gd} that need to be minimized. Reduction of g_{ds} can be achieved by appropriate vertical scaling (to be discussed in the next section). Reduction of R_s and R_g depends mainly on the process technology. Reduction of C_{gd} can be achieved by proper design of the gate-drain region of the FET. The crucial parameter in the design of the gate drain depletion region is the gate-drain separation L_{gd}.[10] Increasing L_{gd} reduces C_{gd} but also increases the effective gate length of the device, reducing the short channel effects. The optimum value of L_{gd} is 2.3 times that of the gate length L_g. Thus it is clear that f_{max} is a better measure of the high-frequency performance of a FET as it is determined not only by the material system used but also by the process technology and device design parameters. Large signal models of HEMTs are essential for designing power amplifiers and are similar to those of MESFETs.

8.3 Scaling Issues in Ultra-High-Speed HEMTs

The frequency at which a HEMT operates is limited by the electron transit time from the source to the drain. Therefore to increase the frequency of operation it is necessary to reduce the gate length. However, as the gate length approaches 0.1 μm it is necessary to reduce the other parasitic delays in the device and take into account short channel effects to maintain the high-frequency performance of the HEMT.

8.3.1 Delay Time Analysis

The reduction of parasitic delays in a FET is essential to improve the high frequency performance as these delays can be as high as 45% of the intrinsic delay.[11] Considering the small-signal model of a FET, the total delay t_T in a FET can be expressed as follows:[12]

$$t_T = t_{pad} + t_{fringe} + t_{channel} + t_{transit} + t_{drain} = 1/(2\pi f_T) \tag{8.20}$$

Here t_{pad} is the charging time for the parasitic pad capacitance and is given by

$$t_{pad} = C_{pad}/g_m \cdot W \tag{8.21}$$

where C_{pad} is the pad capacitance and is typically 10 *fF* per 50 μm × 50 μm bonding pad, g_m is the extrinsic transconductance per unit gate width, and W is the width of the device. To minimize t_{pad} it is necessary to have a high gate width, high transconductance HEMT.

The gate fringe capacitance charging time (t_{fringe}) is given by

$$t_{fringe} = C_{fringe}/g_{mo} \tag{8.22}$$

where g_{mo} is the intrinsic transconductance of the HEMT and is related to the extrinsic transconductance (g_m) and source resistance R_s by the following expression:

$$g_m = g_{mo}/(1 + g_{mo} \cdot R_s) \tag{8.23}$$

The gate fringe capacitance C_{fringe} is typically 0.18 pF/mm, hence for a HEMT with an intrinsic transconductance of 1000 to 1500 mS/mm, t_{fringe} is approximately 0.1 to 0.2 ps.

Channel charging delay $t_{channel}$ is associated with RC delays and is proportional to channel resistance. The channel charging delay is minimum at high current densities. The channel charging delay can be considered as a measure of the effectiveness of a FET operating in the saturated velocity mode.

The transit delay of the FET, $t_{transit}$, can be expressed as the time required to traverse under the gate and is given by

$$t_{transit} = L_g/v_{sat} \tag{8.24}$$

The drain delay (t_{drain}) is the time required by the electron to traverse the depletion region between the gate and the drain and is a function of bias conditions.[13] The drain delay increases with drain bias as the length of the depletion region beyond the gate increases. Drain delay is an important parameter for millimeter-wave power HEMTs. To increase the breakdown voltage of the device, the gate-to-drain spacing has to be increased. When the device is biased at a high drain voltage to maximize the power output, it creates a drain depletion region that is on the order of gate length of the device. Thus the drain delay becomes a major component of the total delay in the device, and can limit the maximum f_T and f_{max}.

8.3.2 Vertical Scaling

Aspect ratio (the ratio between the gate length L_g and the gate-to-channel separation $d_{Barrier}$) needs to be maintained when gate length is reduced. Aspect ratio is a critical factor affecting the operation of the field effect transistor and should be maintained above five. As the gate length is reduced, the distance between the gate and 2DEG (the distance $d_n + d_i$ as seen in Fig. 8.1) has to be reduced so that the aspect ratio of the device is maintained.

However, maintaining the aspect ratio alone does not guarantee improvement in device performance. This is clear if the variation of threshold voltage with the reduction in $d_{Barrier}$ is examined. It is clear that d_i cannot be reduced, as it will result in degradation of mobility in the 2DEG channel due to scattering from the donors in the barrier layers. Therefore, to maintain aspect ratio, the thickness of the doped barrier layer d_n has to be reduced. By examining Eq. (8.3) for threshold voltage, it is clear that this makes the threshold voltage more positive.

At first glance, this does not seem to affect device performance. The effect of the more positive threshold voltage is clear if the access regions of the device are considered. A more positive threshold voltage results in reduction of sheet charge in the access region of the device. This increases the source and drain resistance of the device, which reduces the extrinsic transconductance [see Eq. (8.23)] and also increases the channel charging time (due to increased RC delays). Thus the increased parasitic resistances nullify the improvements in speed in the intrinsic device.

The threshold voltage of the device must be kept constant with the reduction in d_n. From Eq. (8.3) it can be seen that the doping density in the high bandgap donor has to be increased. Since the threshold voltage varies as a square of the doped barrier thickness, a reduction in its thickness by a factor of 2 requires that the doping density be increased by a factor of 4. High doping densities can be difficult to achieve in wide bandgap materials such as AlGaAs due to the presence of DX centers. Increased doping also results in higher gate leakage current, higher output conductance, and a lower breakdown voltage. Utilizing planar or delta doping wherein all the dopants are located in a single plane can alleviate these problems. This leaves most of the higher bandgap layer undoped and enables reduction of its thickness.

The threshold voltage of a planar-doped HEMT is given as follows[14]

$$V_T = \phi_B - \frac{qN_{2D}d_n}{\varepsilon} - \Delta E_c + \Delta E_F \tag{8.25}$$

where N_{2D} is the per unit area concentration of donors in the doping plane and d_n is the distance of the doping plane from the gate. In this case, the 2D doping density has to increase linearly with the reduction in barrier thickness. The transfer efficiency of electrons from the donors to the 2DEG channel also is increased, as all the dopant atoms are close to the 2DEG channel. Hence higher 2DEG sheet densities can be achieved in the channel and thus planar doping enables efficient vertical scaling of devices with reduction in gate length.[15] From a materials point of view, efficient vertical scaling of a HEMT requires a high bandgap donor/barrier semiconductor that can be doped efficiently.

The voltage gain of the device (g_m/g_{ds}) can be considered as a measure of short channel effects in the device. The reduction of gate length and the gate-to-channel separation results in an increase in the transconductance of the device. However, to reduce the output conductance g_{ds} of the device, it is also necessary to reduce the channel thickness, which then increases the carrier confinement in the channel. Enoki et al. have investigated the effect of the donor/barrier and channel layer thickness on the voltage gain of the device.[16] The gate-to-channel separation ($d_{Barrier}$) and the channel thickness ($d_{channel}$) were varied for a 0.08-μm gate length AlInAs/GaInAs HEMT. For a $d_{Barrier}$ of 170 Å and a $d_{channel}$ of 300 Å, the g_m was 790 mS/mm and g_{ds} was 99 mS/mm, resulting in a voltage gain of 8. When $d_{Barrier}$ was reduced to 100 Å and $d_{channel}$ was reduced to 150 Å, the g_m increased to 1100 mS/mm and g_{ds} reduced to 69 mS/mm; this doubled the voltage gain to 16. This illustrates the necessity to reduce the channel thickness to improve charge control in ultra-short gate length devices.

Subthreshold slope is an important parameter to evaluate short channel effects for digital devices. A high value of subthreshold slope is necessary to minimize the off-state power dissipation and to increase the device speed. Two-dimensional simulations performed by Enoki et al. indicate that reduction in channel thickness is more effective than the reduction in barrier thickness, for maintaining the subthreshold slope with reduction in gate length.[16]

The high-frequency performance of a device is a function of the electrical gate length $L_{g,eff}$ of the device, which larger than the metallurgical gate length L_g due to lateral depletion effects near the gate. The relation between $L_{g,eff}$ and L_g is given by,[14]

$$L_{g,eff} = L_g + \beta\left(d_{Barrier} + \Delta d\right) \tag{8.26}$$

where $d_{Barrier}$ is the total thickness of the barrier layers, Δd is the distance of the centroid of the 2DEG from the channel barrier interface and is on the order of 80 Å. The value of parameter β is 2.

Consider a long gate length HEMT (L_g = 1 μm) with a barrier thickness of 300 Å. Using Eq. (8.26), the value of 1.076 μm is obtained for $L_{g,eff}$. Thus the effective gate length is only 7.6% higher than the metallurgical gate length. Now consider an ultra-short gate length HEMT (L_g = 0.05 μm) with an optimally scaled barrier thickness of 100 Å. Using the same analysis, a value of 0.086 μm is obtained for $L_{g,eff}$. In this case the effective gate length is 43% higher than the metallurgical gate length. Hence, to improve the high-frequency performance of a ultra-short gate length HEMT, effective gate length reduction along with vertical scaling is required.

8.3.3 Horizontal Scaling

Reduced gate length is required for the best high-frequency performance. However, it should be kept in mind that the gate series resistance increases with the reduction in gate length. This problem can be solved with a T-shaped gate. This configuration lowers the gate series resistance while maintaining a small footprint. Another advantage of the T-shaped gate is reduced susceptibility to electromigration under large signal RF drive as the large gate cross-section reduces current density. For a 0.1-μm gate length using a T-gate instead of a straight gate, reduces the gate resistance from 2000 Ω/mm to 200 Ω/mm.

The simplified expression for f_T as expressed in Eq. (8.17) does not include the effect of parasitics on the delay time in a FET. A more rigorous expression for f_T, which includes the effects of parasitics on f_T was derived by Tasker and Hughes and is given here,[17]

$$f_T = \frac{g_m/2\pi}{\left[C_{gs} + C_{gd}\right]\left[1 + \left(R_s + R_d\right)/R_{ds}\right] + C_{gd}g_m\left(R_s + R_d\right)} \tag{8.27}$$

It is clear from Eq. (8.27) that it is necessary to reduce source and drain resistances R_s and R_d, respectively, to increase the f_T of a FET. Mishra et al. demonstrated a record f_T of 250 GHz for a 0.15-μm device with a self-aligned gate, which reduces the gate-source and gate-drain spacing and results in the reduction of R_s and R_d.[18] Equation (8.27) can be rearranged as follows,[17]

$$\frac{1}{2\pi f_T} = \frac{\left(C_{gs}+C_{gd}\right)}{g_m} + \frac{\left(C_{gs}+C_{gd}\right)\left(R_s+R_d\right)}{g_m R_{ds}} + C_{gd}\left(R_s+R_d\right) \tag{8.28}$$

where the first term on the right-hand side is the intrinsic delay of the device (τ_{int}) and the rest of the terms contribute to parasitic delay (τ_p). From this equation the ratio of parasitic delay to the total delay ($\tau_t = \tau_p + \tau_{int}$) is given as

$$\frac{\tau_p}{\tau_t} = g_m\left(R_s+R_d\right)\left[\frac{G_{ds}}{g_m} + \frac{1}{\left[1+C_{gs}/C_{gd}\right]}\right] \tag{8.29}$$

Hence to improve the f_T of the device, the parasitic source and drain resistance have to be reduced as the gate length of the device is reduced. This minimizes the contribution of the parasitic delays to the total delay of the device.

8.4 Material Systems for HEMT Devices

The previous portions of this section discussed the various device parameters crucial to high-frequency performance of HEMTs. In this section the relationship between material and device parameters will be discussed. This will enable the selection of the appropriate material system for a particular device application. Table 8.1 illustrates the relationship between the device parameters and material parameters for the various constituent layers of the HEMT, namely the high bandgap donor and buffer layers, and the 2DEG channel. Figure 8.5 shows a schematic diagram of a HEMT, illustrating the material requirements from each component layer.

The first HEMT was implemented in the lattice-matched AlGaAs/GaAs system in 1981.[19] The AlGaAs/GaAs HEMT demonstrated significant improvement in low noise and power performance over GaAs MESFET due to superior electronic transport properties of the 2DEG at the AlGaAs/GaAs interface and better scaling properties. However, the limited band discontinuity at the AlGaAs/GaAs interface limits the 2DEG sheet charge density. Other undesirable effects, such as formation of a parasitic MESFET in the donor layer and real space transfer of electrons from the channel to donor, are prevalent, however.

One way to increase band discontinuity is to increase the Al composition in AlGaAs. However, the presence of deep level centers (DX centers) associated with Si donors in AlGaAs prevents the use of high

TABLE 8.1 Relationship between Device and Material Parameters

Device Type	Device Parameters	Material Parameters: 2DEG Channel Layer	Material Parameters: Barrier/Buffer Layer
Short Gate Length Devices	High Electron Velocity	High Electron Velocity High Electron Mobility	
	High Aspect Ratio		High Doping Efficiency
Power Devices	High Current Density	High 2DEG Density	
	Low Gate Leakage		High Schottky Barrier
	High Breakdown Voltage	High Breakdown Field	High Breakdown Field
	Low Output Conductance		High Quality Buffer
	Good Charge Control	High Modulation Efficiency	
	Low Frequency Dispersion		Low Trap Density
Low Noise Devices	Low Rs	High 2DEG density	
	High Electron Velocity	High Electron Velocity High Electron Mobility	
Digital Devices	Low Gate Leakage Current		High Schottky Barrier
	High Current Drive	High 2DEG Density	

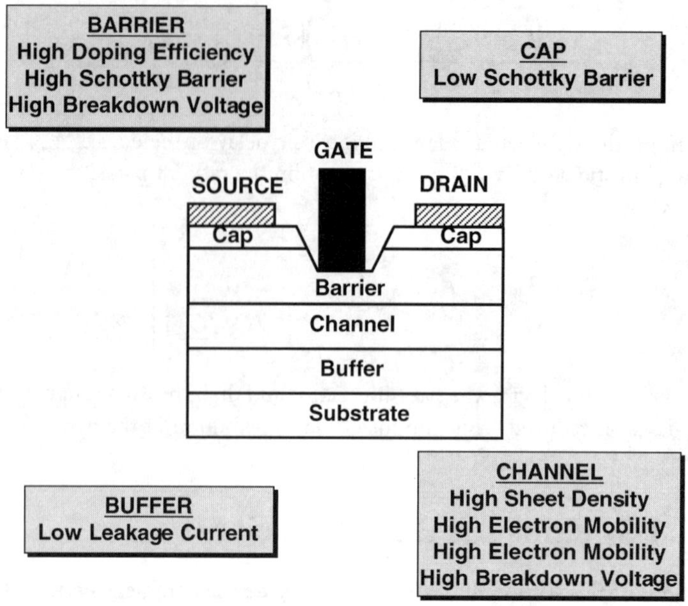

FIGURE 8.5 Material requirements for HEMT devices.

Al composition AlGaAs donor layers to increase the band discontinuity and also limits doping efficiency. Problems relating to low band discontinuity can also be solved by reducing the bandgap of the channel, and by using a material that has higher electron mobility and electron saturation velocity. The first step in this direction was taken by the implementation of an AlGaAs/InGaAs pseudomorphic HEMT (GaAs pHEMT).[20] In an AlGaAs/InGaAs pHEMT the electron channel consists of a thin layer of narrow bandgap InGaAs that is lattice mismatched to GaAs by 1 to 2%. The thickness of the InGaAs channel is thin enough (~200 Å) so that the mismatch strain is accommodated coherently in the quantum well, resulting in a dislocation free "pseudomorphic" material. However the indium content in the InGaAs channel can be increased only up to 25%. Beyond this limit the introduction of dislocations due to high lattice mismatch degrades the electronic properties of the channel. The maximum Al composition that can be used in the barrier is 25% and the maximum indium composition that can be used in the channel is 25%.

Using the $Al_{0.48}In_{0.52}As/Ga_{0.47}In_{0.53}As$ material system lattice matched to InP can simultaneously solve the limitations of the high bandgap barrier material and the lower bandgap channel material. The AlInAs/GaInAs HEMT (InP HEMT) has demonstrated excellent low-noise and power performance that extends well into the millimeter-wave range; they currently hold all the high-frequency performance records for FETs. The GaInAs channel has high electron mobility (>10,000 cm^2/Vs at room temperature), high electron saturation velocity (2.6×10^7 cm/s) and higher intervalley (Γ-L) energy separation. The higher conduction band offset at the AlInAs/GaInAs interface ($\Delta E_c = 0.5$ eV) and the higher doping efficiency of AlInAs (compared to AlGaAs) results in a sheet charge density that is twice that of the AlGaAs/InGaAs material system. Higher doping efficiency of AlInAs also enables efficient vertical scaling of short gate length HEMTs. The combination of high sheet charge and electron mobility in the channel results in low source resistance, which is necessary to achieve high transconductance. However, the low bandgap of the InGaAs channel results in low breakdown voltage due to high impact ionization rates.

Table 8.2 summarizes the material properties of the three main material systems used for the fabrication of HEMTs. The emergence of growth techniques like Metal Organic Chemical Vapor Deposition (MOCVD) and Gas Source Molecular Beam Epitaxy (GSMBE) and continuing improvement in the existing growth techniques like molecular beam epitaxy (MBE) have enabled a new class of phosphorus-based material systems for fabrication of HEMTs. On the GaAs substrate, the GaInP/InGaAs has emerged as an alternative to the AlGaAs/InGaAs material system. GaInP has a higher bandgap than AlGaAs and

TABLE 8.2 Material Parameters of AlGaAs/GaAs, AlGaAs/InGaAs, and AlInAs/GaInAs Material Systems

Material Parameter	AlGaAs/GaAs	AlGaAs/InGaAs	AlInAs/GaInAs
ΔE_c	0.22 eV	0.42 eV	0.51 eV
Maximum donor doping	$5 \times 10^{18}/cm^3$	$5 \times 10^{18}/cm^3$	$1 \times 10^{19}/cm^3$
Sheet charge density	$1 \times 10^{12}/cm^2$	$1.5 \times 10^{12}/cm^2$	$3 \times 10^{12}/cm^2$
Mobility	8000 cm^2/Vs	6000 cm^2/Vs	12,000 cm^2/Vs
Peak Electron Velocity	2×10^7 cm/s		2.7×10^7 cm/s
Γ-L valley separation	0.33 eV		0.5 eV
Schottky Barrier	1.0 eV	1.0 eV	0.45 eV

hence enables high 2DEG densities due to the increased conduction band discontinuity (ΔE_c) at the GaInP/InGaAs interface. As GaInP has no aluminum it is less susceptible to environmental oxidation. The availability of high selectivity etchants for GaAs and GaInP simplifies device processing. However, the high conduction band discontinuity is achieved only for disordered GaInP, which has a bandgap of 1.9 eV. Using graded GaInP barrier layers and an $In_{0.22}Ga_{0.78}As$ channel, 2DEG density as high as $5 \times 10^{12}/cm^2$ and a mobility of 6000 cm^2/Vs was demonstrated.[21]

On InP substrates, the InP/InGaAs material system can be used in place of the AlInAs/GaInAs material system. The presence of deep levels and traps in AlInAs degrades the low frequency noise performance of AlInAs/GaInAs HEMT. Replacing the AlInAs barrier by InP or pseudomorphic InGaP can solve this problem. One disadvantage of using the InP-based barrier is the reduced band discontinuity (0.25 eV compared to 0.5 eV for AlInAs/GaInAs) at the InP/InGaAs interface. This reduces 2DEG density at the interface and modulation efficiency. Increasing the indium content up to 75% in the InGaAs channel can increase the band discontinuity at the InP/InGaAs interface. The poor Schottky characteristics on InP necessitate the use of higher bandgap InGaP barrier layers or depleted p-type InP layers. A sheet density of $3.5 \times 10^{12}/cm^2$ and mobility of 11,400 cm^2/Vs was demonstrated in an $InP/In_{0.75}Ga_{0.25}As/InP$ double heterostructure.[22]

Despite the large number of material systems available for fabrication of HEMTs, the GaAs pHEMT implemented in the $Al_xGa_{1-x}As/In_yGa_{1-y}As$ ($x \sim 0.25$; $y \sim 0.22$) material system and the InP HEMT implemented in the $Al_{0.48}In_{0.52}As/Ga_{0.47}In_{0.53}As$ material system have emerged as industry vehicles for implementation of millimeter-wave analog and ultra high-speed digital circuits. The next two portions of this section will discuss the various performance aspects of GaAs pHEMT and InP HEMT.

8.5 AlGaAs/InGaAs/GaAs Pseudomorphic HEMT (GaAs pHEMT)

The first AlGaAs/InGaAs pseudomorphic HEMT was demonstrated in 1985.[23] Significant performance improvement over AlGaAs/GaAs HEMT was observed. Devices with a 1-μm gate length had peak transconductance of 270 mS/mm and maximum drain current density of 290 mA/mm.[20] The current gain cutoff frequency (f_T) was 24.5 GHz and the power gain cutoff frequency (f_{max}) was 40 GHz. An f_T of 120 GHz was reported for 0.2-μm gate length devices with $In_{0.25}Ga_{0.75}As$ channel.[24] Devices with a 0.1-μm gate length with an f_{max} of 270 GHz were demonstrated in 1989.[14]

8.5.1 Millimeter-Wave Power GaAs pHEMT

In the past few years, the GaAs pHEMT has emerged as a device of choice for implementing microwave and millimeter-wave power amplifiers. To achieve a high output power density, device structures with higher current density and consequently higher sheet charge are required. As the sheet charge density in a single heterojunction AlGaAs/InGaAs pHEMT is limited to $2.3 \times 10^{12}/cm^2$, a double heterojunction (DH) device structure must be used to increase the sheet charge. In a DH GaAs pHEMT, carriers are introduced in the InGaAs channel by doping the AlGaAs barriers on both sides of the InGaAs channel.

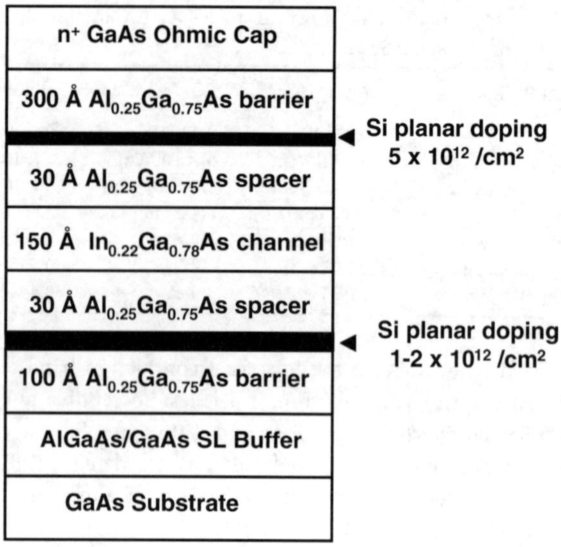

FIGURE 8.6 Layer structure of a GaAs power pHEMT.

The AlGaAs barriers are doped with silicon using atomic planar doping to increase the electron transfer efficiency. A typical charge density of $3.5 \times 10^{12}/cm^2$ and a mobility of 5000 cm^2/Vs is obtained for a double heterojunction GaAs pHEMT structure. The high sheet charge thus obtained enables higher current drive and power handling capability. Figure 8.6 shows the layer structure of a typical millimeter-wave power GaAs pHEMT. In some cases a doped InGaAs channel is also used to increase sheet charge density.[25,26]

Breakdown voltage is an important parameter for power devices. A device with high breakdown voltage can be biased at high drain voltages, which increases the drain efficiency, voltage gain, and power added efficiency (PAE). Typical breakdown voltages of GaAs pHEMTs range from 8 to 15 V. The breakdown mechanism of a GaAs pHEMT can be either at the surface in the gate-drain of the device or in the channel (due to impact ionization).

There are several approaches used to increase the breakdown voltage of a GaAs pHEMT. The planar doping of AlGaAs barriers (as already described) helps in maintaining a high breakdown voltage, as most of the AlGaAs barrier is undoped. Another approach to increase the breakdown voltage uses a low-temperature grown (LTG) GaAs buffer below the channel. Using this approach, a 45% increase in channel breakdown voltage with a 12% increase in output power was demonstrated.[27] Using a double recessed gate structure to tailor the electric field in the gate drain depletion region can also increase breakdown voltage. The increase in breakdown voltage is mainly due to reduction in the electric field at the gate edge by surface states in the exposed recess region.[28]

The output power obtained from a HEMT also depends on the biasing conditions. To achieve high efficiency devices (as in Class B operation), the device is biased near pinch off, and therefore, high gain is required near pinch off. The mode of operation is ideally suited for pHEMTs, which typically have high transconductance near pinch off due to their superior charge-control properties. The effect for gate bias on the power performance of HEMT has been investigated.[29] Higher gain is achieved under Class A conditions. Biasing the device at higher drain voltages can increase the output power. Table 8.3 presents a summary of power performance of GaAs pHEMTs at various microwave and millimeter-wave frequencies.

From Table 8.3 it can be seen that at a given frequency, the device power output and gain increases with a decrease in gate length due to better high-frequency operation. Reduction in power gain is also observed for wider devices. This is due to the increase in source inductance, which increases with gate width and frequency and is due to the increase in gate resistance as a square of gate width. Low inductance via hole source grounding and proper gate layout is required to reduce these parasitics.

TABLE 8.3 Summary of Power Performance of GaAs pHEMTs

Frequency	Gate Length (µm)	Gate Width	Power Density (W/mm)	Output Power[a]	Gain[b] (dB)	PAE[b] (%)	Device, Drain Bias (Reference)
12 GHz	0.45	1.05 mm	0.77	0.81 W	10.0	60	Double HJ[c] V_{ds} = 7 V[32]
(Ku-Band)	0.25	1.6 mm	1.37	2.2 W	14.0	39	Double HJ[33]
20 GHz	0.25	600 µm	0.51	306 mW	7.4	45	Prematched, V_{ds} = 7 V [34]
(K-Band)	0.15	400 µm	1.04	416 mW	10.5	63	LTG Buffer, V_{ds} = 5.9 V[27]
	0.15	600 µm	0.12	72 mW	8.6	68	V_{ds} = 2 V[35]
	0.15	600 µm	0.84	501 mW	11	60	V_{ds} = 8 V[35]
35 GHz	0.25	500 µm	0.62	310 mW	6.8	40	Double HJ, V_{ds} = 5 V[36]
(Ku-Band)	0.15	150 µm	0.63	95 mW	9.0	51	Double HJ[37]
			0.91	*137 mW*	7.6	40	
44 GHz	0.25		0.79		5.1	41	Doped Channel[d] [38]
(Q-Band)	0.15	400 µm	0.5	200 mW	9.0	41	Double HJ, V_{ds} = 5 V[39]
	0.2	600 µm	0.53	318 mW	5.0	30	Double HJ, V_{ds} = 5 V[40]
	0.15	1.8 mm	*0.44*	*800 mW*	5.8	25	Double HJ, V_{ds} = 5 V[41]
55 GHz	0.25	400 µm	*0.46*	*184 mW*	4.6	25	Doped Channel, V_{ds} = 5.5 V[42]
(V-Band)	0.2	50 µm	*0.85*	*42 mW*	3.3	22	Doped Channel, V_{ds} = 4.3[43]
60 GHz	0.15	150 µm	0.83	125 mW	4.5	32	Double HJ[37]
(V-Band)			0.55	82 mW	4.7	38	
	0.15	400 µm	*0.55*	*225 mW*	4.5	25	Double HJ, V_{ds} = 5 V[44]
			0.44	174 mW	4.4	*28*	
94 GHz	0.25	75 µm	*0.43*	*32 mW*	3.0	15	Doped Channel, V_{ds} = 4.3 V[25]
W-Band				25 mW	4.0	14	
	0.15	150 µm	*0.38*	*57 mW*	2.0	16	Double HJ[37]
			0.30	45 mW	3.0	16	
	0.1	40 µm	*0.31*	*13 mW*	6.0	13	Doped Channel V_{ds} = 3.4 V[26]
	0.1	160 µm	*0.39*	*63 mW*	4.0	13	Doped Channel V_{ds} = 4.3 V[26]

[a] *Output Power* in italics indicates device biased for maximum power output.
[b] *PAE/Gain* in italics indicates device biased for maximum PAE/Gain.
[c] Double HJ — Double heterojunction GaAs pHEMT.
[d] Doped channel — Doped channel GaAs pHEMT.

Reliability is important for space applications, typically a mean-time-to-failure (MTTF) of 10^7 h (1142 yr) is required for space applications. GaAs pHEMTs have demonstrated a MTTF of 1×10^7 h at a channel temperature of 125°C. The main failure mechanism is the atmospheric oxidation of the exposed AlGaAs barrier layers and interdiffusion of the gate metallization with the AlGaAs barrier layers (gate sinking).[30] Using dielectric passivation layers to reduce the oxidation of AlGaAs can solve these problems. Using refractory metal for gate contacts will minimize their interaction with the AlGaAs barrier layers. A MTTF of 1.5×10^7 hours at a channel temperature of 150°C was achieved using Molybdenum based gate contacts.[31]

Traveling wave tubes (TWT) have been traditionally used as multiwatt power sources for microwave applications up to K-band (20 GHz). Using GaAs pHEMT in place of TWT for these applications has many advantages, including lower cost, smaller size, smaller weight, and higher reliability. However, the typical power density of a GaAs pHEMT at 20 GHz is on the order of 1 W/mm. Hence, the output power from a large number of devices has to be combined. To minimize combining losses, it is desirable to maximize power output of a single device. When large devices (gate width on the order of mm) are used to increase the total power output, other factors such as device layout, input signal distribution, output power combining networks, and substrate thickness are of critical importance. Several multiwatt GaAs pHEMT power modules have been demonstrated recently. A power module with 9.72-mm wide GaAs pHEMTs that delivered an output power of 4.7 W in the 18 to 21.2 GHz band with a PAE of 38% was demonstrated.[45] Table 8.4 summarizes the recent results of multiwatt GaAs pHEMT power modules.

TABLE 8.4 Summary of Power Performance of Multiwatt pHEMT Power Modules

Width (mm)	Frequency (GHz)	Power (W)	Gain (dB)	PAE (%)	Ref.
16.8	2.45	10.0	13.5	63.0	V_{ds} = 7 V[46]
24	2.45	11.7	14.0	58.2	V_{ds} = 8 V[47]
32.4	8.5–10.5	12.0	7.2	40.0	V_{ds} = 7 V[48]
16.8	12	12.0	10.1	48.0	[32]
25.2	12	15.8	9.6	36.0	V_{ds} = 9 V[49]
8	12	6	10.8	53.0	V_{ds} = 9 V[50]
9.72	18–21.2	4.7	7.5	41.4	V_{ds} = 5.5 V[45]

Note: V_{ds} = Drain bias for power measurements.

8.5.2 Low Noise GaAs pHEMTs

Figure 8.7 shows the structure of a generic low-noise GaAs pHEMT. As the drain current requirement for a low-noise bias is low, single-side doped heterojunctions are sufficient for low-noise devices. The emphasis here is on achieving higher mobility to reduce the parasitic source resistance. As already discussed, the AlInAs/GaInAs material system is the ideal choice for fabrication of low-noise microwave and millimeter-wave devices. However, GaAs pHEMTs also find significant use in millimeter-wave low noise applications due to wafer size, cost, and process maturity related advantages.

Henderson et al. first reported on the low noise performance of GaAs pHEMTs in 1986. Devices with 0.25-μm gate length had a noise figure of 2.4 dB and an associated gain of 4.4 dB at 62 GHz.[51] A 0.15-μm gate length $Al_{0.25}Ga_{0.75}As/In_{0.28}Ga_{0.72}As$ pHEMT with a noise figure of 1.5 dB and an associated gain of 6.1 dB at 61.5 GHz was demonstrated in 1991.[52] The reduction in noise figure was a direct result of reducing the gate length, which increased the f_T of the device. Low noise operation of a GaAs pHEMT was also demonstrated at 94 GHz.[53] For a 0.1-μm gate length device a noise figure of 3.0 dB and associated gain of 5.1 dB was achieved. A noise figure of 2.1 dB and an associated gain of 6.3 dB was reported for a 0.1 μm gate length GaAs pHEMT at 94 GHz.[54] The improvement in noise figure is attributed to the use of a T-shaped gate with end-to-end resistance of 160 Ω/mm by Tan et al.,[52] compared to a trapezoidal gate with end-to-end resistance of 1700 Ω/mm used by Chao et al.[53] This further emphasizes the need to reduce parasitic resistances in low-noise devices.

One of the main system applications of low-noise GaAs pHEMTs is satellite direct broadcasting receiver systems (DBS) that are in increasing demand worldwide. Low-noise amplifiers operating at 12 GHz are a critical component in these systems. The low-noise performance of GaAs pHEMTs is more than adequate for these applications. A 0.25-μm gate length GaAs pHEMT with a noise figure of 0.6 dB and an associated gain of 11.3 dB at 12 GHz was reported by Tokue et al.[55] Performance coupled

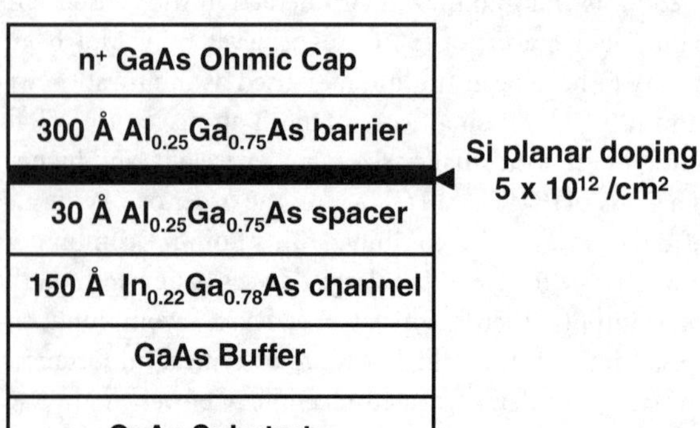

FIGURE 8.7 Layer structure of a low noise GaAs pHEMT.

TABLE 8.5 Summary of Low Noise Performance of GaAs pHEMTs

Frequency (GHz)	Gate Length (μm)	F_{min} (dB)	Ga (dB)	Ref.	Comments
12	0.25	0.6	11.3	55	
12	0.17	0.35	12.5	57	Packaged Device
18	0.25	0.9	10.4	51	
62	0.25	2.4	4.4	51	Improvement by
60	0.15	1.5	6.1	52	Reduction in L_g
94	0.10	3.0	5.1	53	Improvement by
94	0.10	2.1	6.3	54	Reduction in R_g

with low cost packaging is one of the crucial factors in the high volume DBS market. Hwang et al. have demonstrated a 0.2-μm gate length GaAs pHEMT in plastic packaging with a 1.0 dB noise figure and 9.9 dB associated gain at 12 GHz.[56] A plastic packaged GaAs pHEMT device with a gate length of 0.17 μm demonstrated a noise figure of 0.35 dB, and 12.5 dB associated gain at 12 GHz.[57] Table 8.5 summarizes the low-noise performance of GaAs pHEMTs at various microwave and millimeter-wave frequencies.

8.5.3 GaAs pHEMT for Wireless Applications

The explosive growth of the wireless communication industry has opened up a new area of application for GaAs pHEMTs. Unlike the millimeter-wave military and space applications, the frequencies of operation of these applications are much lower. The frequencies used in typical cellular phones range from 850 MHz for the American Mobile Phone System (AMPS) to 1.9 GHz for the Japanese Personal Handy Phone System (PHS) and the Digital European Cordless Telephone (DECT). The device parameters of interest when considering device technologies for wireless applications are operating voltages, which must be positive, power density, output match and linearity requirements, and gate leakage current.[58] Using enhancement mode devices (threshold/pinch-off voltage > 0) eliminates the negative supply voltage generator and power-cutoff switch. The high gate turn-on voltage of an enhancement mode GaAs pHEMT, as compared to the enhancement mode GaAs MESFET, enables higher input voltage swing.

The power performance of MESFET and GaAs pHEMT for wireless applications has been compared.[59] For the same saturation drain current density, at a frequency of 950 MHz, the saturated power output from the pHEMT is 2.5 W, whereas it is 1.8 W from the MESFET. The power-added efficiency of the pHEMT is 68%, which is 8% higher than that of the MESFET. This difference is due to the transfer characteristics of the two devices. The pHEMT performs as a better power amplifier than the MESFET because the input power is effectively amplified with higher g_m near the pinch-off voltage. This is a direct consequence of better charge control properties of the HEMT when compared to the MESFET. The pHEMT also has a lower gate leakage current than the MESFET due to a higher Schottky barrier on AlGaAs.

The power performance of enhancement mode GaAs pHEMT with a threshold voltage of 0.05 V for wireless applications has also been investigated.[60] A device with a gate width of 3.2 mm delivered an output power of 22 dBm with power-added-efficiency of 41.7%. The standby current at a gate bias of 0 V was 150 μA. Enhancement mode GaAs HFET with a higher threshold voltage of 0.5 V has also been demonstrated.[61] A 1-μm gate length device with a gate width of 12 mm delivered an output power of 31.5 dBm with a PAE of 75% at 850 MHz and at a drain bias of 3.5 V. The standby current at a gate bias of 0 V was 1 μA. This eliminates the need for a switch in the drain current of the power amplifier. The device was manufactured using Motorola's CGaAs™ process, which is cost effective as it uses processes that are similar to standard silicon MOS and bipolar processes. Table 8.6 shows a summary of power performance of GaAs pHEMTs for cellular phones.

TABLE 8.6 Summary of Power Performance of GaAs pHEMTs for Cellular Phones

Frequency	Device Width (μm)	Drain Bias (V)	Power Output	PAE (%)	ACPL	Device/Ref.
850 MHz	5	1.2	19.6 dBm	65.2		AlGaInP/InGaAs pHEMT[62]
	10	1.3	21.5 dBm	57.4		InGaP/InGaAs pHEMT[63]
	12	3.5	31.5 dBm	75		Enhancement Mode HFET[61]
	12	3.5	33.1 dBm	84.8		GaAs pHEMT[64]
	30	3.7	31.0 dBm	59.0	–30.1 dBc @ 30 kHz	1 μm GaAs MESFET[65]
900 MHz	12	3.5	31.5 dBm	75		Enhancement Mode CGaAs™[66]
	21	2.3	31.3 dBm	68		0.8 μm MESFET[67]
	14	3.0	32.3 dBm	71		pHEMT[68]
	40	1.5	31.5 dBm	65		MESFET[69]
950 MHz	28	1.2	1.1 W	54		pHEMT[70]
	21	2.2	32.7 dBm	62.8	–50.5 dBc	PHEMT[71]
	12	3.0	1.4 W	60.0	—	pHEMT[72]
	7	3.4	30.9 dBm	56.3	–51.5 dBc @ 50 kHz	pHEMT[73]
	16	3.4	1.42 W	60.0	–48.2 dBc @ 50 kHz	pHEMT[74]
	12	4.7	2.5 W	68.0		pHEMT[59]
	12	4.7	1.8 W	60.0		MESFET[59]
1.9 GHz	1	2.0	20.2 dBm	45.3	–55.2 dBc @ 600 kHz	pHEMT[75]
	2.4	2.0	21.1 dBm	54.4	–55 dBc @ 600 kHz	MESFET[76]
	3.2	3.0	22.0 dBm	41.7	–58.2 dBc @ 600 kHz	Enhancement Mode pHEMT[60]
	12	3.5	30 dBm	50	–30 dBc	Enhancement Mode pHEMT[61]
	5	2.0	25.0 dBm	53.0		AlGaInP/InGaAs pHEMT[62]

8.6 AlInAs/GaInAs/InP (InP HEMT)

Future military and commercial electronic applications will require high-performance microwave and millimeter-wave devices. Important applications include low-noise amplifiers for receiver front ends, power amplifiers for phased-array radars, ultra high-speed digital circuits for prescalers, and MUX/DEMUX electronics for high-speed (> 40 Gb/s) optical links.

A HEMT device capable of operating at millimeter-wave frequency requires a channel with high electron velocity, high current density, and minimal parasitics. As discussed, the $Al_{0.48}In_{0.52}As/Ga_{0.47}In_{0.53}As$ material system lattice matched to InP satisfies these criteria. A 1-μm gate length AlInAs/GaInAs HEMT with extrinsic transconductance as high as 400 mS/mm was demonstrated.[77] The microwave performance of 1-μm gate-length devices showed an improvement of 20 to 30% over the AlGaAs/GaAs HEMT.[78] In 1988 Mishra et al. demonstrated a 0.1 μm InP HEMT with a f_T of 170 GHz.[79] Using a T-gate to self-align the source and drain contacts results in reduction of source-gate and source-drain spacing. This not only reduces the parasitic source and drain resistances, but also the drain delay. Using the preceding technique an f_T of 250 GHz was achieved in a 0.13-μm gate length self-aligned HEMT.[18] Recently, a 0.07-μm AlInAs/GaInAs HEMT with an f_T of 300 GHz and an f_{max} of 400 GHz was reported.[80]

The high-frequency performance of the InP HEMT can be further improved by using a pseudomorphic InGaAs channel with an indium content as high as 80%. The f_T of a 0.1-μm InP HEMT increased from 175 to 205 GHz when the indium content in the channel was increased from 53 to 62%.[81] Although devices with high indium content channels have low breakdown voltages, they are ideal for low noise applications and ultra high-speed digital applications. An f_T of 340 GHz was achieved in a 0.05-μm gate length psuedomorphic InP HEMT with a composite $In_{0.8}Ga_{0.2}As/In_{0.53}Ga_{0.47}As$ channel.[82] This is the highest reported f_T of any 3-terminal device.

Compared to the GaAs pHEMT, the AlInAs/GaInAs HEMT has a higher current density that makes it suitable for ultra high-speed digital applications. The high current gain cutoff frequency and low parasitics make the AlInAs/GaInAs HEMT the most suitable choice for low- noise applications extending well beyond 100 GHz. The high current density and superior high-frequency performance can be utilized

TABLE 8.7 Summary of Low-Noise Performance of AlInAs/GaInAs HEMTs

Frequency (GHz)	Gate Length (μm)	F_{min} (dB)	G_a (dB)	Comments/Ref.
12	0.15	0.39	16.5	$In_{0.7}Ga_{0.3}As$ channel[87]
18	0.25	0.5	15.2	[85]
18	0.15	0.3	17.2	[86]
26	0.18	0.43	8.5	Passivated device [88]
57	0.25	1.2	8.5	[85]
60	0.1	0.8	8.9	[89]
63	0.1	0.8	7.6	Passivated[90]
		0.7	8.6	Unpassivated
94	0.15	1.4	6.6	[86]
94	0.1	1.2	7.2	[89]

for high-performance millimeter-wave power applications provided the breakdown voltage is improved. Some of the state-of-the-art millimeter-wave analog circuits and ultra high-speed digital circuits have been implemented using InP HEMTs. A low-noise amplifier with 12 dB gain at a frequency of 155 GHz using a 0.1-μm InP HEMT with a $In_{0.65}Ga_{0.35}As$ psuedomorphic channel was demonstrated.[83] Pobanz et al. demonstrated an amplifier with 5 dB gain at 184 GHz using a 0.1 μm gate $In_{0.8}Ga_{0.2}As$/InP composite channel HEMT.[84]

8.6.1 Low-Noise AlInAs/GaInAs HEMT

The superior electronic properties of the GaInAs channel enable the fabrication of extremely high f_T and $f_{\max}$ devices. The superior carrier confinement at the AlInAs/GaInAs interface results in a highly linear transfer characteristic. High transconductance is also maintained very close to pinch off. This is essential because the noise contribution of the FET is minimized at low drain current levels. Hence high gain can be achieved at millimeter-wave frequencies under low-noise bias conditions. The high mobility at the AlInAs/GaInAs interface also results in reduced parasitic source resistance of the device. AlInAs/GaInAs HEMTs with 0.25-μm gate length exhibited a noise figure of 1.2 dB at 58 GHz.[85] At 95 GHz a noise figure of 1.4 dB with associated gain of 6.6 dB was achieved in a 0.15-μm gate length device.[86] Table 8.7 summarizes the low-noise performance of AlInAs/GaInAs HEMTs.

8.6.2 Millimeter-Wave AlInAs/GaInAs Power HEMT

The millimeter-wave power capability of single heterojunction AlInAs/GaInAs HEMTs has been demonstrated.[91,92] The requirements for power HEMT are high gain, high current density, high breakdown voltage, low access resistance, and low knee voltage to increase power output and power-added efficiency. The AlInAs/GaInAs HEMT satisfies all of these requirements with the exception of breakdown voltage. This limitation can be overcome by operating at a lower drain bias. In fact, the high gain and PAE characteristics of InP HEMTs at low drain bias voltages make them ideal candidates for battery-powered applications.[93] Another advantage is the use of InP substrate that has a 40% higher thermal conductivity than GaAs. This allows higher dissipated power per unit area of the device or lower operating temperature for the same power dissipation. As low breakdown voltage is a major factor that limits the power performance of InP HEMTs, this section will discuss in detail the various approaches used to increase breakdown voltage.

Breakdown in InP HEMT is a combination of electron injection from the gate contact and impact ionization in the channel.[94] The breakdown mechanism in the off-state (when the device is pinched off) is electron injection from the gate. The low Schottky barrier height of AlInAs results in increased electron injection from the gate and, consequently, higher gate leakage current compared to the GaAs pHEMT. These injected hot electrons cause impact ionization in the high-field drain end of the GaInAs channel. The high impact ionization rate in the low bandgap GaInAs channel is the main mechanism that determines the on-state breakdown. Some of the holes generated by impact ionization are collected by the negatively biased gate and result in increased gate leakage. The potential at the source end of the channel is modulated by holes collected by the source. This results in increased output conductance.

Lowering the electric field in the gate-drain region can reduce the impact ionization rate. This is achieved by using a double recess gate fabrication process that increases the breakdown voltage from 9 to 16 V.[95] A gate-drain breakdown voltage of 11.2 V was demonstrated for 0.15-μm gate length devices with a 0.6-μm recess width.[96] In addition, reduction in output conductance (g_{ds}) and gate-drain feedback capacitance (C_{gd}) was observed when compared to single recessed devices. The f_{max} of a double recessed device increased from 200 to 300 GHz.[97] Hence, it is desirable for power devices. Another approach to reduce electric field in the gate-drain region is to use an undoped GaInAs cap instead of a doped GaInAs cap.[98] The output conductance can be reduced from 50 to 20 mS/mm for a 0.15-μm gate length device by replacing the doped GaInAs cap by an undoped cap.[99] This also improved the breakdown voltage from 5 to 10 V. The reduction in C_{gd} and g_{ds} resulted in an f_{max} as high as 455 GHz. Redistribution of the dopants in the AlInAs barrier layers can also increase breakdown voltage. An increase in breakdown voltage from 4 to 9 V is achieved by reducing doping in the top AlInAs barrier layer and transferring it to the AlInAs barrier layers below the channel.[92]

The gate leakage current can be reduced and the breakdown voltage increased by using a higher bandgap strained AlInAs barrier.[100,101] By increasing the Al composition in the barrier layers from 48 to 70%, the gate-to-drain breakdown voltage was increased from 4 to 7 V. This also results in reduction of gate leakage as the Schottky barrier height increases from 0.5 to 0.8 eV. The use of $Al_{0.25}In_{0.75}P$ as an Schottky barrier improves the breakdown voltage from –6 to –12V.[102]

The on-state breakdown can be improved in two ways. The first is to reduce the gate leakage current by the impact ionization generated holes by increasing the barrier height for holes. This was achieved by increasing the valence band discontinuity at the channel-barrier interface. The use of a strained 25-Å $In_{0.5}Ga_{0.5}P$ spacer instead of AlInAs increases the valence band discontinuity at the interface from 0.2 to 0.37 eV. An on-state breakdown voltage of 8 V at a drain current density of 400 mA/mm for a 0.7-μm gate length InP HEMT was achieved by using a strained InGaP barrier.[103]

The various approaches to increase the breakdown voltage, as already discussed here, concentrate on reducing the electron injection from the Schottky gate and reducing the gate leakage current. These approaches also have an inherent disadvantage as Al-rich barriers result in high source resistance and are more susceptible to atmospheric oxidation. Additionally, these approaches do not address the problem of a high impact ionization rate in the GaInAs channel and carrier injection from contacts. In the recent past, various new approaches have been investigated to increase breakdown voltage without compromising the source resistance or atmospheric stability of the device. These include the junction-modulated AlInAs/GaInAs HEMT (JHEMT), the composite GaInAs/InP channel HEMT, and the use of regrown contacts. Table 8.8 summarizes the power performance of AlInAs/GaInAs HEMTs.

8.6.3 GaInAs/InP Composite Channel HEMT

The high speed and power performance of InP HEMT can be improved by the use of composite channels that are composed of two materials with complementing electronic properties. The high-speed performance of an InP HEMT can be improved by inserting InAs layers in the InGaAs channel. The current gain cutoff frequency of a 0.15-μm gate length device increased from 179 to 209 GHz due to improved electron properties.[111] An f_T as high as 264 GHz was achieved for a 0.08-μm gate length device.

The GaInAs channel has excellent electronic properties at a low electric field, but suffers from high impact ionization at high electric fields. On the other hand, InP has excellent electronic transport properties at high fields but has lower electron mobility. In a composite InGaAs/InP channel HEMT, the electrons are in the InGaAs channel at the low field source end of the channel and are in the InP channel at the high field drain end of the channel. This improves the device characteristics at high drain bias while still maintaining the advantages of the GaInAs channel at low bias voltages.[112] A typical submicron gatelength AlInAs/GaInAs HEMT has an off-state breakdown voltage (BV_{dsoff}) of 7 V, and on-state

TABLE 8.8 Summary of Power Performance of InP HEMTs

Frequency (GHz)	Gate Length (μm)	Gate Width	Power Density (W/mm)	Power Output (mW)	Gain (dB)	PAE (%)	Device/Drain Bias (Ref.)
4	0.5	2 mm	0.13	269	18	66	V_{ds} = 2.5 V[93]
	0.15	0.8mm	0.4	320	18	57	V_{ds} = 3 V[93]
12	0.22	150 μm	0.78	117	8.4	47	V_{ds} = 4 V[92]
			0.47	70	11.3	59	V_{ds} = 3 V, Double HJ
18	0.15	600 μm	0.74	446	13	59	Double recessed V_{ds} = 7 V[104]
20	0.15	50 μm	0.78	39	10.2	44	V_{ds} = 4.9 V[91]
(K-Band)			0.41	21	10.5	52	V_{ds} = 2.5 V Single heterojunction
	0.15	50 μm	0.61	30	12.2	44	Single heterojunction $In_{0.69}Ga_{0.31}As$ channel V_{ds} = 4.1 V[91]
	0.15	800 μm	0.65	516	7.1	47	70% AlInAs, V_{ds} = 4 V[105]
44	0.15	450 μm	0.55	251	8.5	33	$Al_{0.6}In_{0.4}As$ barrier
(Q-Band)			0.88	398	6.7	30	Doped Channel[100]
	0.2	600 μm	0.37	225	5	39	$Al_{0.6}In_{0.4}As$ barrier Single heterojunction, V_{ds} = 4 V[106]
57	0.22	450 μm	0.33	150	3.6	20	$Al_{0.6}In_{0.4}As$ barrier
(V-Band)			0.44	200		17	Doped channel, V_{ds} = 3.5 V[101]
60	0.15	50 μm	0.35		7.2	41	V_{ds} = 2.6 V[91]
(V-Band)			0.52	26	5.9	33	V_{ds} = 3.6 V Single heterojunction
	0.1	50 μm	0.30	15	8.6	49	V_{ds} = 3.35 V[107]
			0.41	21	8.0	45	
	0.1	400 μm	0.48	192	4.0	30	V_{ds} = 4.12 V, 67% In[107]
94	0.15	50 μm	0.30	15	4.6	21	Single HJ, Passivated device V_{ds} = 2.6 V[108]
(W-Band)							
	0.15	640 μm	0.20	130	4.0	13	Double HJ, V_{ds} = 2.7[109]
	0.1	200 μm	0.29	58		33	$In_{0.68}Ga_{0.32}As$ channel[110]

breakdown voltage (BV_{dson}) of 3.5 V. Using a composite channel, (30 Å GaInAs/50 Å InP/100 Å n^+ InP), Matloubian et al. demonstrated a BV_{dsoff} of 10 V and BV_{dson} of 8 V for a 0.15-μm gate length device.[113] A 0.25-μm GaInAs/InP composite channel HEMT with a two-terminal gate drain voltage of 18 V was also demonstrated.[114]

The increased breakdown voltage of a composite channel HEMT enables operation at a higher drain bias. This increases the drain efficiency and the PAE of the device. An output power of 0.9 W/mm with a PAE of 76% at 7 GHz was demonstrated for a 0.15-μm GaInAs/InP composite channel HEMT at a drain bias of 5 V.[115] At 20 GHz, an output power density of 0.62 W/mm (280 mW) and a PAE 46% was achieved for a 0.15 μm gate length device at a drain bias of 6 V.[113] At 60 GHz, a 0.15-μm GaInAs/InP composite channel HEMT demonstrated an output power of 0.35 W/mm, and a power gain of 6.2 dB with a PAE of 12% at a drain bias of 2.5 V.[116]

8.7 Technology Comparisons

8.7.1 Comparison between FETs and HBTs for RF/Microwave Applications

In recent years hetero-structure bipolar transistors (HBTs) have emerged as strong contenders for wireless, microwave, and millimeter-wave applications. HBTs have similar high frequency performance with modest lateral dimensions as transport is along the vertical direction. This also offers higher current drivability per unit chip area. On the other hand, submicron gate lengths are required to achieve microwave and millimeter-wave operation in FETs. In addition, HBTs have high transconductance due to an exponential relationship between the input (base) voltage and output (collector) current. Device uniformity over a wafer can be easily achieved in a HBT as the turn-on voltage depends on the built-in voltage of a pn junction (and is governed by the uniformity of the epitaxial growth technique). The threshold voltage of a HEMT (which is a measure of device uniformity) is mainly determined by uniformity in gate recess etching. The FET is mainly a surface device and therefore has a higher 1/f noise due generation-recombination processes in the surface traps. On the other hand, current transport in a HBT is in the bulk, hence HBTs have lower 1/f noise which makes them ideal for low-phase-noise microwave and millimeter-wave oscillators.

Semiconductor processing of HBTs is more complex than FETs owing to the vertical structure. Another advantage of FETs when compared to HBTs is lower operating voltage. The turn-on voltage of GaAs HBTs is significantly higher (1.4 V) than that of HEMTs (0.5 V). This makes the HEMT an ideal candidate for low-voltage battery-operated applications. Owing to low parasitics, HEMTs have a lower noise figure than HBTs at any frequency and therefore are preferred for low-noise receivers. For microwave and millimeter-wave power applications gain, HEMTs are again the preferred choice. This is because at millimeter-wave frequency (> 50 GHz) the gain in a HBT is mainly limited by the high collector base feedback capacitance. Another disadvantage of large area HBTs is thermal runaway due to nonuniform junction temperatures. This can be solved by the use of emitter ballast resistors, but this reduces device gain and processing complexity. In the following sections, the various HEMT technologies are compared for specific applications.

8.7.2 Power Amplifiers for Wireless Phones

The two main aspects that govern the choice of technology for wireless power amplifiers are the operating voltage and the cost. As far as operating voltages are concerned, the use of GaAs HBTs is limited to 3.6 V supply voltages. This is due to the high turn-on voltage (1.4 V) for the emitter-base junction in the GaAs HBT. On the other hand, GaAs MESFETs and pHEMT can be used in 1.2 V systems due to their low knee voltages. As far as cost is concerned, the major contenders are ion-implanted GaAs MESFETs and GaAs HBTs. However, GaAs MESFETs require a negative gate voltage, which potentially increases circuit complexity and cost. The enhancement mode GaAs pHEMT, which has a low knee voltage (like a GaAs MESFET) and positive voltage operation (like a GaAs HBT) is therefore a viable technology option for high performance, low voltage wireless phones.

8.7.3 Microwave Power Amplifiers

The GaAs pHEMT and the InP HEMT are the two major competing technologies for microwave and millimeter-wave power amplifiers. The relation between output power density (P_{out}) and power-added efficiency (*PAE*) and device parameters is given by the following expressions,

$$P_{out} = \frac{1}{8}\left(I_{max}\right)\left(BV_{gd} - V_{knee}\right) \tag{8.30}$$

$$PAE = \alpha\left(\frac{V_{DD} - V_{knee}}{V_{DD}}\right)\left(1 - \frac{1}{G_a}\right) \tag{8.31}$$

In Eq. (8.31), α is ½ for Class A operation and $\pi/4$ for Class B operation. Figure 8.8 compares the P_{out} and *PAE* of GaAs pHEMTs and InP HEMTs as a function of operating frequency. It can be seen that GaAs pHEMTs have a higher power density than InP HEMTs at all frequencies than InP HEMT. This is due to higher breakdown voltages (BV_{gd}) in GaAs pHEMTs that enables higher operating voltage (V_{DD}). The InP HEMT operating voltage is limited by the low breakdown voltage, hence the power output is low. However, InP HEMTs have comparable power output at millimeter-wave frequencies. This is enabled

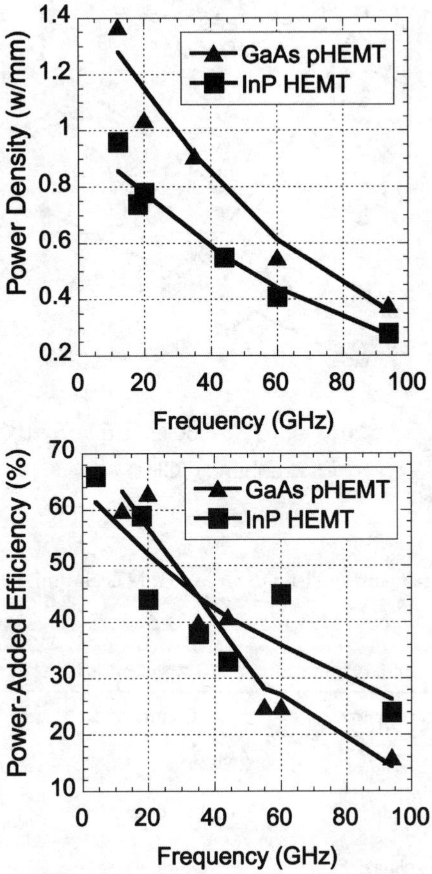

FIGURE 8.8 Comparison of output power density and power-added-efficiency of GaAs pHEMTs and InP HEMTs as a function of frequency.

by the low knee voltage (V_{knee}) and high current drive (I_{max}). On the other hand, due to their low knee voltage and high gain (G_a), InP HEMTs have higher gain and PAE at frequencies exceeding 60 GHz. The potential of InP HEMTs as millimeter-wave power devices is evident in the fact that comparable power performance is achieved at drain biases 2 to 3 V lower than those for GaAs pHEMTs. Hence a high voltage InP HEMT technology should be able to outperform the GaAs pHEMT for power applications at all frequencies.

8.7.4 Low Noise Amplifiers

Figure 8.9 compares the gain and noise figure of GaAs pHEMTs and InP HEMTs as a function of frequency. At any given frequency, the InP HEMT has a noise figure about 1 dB lower than the GaAs pHEMT and gain 2 dB higher than the GaAs pHEMT. This is mainly due to the excellent transport properties of the AlInAs/GaInAs material system.

Digital applications — High current driving capability with low voltage operation is essential for high speed, low power digital circuits. High f_T devices are also required for increasing the frequency of operation. In this area, GaAs MESFETs and GaAs pHEMTs are sufficient for 10 Gb/s digital circuits. However, for digital circuits operating at bit rates exceeding 40 Gb/s, the InP HEMT is the preferred technology option.

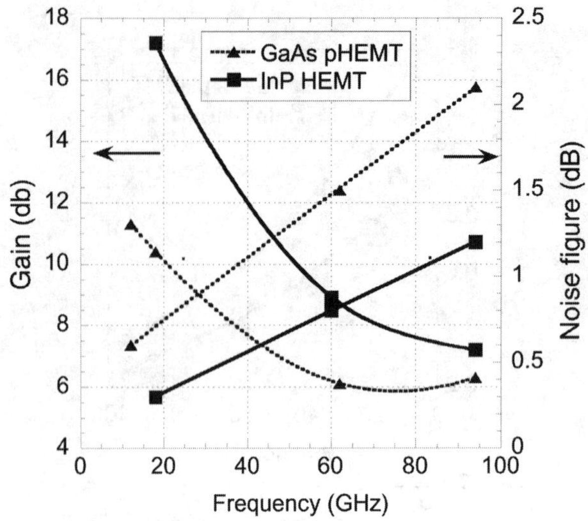

FIGURE 8.9 Comparison of noise figure and gain of GaAs pHEMTs and InP HEMTs as a function of frequency.

TABLE 8.9 Frequency Bands and Military and Commercial Applications

Frequency	Military/Space	Commercial	Device Technology
850 MHz–1.9 GHz		Wireless	Low noise — GaAs pHEMT Power — GaAs pHEMT
12 GHz (Ku-Band)	Phased array radar	Direct broadcast satellite	Low noise — GaAs pHEMT Power — GaAs pHEMT
20 GHz (K-Band)	Satellite downlinks		
27–35 GHz (Ka-Band)	Missile seekers	LMDS — Local multipoint distribution system	Low noise — GaAs pHEMT Power — GaAs pHEMT
44 GHz (Q-Band)	SATCOM ground terminals	MVDS — Multipoint video distribution system	Low noise — InP HEMT Power — GaAs pHEMT
60 GHz (V-Band)	Satellite crosslinks	Wireless LAN	Low noise — InP HEMT Power — GaAs pHEMT/InP HEMT
77 GHz		Collision avoidance radar	Low noise — InP HEMT Power — GaAs pHEMT/InP HEMT
94 GHz (W-Band)	FMCW radar		Low noise — InP HEMT Power — InP HEMT
100–140 GHz	Radio astronomy		Low noise — InP HEMT
Digital 10 Gb/s		Fiber-optic communication	GaAs pHEMT
Digital 40 Gb/s		Fiber-optic communication	InP HEMT

Table 8.9 lists the various frequency bands and military and space applications in each band with the appropriate HEMT device technology for each application.

8.8 Conclusion

GaAs- and InP-based high electron mobility transistors have emerged as premier devices for the implementation of millimeter-wave analog circuits and ultra high-speed digital circuits. In this section, principles of HEMT operation were discussed. The design aspects of HEMT for both low-noise and high-power applications were discussed. Reduction in gate length is essential for improved performance at high frequencies. Appropriate device scaling with gate length reduction is necessary to minimize the effect of parasitics on device performance.

Millimeter-wave power modules have been demonstrated using GaAs pHEMT devices. The superior device performance of GaAs pHEMTs is being used to improve the performance of power amplifiers for wireless phone systems. The superior material characteristics of the AlInAs/GaInAs material system have been used to achieve record low-noise performance at millimeter-wave frequencies using InP HEMTs. Despite their low breakdown voltage, InP HEMTs have demonstrated superior power performance at millimeter-wave frequency. Improving the breakdown voltage using approaches that include composite channel GaInAs/InP HEMT and junction-modulated HEMT will further improve power performance. The development of the GaAs pHEMT and InP HEMT technology was traditionally supported by low-volume, high-cost military and space applications. The recent emergence of high-volume commercial applications such as wireless and optical communications systems has new constraints that include manufacturability and low-voltage operation.

References

1. Dingle, R., Stromer, H. L., Gossard, A. C., and Wiemann, W., Electron mobility in modulation doped semiconductor superlattices, *Applied Physics Letters*, 33, 665, 1978.
2. Delagebeaudeuf, D. and Linh, N. T., Metal-(n) AlGaAs-GaAs two-dimensional electron gas FET, *IEEE Transactions on Electron Devices*, 29, 955, 1982.
3. Drummond, T. J., Masselink, W. T., and Morkoc, H., Modulation Doped GaAs/(Al,Ga)As heterojunction field-effect transistors: MODFETs, *Proceedings of the IEEE*, 74, 773, 1986.
4. Vinter, B., Subbands and charge control in two-dimensional electron gas field effect transistor, *Applied Physics Letters*, 44, 307, 1984.
5. Foisy, M. C., Tasker, P. J., Hughes, B., and Eastman, L. F., The role of insufficient charge modulation in limiting the current gain cutoff frequency of the MODFET, *IEEE Transactions of Electron Devices*, 35, 871, 1988.
6. Nguyen, L. D., Larson, L. E., and Mishra, U. K., Ultra-high-speed modulation-doped field-effect transistors: A Tutorial Review, *Proceedings of the IEEE*, 80, 494, 1992.
7. Dambrine, G., Cappy, A., Heliodore, F., and Playez, E., A new method for determining the FET small-signal equivalent circuit, *IEEE Transactions on Microwave Theory and Techniques*, 36, 1151, 1988.
8. Berroth, M. and Bosch, R., Broad-band determination of the FET small-signal equivalent circuit, *IEEE Transactions on Microwave Theory and Techniques*, 38, 891, 1990.
9. Das, M. B., A high aspect ratio design approach to millimeter-wave HEMT structures, *IEEE Transactions on Electron Devices*, 32, 11, 1985.
10. Lester, L. F., Smith, P. M., Ho, P., Chao, P. C., Tiberio, R. C., Duh, K. H. G., and Wolf, E. D., 0.15 μm gate length double recess psuedomorphic HEMT with f_{max} of 350 GHz, *IEDM Technical Digest*, 172, 1988.
11. Nguyen, L. D., Tasker, P. J., Radulescu, D. C., and Eastman, L. F., Characterization of ultra-high speed pseudomorphic AlGaAs/InGaAs (on GaAs) MODFET's, *IEEE Transactions on Electron Devices*, 36, 2243, 1989.
12. Nguyen, L. D. and Tasker, P. J., Scaling issue of ultra-high speed HEMTs, *SPIE Conf. on High Speed Electronics and Device Scaling*, 1288, 251, 1990.
13. Moll, N., Hueschen, M. R., and Fischer-Colbrie, A., Pulse doped AlGaAs/InGaAs pseudomorphic MODFETs, *IEEE Transactions on Electron Devices*, 35, 879, 1988.
14. Chao, P. C., Shur, M. S., Tiberio, R. C., Duh, K. H. G., Smith, P. M., Ballingall, J. M., Ho, P., and Jabra, A. A., DC and microwave characteristics of Sub 0.1 μm gate length planar doped pseudomorphic HEMTs, *IEEE Transactions on Electron Devices*, 36, 461, 1989.

15. Nguyen, L. D., Brown, A., Delaney, M., Mishra, U., Larson, L., Jelloian, L., Melendes, M., Hooper, C., and Thompson, M., Vertical scaling of ultra-high-speed AlInAs-GaInAs HEMTs, *IEDM Technical Digest*, 105, 1989.
16. Enoki, T., Tomizawa, M., Umeda, Y., and Ishii, Y., 0.05 μm Gate InAlAs/InGaAs High Electron mobility transistor and reduction of its short channel effects, *Japanese Journal of Applied Physics*, 33, 798, 1994.
17. Tasker, P. J. and Hughes, B., Importance of source and drain resistance to the maximum f_T of millimeter-wave MODFETs, *IEEE Electron Device Letters*, 10, 291, 1989.
18. Mishra, U. K., Brown, A. S., Jelloian, L. M., Thompson, M., Nguyen, L. D., and Rosenbaum, S. E., Novel high performance self aligned 0.15 μm long T-gate AlInAs-GaInAs HEMTs, *IEDM Technical Digest*, 101, 1989.
19. Mimura, T., Hiyamizu, S., and Hikosak, S., Enhancement mode high electron mobility transistors for logic applications, *Japanese Journal of Applied Physics*, L317, 1981.
20. Ketterson, A. A., Masselink, W. T., Gedymin, J. S., Klem, J., Peng, C.-K., Kopp, W. F., Morkoç, H., and Gleason, K. R., Characterization of InGaAs/AlGaAs pseudomorphic modulation-doped field-effect transistors, *IEEE Transactions on Electron Devices*, 33, 564, 1986.
21. Pereiaslavets, B., Bachem, K. H., Braunstein, J., and Eastman, L. F., GaInP/InGaAs/GaAs graded barrier MODFET grown by OMVPE: design, fabrication, and device results, *IEEE Transactions on Electron Devices*, 43, 1659, 1996.
22. Mesquida-Kusters, A. and Heime, K., Al-Free InP-based high electron mobility transistors: design, fabrication and performance, *Solid State Electronics*, 41, 1159, 1997.
23. Ketterson, A., Moloney, M., Masselink, W. T., Klem, J., Fischer, R., Kopp, W., and Morkoc, H., High transconductance InGaAs/AlGaAs pseudomorphic modulation doped field effect transistors, *IEEE Electron Device Letters*, 6, 628, 1985.
24. Nguyen, L. D., Radulescu, D. C., Tasker, P. J., Schaff, W. J., and Eastman, L. F., 0.2 μm gate-length atomic-planar doped pseudomorphic $Al_{0.3}Ga_{0.7}As/In_{0.25}Ga_{0.75}As$ MODFET's with f_τ over 120 GHz, *IEEE Electron Device Letters*, 9, 374, 1988.
25. Smith, P. M., Lester, L. F., Chao, P.-C., Ho, P., Smith, R. P., Ballingall, J. M., and Kao, M.-Y., A 0.25 μm gate length pseudomorphic HFET with 32 mW Output power at 94 GHz, *IEEE Electron Device Letters*, 10, 437, 1989.
26. Streit, D. C., Tan, K. L., Dia, R. M., Liu, J. K., Han, A. C., Velebir, J. R., Wang, S. K., Trinh, T. Q., Chow, P. D., Liu, P. H., and Yen, H. C., High gain W-band pseudomorphic InGaAs power HEMTs, *IEEE Electron Device Letters*, 12, 149, 1991.
27. Actis, R., Nichols, K. B., Kopp, W. F., Rogers, T. J., and Smith, F. W., High performance 0.15 μm gate length pHEMTs enhanced with a low temperature grown GaAs buffer, *IEEE MTT-S Int. Microwave Symp. Dig.*, 445, 1995.
28. Huang, J. C., Boulais, W., Platzker, A., Kazior, T., Aucoin, L., Shanfield, S., Bertrand, A., Vafiades, M., and Niedzwiecki, M., The effect of channel dimensions on the millimeter wave performance of pseudomorphic HEMT, *Proceedings of GaAs IC Symposium*, 177, 1993.
29. Danzilio, D., White, P., Hanes, L. K., Lauterwasser, B., Ostrowski, B., and Rose, F., A high efficiency 0.25 μm pseudomorphic HEMT power process, *Proceedings of GaAs IC Symposium*, 255, 1992.
30. Chen, C. H., Saito, Y., Yen, H. C., Tan, K., Onak, G., and Mancini, J., Reliability study on pseudomorphic InGaAs power HEMT devices at 60 GHz, *IEEE MTT-S Int. Microwave Symp. Dig.*, 817, 1994.
31. Hori, Y., Onda, K., Funabashi, M., Mizutani, H., Maruhashi, K., Fujihara, A., Hosoya, K., Inoue, T., and Kuzuhara, M., Manufacturable and reliable millimeter wave HJFET MMIC technology using novel 0.15 μm MoTiPtAu gates, *IEEE MTT-S Int. Microwave Symp. Dig.*, 431, 1995.
32. Matsunaga, K., Okamoto, Y., and M. Kuzuhara, A 12-GHz, 12-W HJFET amplifier with 48% peak power added-efficiency, *IEEE Microwave and Guided Wave Letters*, 5, 402, 1995.

33. Helms, D., Komiak, J. J., Kopp, W. F., Ho, P., Smith, P. M., Smith, R. P., and Hogue, D., Ku band power amplifier using pseudomorphic HEMT devices for improved efficiency, *IEEE MTT-S Int. Microwave Symp. Dig.*, 819, 1991.
34. Yarborough, R., Heston, D., Saunier, P., Tserng, H. Q., Salzman, K., and B. Smith, Four watt, Kt-band MMIC amplifier, *IEEE MTT-S Int. Microwave Symp. Dig.*, 797, 1994.
35. Kao, M.-Y., Saunier, P., Ketterson, A. A., Yarborough, R., and Tserng, H. Q., 20 GHz power PHEMTs with power-added efficiency of 68% at 2 volts, *IEDM Technical Digest*, 931, 1996.
36. Dow, G. S., Tan, K., Ton, N., Abell, J., Siddiqui, M., Gorospe, B., Streit, D., Liu, P., and Sholley, M., Ka-band high efficiency 1 watt power amplifier, *IEEE MTT-S Int. Microwave Symp. Dig.*, 579, 1992.
37. Kao, M.-Y., Smith, P. M., Ho, P., Chao, P.-C., Duh, K. H. G., Abra, A. A. J., and Ballingall, J. M., Very high power-added efficiency and low noise 0.15 μm gate-length pseudomorphic HEMTs, *IEEE Electron Device Letters*, 10, 580, 1989.
38. Ferguson, D. W., Smith, P. M., Chao, P. C., Lester, L. F., Smith, R. P., Ho, P., Jabra, A., and Ballingall, J. M., 44 GHz hybrid HEMT power amplifiers, *IEEE MTT-S Int. Microwave Symp. Dig.*, 987, 1989.
39. Kasody, R., Wang, H., Biedenbender, M., Callejo, L., Dow, G. S., and Allen, B. R., Q Band high efficiency monolithic HEMT power prematch structures, *Electronics Letters*, 31, 505, 1995.
40. Boulais, W., Donahue, R. S., Platzker, A., Huang, J., Aucoin, L., Shanfield, S., and Vafiades, M., A high power Q-band GaAs pseudomorphic HEMT monolithic amplifier, *IEEE MTT-S Int. Microwave Symp. Dig.*, 649, 1994.
41. Smith, P. M., Creamer, C. T., Kopp, W. F., Ferguson, D. W., Ho, P., and Willhite, J. R., A high power Q-band HEMT for communication terminal applications, *IEEE MTT-S Int. Microwave Symp. Dig.*, 809, 1994.
42. Tan, K. L., Streit, D. C., Dia, R. M., Wang, S. K., Han, A. C., Chow, P. D., Trinh, T. Q., Liu, P. H., Velebir, J. R., and Yen, H. C., High power V-band pseudomorphic InGaAs HEMT, *IEEE Electron Device Letters*, 12, 213, 1991.
43. Saunier, P., Matyi, R. J., and Bradshaw, K., A double-heterojunction doped-channel pseudomorphic power HEMT with a power density of 0.85 W/mm at 55 GHz, *IEEE Electron Device Letters*, 9, 397, 1988.
44. Lai, R., Wojtowicz, M., Chen, C. H., Biedenbender, M., Yen, H. C., Streit, D. C., Tan, K. L., and Liu, P. H., High power 0.15 μm V-band pseudomorphic InGaAs-AlGaAs-GaAs HEMT, *IEEE Microwave and Guided Wave Letters*, 3, 363, 1993.
45. Kraemer, B., Basset, R., Baughman, C., Chye, P., Day, D., and Wei, J., Power PHEMT module delivers 4 Watts, 38% PAE over the 18.0 to 21.2 GHz band, *IEEE MTT-S Int. Microwave Symp. Dig.*, 801, 1994.
46. Aucoin, L., Bouthillette, S., Platzker, A., Shanfield, S., Bertrand, A., Hoke, W., and Lyman, P., Large periphery, high power pseudomorphic HEMTs, *Proceedings of GaAs IC Symposium*, 351, 1993.
47. Bouthillette, S., Platzker, A., and Aucoin, L., High efficiency 40 Watt PsHEMT S-band MIC power amplifiers, *IEEE MTT-S Int. Microwave Symp. Dig.*, 667, 1994.
48. Kraemer, B., Basset, R., Chye, P., Day, D., and Wei, J., Power PHEMT module delivers 12 watts, 40% PAE over the 8.5 to 10.5 GHz band, *IEEE MTT-S Int. Microwave Symp. Dig.*, 683, 1994.
49. Matsunaga, K., Okamoto, Y., Miura, I., and Kuzuhara, M., Ku-band 15 W single chip HJFET power amplifier, *IEEE MTT-S Int. Microwave Symp. Dig.*, 697, 1996.
50. Fu, S. T., Lester, L. F., and Rogers, T., Ku band high power high efficiency pseudomorphic HEMT, *IEEE MTT-S Int. Microwave Symp. Dig.*, 793, 1994.
51. Henderson, T., Aksun, M. I., Peng, C. K., Morkoc, H., Chao, P. C., Smith, P. M., Duh, K.-H. G., and Lester, L. F., Microwave performance of quarter micrometer gate low noise pseudomorphic InGaAs/AlGaAs modulation doped field effect transistor, *IEEE Electron Device Letters*, 7, 649, 1986.
52. Tan, K. L., Dia, R. M., Streit, D. C., Shaw, L. K., Han, A. C., Sholley, M. D., Liu, P. H., Trinh, T. Q., Lin, T., and Yen, H. C., 60 GHz pseudomorphic $Al_{0.25}Ga_{0.75}As/In_{0.28}Ga_{0.72}As$ low- noise HEMTs, *IEEE Electron Device Letters*, 12, 23, 1991.

53. Chao, P. C., Duh, K. H. G., Ho, P., Smith, P. M., Ballingall, J. M., Jabra, A. A., and Tiberio, R. C., 94 GHz low noise HEMT, *Electronics Letters*, 25, 504, 1989.
54. Tan, K. L., Dia, R. M., Streit, D. C., Lin, T., Trinh, T. Q., Han, A. C., Liu, P. H., Chow, P. D., and Yen, H. C., 94 GHz 0.1 μm T-gate low noise pseudomorphic InGaAs HEMTs, *IEEE Electron Device Letters*, 11, 585, 1990.
55. Tokue, T., Nashimoto, Y., Hirokawa, T., Mese, A., Ichikawa, S., Negishi, H., Toda, T., Kimura, T., Fujita, M., Nagasako, I., and Itoh, T., Ku band super low noise pseudomorphic heterojunction field effect transistor (HJFET) with high producibility and high reliability, *IEEE MTT-S Int. Microwave Symp. Dig.*, 705, 1991.
56. Hwang, T., Kao, T. M., Glajchen, D., and Chye, P., Pseudomorphic AlGaAs/InGaAs/GaAs HEMTs in low-cost plastic packaging for DBS application, *Electronics Letters*, 32, 141, 1996.
57. Hirokawa, T., Negishi, H., Nishimura, Y., Ichikawa, S., Tanaka, J., Kimura, T., Watanbe, K., and Nashimoto, Y., A Ku-band ultra super low-noise pseudomorphic heterojunction FET in a hollow plastic PKG, *IEEE MTT-S Int. Microwave Symp. Dig.*, 1603, 1996.
58. Halchin, D. and Golio, M., Trends for portable wireless applications, *Microwave Journal*, 40, 62, 1997.
59. Ota, Y., Adachi, C., Takehara, H., Yangihara, M., Fujimoto, H., Masato, H., and Inoue, K., Application of heterojunction FET to power amplifier for cellular telephone, *Electronics Letters*, 30, 906, 1994.
60. Kunihisa, T., Yokoyama, T., Nishijima, M., Yamamoto, S., Nishitsuji, M., K. Nishii, Nakayama, and Ishikawa, O., A high-efficiency normally-off MODFET power MMIC for PHS operating under 3.0V single-supply condition, *Proceedings of GaAs IC Symposium*, 37, 1997.
61. Glass, E., Huang, J.-H., Abrokwah, J., Bernhradt, B., Majerus, M., Spears, E., Droopad, R., and Ooms, B., A true enhancement mode single supply power HFET for portable applications, *IEEE MTT-S Int. Microwave Symp. Dig.*, 1399, 1997.
62. Wang, Y. C. and J.M. Kuo, J. R. L., F. Ren, H.S. Tsai, J.S. Weiner, J. Lin, A. Tate, Y.K. Chen, W.E. Mayo, An $In_{0.5}(Al_{0.3}Ga_{0.7})_{0.5}P/In_{0.2}Ga_{0.8}As$ power HEMT with 65.2% power-added efficiency under 1.2 V operation, *Electronics Letters*, 34, 594, 1998.
63. Ren, F., Lothian, J. R., Tsai, H. S., Kuo, J. M., Lin, J., Weiner, J. S., Ryan, R. W., Tate, A., and Chen, Y. K., High performance pseudomorphic InGaP/InGaAs power HEMTs, *Solid State Electronics*, 41, 1913, 1997.
64. Martinez, M. J., Schirmann, E., Durlam, M., Halchin, D., Burton, R., Huang, J.-H., Tehrani, S., Reyes, A., Green, D., and Cody, N., P-HEMTs for low-voltage portable applications using filled gate fabrication process, *Proceedings of GaAs IC Symposium*, 241, 1996.
65. Masato, H., Maeda, M., Fujimoto, H., Morimoto, S., Nakamura, M., Yoshikawa, Y., Ikeda, H., Kosugi, H., and Ota, Y., Analogue/digital dual power module using ion-implanted GaAs MESFETs, *IEEE MTT-S Int. Microwave Symp. Dig.*, 567, 1995.
66. Huang, J. H. and E. Glass, J. A., B. Bernhardt, M. Majerus, E. Spears, J.M. Parsey Jr., D. Scheitlin, R. Droopad, L.A. Mills, K. Hawthorne, J. Blaugh, Device and process optimization for a low voltage enhancement mode power heterojunction FET for portable applications, *Proceedings of GaAs IC Symposium*, 55, 1997.
67. Lee, J.-L., Mun, J. K., Kim, H., Lee, J.-J., and Park, H.-M., A 68% PAE, GaAs power MESFET operating at 2.3 V drain bias for low distortion power applications, *IEEE Transactions on Electron Devices*, 43, 519, 1996.
68. Inosako, K., Matsunaga, K., Okamoto, Y., and Kuzuhara, M., Highly efficient double doped heterojunction FET's for battery operated portable power applications, *IEEE Electron Device Letters*, 15, 248, 1994.
69. Tanaka, T., Furukawa, H., Takenaka, H., Ueda, T., Noma, A., Fukui, T., Tateoka, K., and Ueda, D., 1.5 V Operation GaAs spike-gate power FET with 65% power-added efficiency, *IEDM Technical Digest*, 181, 1995.
70. Inosako, K., Iwata, N., and Kuzuhara, M., 1.2 V operation 1.1 W heterojunction FET for portable radio applications, *IEEE Electron Devices Meeting*, 185, 1995.

71. Iwata, N., Inosako, K., and Kuzuhara, M., 2.2 V operation power heterojunction FET for personal digital cellular telephones, *Electronics Letters*, 31, 2213, 1995.
72. Iwata, N., Inosako, K., and Kuzuhara, M., 3V operation L-band power double-doped heterojunction FETs, *IEEE MTT-S Int. Microwave Symp. Dig.*, 1465, 1993.
73. Iwata, N., Tomita, M., Yamaguchi, K., Oikawa, H., and Kuzuhara, M., 7 mm gate width power heterojunction FETs for Li-ion battery operated personal digital cellular phones, *Proceedings of GaAs IC Symposium*, 119, 1996.
74. Bito, Y., Iwata, N., and Tomita, M., Single 3.4 V operation power heterojunction FET with 60% efficiency for personal digital cellular phones, *Electronics Letters*, 34, 600, 1998.
75. Lai, Y.-L., Chang, E. Y., Chang, C.-Y., Liu, T. H., Wang, S. P., and Hsu, H. T., 2-V-operation δ-doped power HEMT's for personal handy-phone systems, *IEEE Microwave and Guided Wave Letters*, 7, 219, 1997.
76. Choumei, K., Yamamoto, K., Kasai, N., Moriwaki, T., Y, Y., Fujii, T., Otsuji, J., Miyazaki, Y., Tanino, N., and Sato, K., A high efficiency, 2 V single-supply voltage operation RF front-end MMIC for 1.9 GHz personal hand phone systems, *Proceedings of GaAs IC Symposium*, 73, 1998.
77. Hirose, K., Ohata, K., Mizutani, T., Itoh, T., and Ogawa, M., 700 mS/mm 2DEGFETs fabricated from high electron mobility MBE-grown n-AlInAs/GaInAs heterostructures, *GaAs and Related Compounds*, 529, 1985.
78. Palamateer, L. F., Tasker, P. J., Itoh, T., Brown, A. S., Wicks, G. W., and Eastman, L. F., Microwave characterization of 1 μm gate $Al_{0.48}In_{0.52}As/Ga_{0.47}In_{0.53}As$/InP MODFETs, *Electronics Letters*, 23, 53, 1987.
79. Mishra, U. K., Brown, A. S., Rosenbaum, S. E., Hooper, C. E., Pierce, M. W., Delaney, M. J., Vaughn, S., and White, K., Microwave performance of AlInAs-GaInAs HEMT's with 0.2 μm and 0.1 μm gate length, *IEEE Electron Device Letters*, 9, 647, 1988.
80. Suemitsu, T., Enoki, T., Yokoyama, H., Umeda, Y., and Ishii, Y., Impact of two-step-recessed gate structure on RF performance of InP-based HEMTs, *Electronics Letters*, 34, 220, 1998.
81. Mishra, U. K., Brown, A. S., and Rosenbaum, S. E., DC and RF performance of 0.1 μm gate length $Al_{0.48}In_{0.52}As\text{-}Ga_{0.38}In_{0.62}As$ pseudomorphic HEMTs, *IEDM Technical Digest*, 180, 1988.
82. Nguyen, L. D., Brown, A. S., Thompson, M. A., and Jelloian, L. M., 50-nm self-aligned-gate pseudomorphic AlInAs/GaInAs high electron mobility transistors, *IEEE Transactions on Electron Devices*, 39, 2007, 1992.
83. Lai, R., Wang, H., Chen, Y. C., Block, T., Liu, P. H., Streit, D. C., Tran, D., Barsky, M., Jones, W., Siegel, P., and Gaier, T., 155 GHz MMIC LNAs with 12 dB gain fabricated using a high yield InP HEMT MMIC process, *Microwave Journal*, 40, 166, 1997.
84. Pobanz, C., Matloubian, M., Lui, M., Sun, H.-C., Case, M., Ngo, C., Janke, P., Gaier, T., and Samoska, L., A high-gain monolithic D-band InP HEMT amplifier, *Proceedings of GaAs IC Symposium*, 41, 1998.
85. Ho, P., Chao, P. C., Du, K. H. G., Jabra, A. A., Ballingall, J. M., and Smith, P. M., Extremely high gain, low noise InAlAs/InGaAs HEMTs grown by molecular beam epitaxy, *IEDM Technical Digest*, 184, 1988.
86. Chao, P. C., Tessmer, A. J., Duh, K. H. G., Ho, P., Kao, M.-Y., Smith, P. M., Ballingall, J. M., Liu, S.-M. J., and Jabra, A. A., W-band low-noise InAlAs/InGaAs lattice matched HEMTs, *IEEE Electron Device Letters*, 11, 59, 1990.
87. Onda, K., Fujihara, A., Miyamoto, H., Nakayama, T., Mizuki, E., Samoto, N., and Kuzuhara, M., Low noise and high gain InAlAs/InGaAs heterojunction FETs with high indium composition channels, *GaAs and Related Compounds*, 139, 1993.
88. Umeda, Y., Enoki, T., Arai, K., and Ishii, Y., Silicon nitride passivated ultra low noise InAlAs/InGaAs HEMT's with n^+ InGaAs/n^+-InAlAs cap layer, *IEICE Transactions on Electronics*, E75-C, 649, 1992.
89. Duh, K. H. G., Chao, P. C., Liu, S. M. J., Ho, P., Kao, M. Y., and Ballingall, J. M., A super low noise 0.1 μm T-gate InAlAs-InGaAs-InP HEMT, *IEEE Microwave and Guided Wave Letters*, 1, 114, 1991.

90. Kao, M.-Y., Duh, K. H. G., Ho, P., and Chao, P.-C., An extremely low noise InP-based HEMT with silicon nitride passivation, *IEDM Technical Digest*, 907, 1994.
91. Kao, M. Y., Smith, P. M., Chao, P. C., and Ho, P., Millimeter wave power performance of InAlAs/InGaAs/InP HEMTs, *IEEE/Cornell Conference on Advanced Concepts in High Speed Semiconductor Devices and Circuits*, 469, 1991.
92. Matloubian, M., Nguyen, L. D., Brown, A. S., Larson, L. E., Melendes, M. A., and Thompson, M. A., High power and high efficiency AlInAs/GaInAs on InP HEMTs, *IEEE MTT-S Int. Microwave Symp. Dig.*, 721, 1991.
93. Larson, L. E., Matloubian, M., Brown, J. J., Brown, A. S., Rhodes, R., Crampton, D., and Thompson, M., AlInAs/GaInAs on InP HEMTs for low power supply voltage operation of high power-added efficiency microwave amplifiers, *Electronics Letters*, 29, 1324, 1993.
94. Bahl, S. R., Azzam, W. J., delAlamo, J. A., Dickmann, J., and Schildberg, S., Off-state breakdown in InAlAs/InGaAs MODFET's, *IEEE Transactions on Electron Devices*, 42, 15, 1995.
95. Boos, J. B. and Kruppa, W., InAlAs/InGaAs/InP HEMTs with High Breakdown voltages using double recess gate process, *Electronics Letters*, 27, 1909, 1991.
96. Hur, K. Y., McTaggart, R. A., LeBlanc, B. W., Hoke, W. E., Lemonias, P. J., Miller, A. B., Kazior, T. E., and Aucoin, L. M., Double recessed AlInAs/GaInAs/InP HEMTs with high breakdown voltages, *Proceedings of GaAs IC Symposium*, 101, 1995.
97. Hur, K. Y., Mctaggart, R. A., Miller, A. B., Hoke, W. E., Lemonias, P. J., and Aucoin, L. M., DC and RF characteristics of double recessed and double pulse doped AlInAs/GaInAs/InP HEMTs, *Electronics Letters*, 31, 135, 1995.
98. Pao, Y. C., Nishimoto, C. K., Majidi-Ahy, R., Archer, J., Betchel, N. G., and Harris, J. S., Characterization of surface undoped $In_{0.52}Al_{0.48}As/In_{0.53}Ga_{0.47}As$ high electron mobility transistors, *IEEE Transactions on Electron Devices*, 37, 2165, 1990.
99. Ho, P., Kao, M. Y., Chao, P. C., Duh, K. H. G., Ballingall, J. M., Allen, S. T., Tessmer, A. J., and Smith, P. M., Extremely high gain 0.15 μm gate-length InAlAs/InGaAs/InP HEMTs, *Electronics Letters*, 27, 325, 1991.
100. Matloubian, M., Larson, L., Brown, A., Jelloian, L., Nguyen, L., Lui, M., Liu, T., Brown, J., Thompson, M., Lam, W., Kurdoghlian, A., Rhodes, R., Delaney, M., and Pence, J., InP based HEMTs for the realization of ultra high efficiency millimeter wave power amplifiers, *IEEE/Cornell Conference on Advanced Concepts in High Speed Semiconductor Devices and Circuits*, 520, 1993.
101. Matloubian, M., Brown, A. S., Nguyen, L. D., Melendes, M. A., Larson, L. E., Delaney, M. J., Pence, J. E., Rhodes, R. A., Thompson, M. A., and Henige, J. A., High power V-band AlInAs/GaInAs on InP HEMT's, *IEEE Electron Device Letters*, 14, 188, 1993.
102. Brown, J. J., Matloubian, M., Liu, T. K., Jelloian, L. M., Schmitz, A. E., Wilson, R. G., Lui, M., Larson, L., Melendes, M. A., and Thompson, M. A., InP Based HEMTs with $Al_xIn_{1-x}P$ Schottky layers grown by gas source MBE, *Proceedings of International Conference on InP and Related Materials*, 419, 1994.
103. Scheffer, F., Heedt, C., Reuter, R., Lindner, A., Liu, Q., Prost, W., and Tegude, F. J., High breakdown voltage InGaAs/InAlAs HFET using $In_{0.5}Ga_{0.5}$ P spacer layer, *Electronics Letters*, 30, 169, 1994.
104. Hur, K. Y., McTaggart, R. A., Lemonias, P. J., and Hoke, W. E., Development of double recessed AlInAs/GaInAs/InP HEMTs for millimeter wave power applications, *Solid State Electronics*, 41, 1581, 1997.
105. Matloubian, M., Brown, A. S., Nguyen, L. D., Melendes, M. A., Larson, L. E., Delaney, M. J., Thompson, M. A., Rhodes, R. A., and Pence, J. A., 20 GHz high efficiency AlInAs-GaInAs on InP power HEMT, *IEEE Microwave and Guided Wave Letters*, 3, 142, 1993.
106. Hur, K. Y., McTaggart, R. A., Ventresca, M. P., Wohlert, R., Hoke, W. E., Lemonias, P. J., Kazior, T. E., and Aucoin, L. M., High efficiency single pulse doped $Al_{0.60}In_{0.40}As$/GaInAs HEMTs for Q band power applications, *Electronics Letters*, 31, 585, 1995.

107. Ho, P., Smith, P. M., Hwang, K. C., Wang, S. C., Kao, M. Y., Chao, P. C., and Liu, S. M. J., 60 GHz power performance of 0.1 μm gate length InAlAs/GaInAs HEMTs, *Proceedings of International Conference on InP and Related Materials*, 411, 1994.
108. Hwang, K. C., Ho, P., Kao, M. Y., Fu, S. T., Liu, J., Chao, P. C., Smith, P. M., and Swanson, A. W., W Band high power passivated 0.15 μm InAlAs/InGaAs HEMT device, *Proceedings of International Conference on InP and Related Materials*, 18, 1994.
109. Chen, Y. C., Lai, R., Lin, E., Wang, H., Block, T., Yen, H. C., Streit, D., Jones, W., Liu, P. H., Dia, R. M., Huang, T.-W., Huang, P.-P., and Stamper, K., A 94-GHz 130-mW InGaAs/InAlAs/InP HEMT High-Power MMIC Amplifier, *IEEE Microwave and Guided Wave Letters*, 7, 133, 1997.
110. Smith, P. M., Liu, S. M. J., Kao, M. Y., Ho, P., Wang, S. C., Duh, K. H. G., Fu, S. T., and Chao, P. C., W-Band high efficiency InP-based power HEMT with 600 GHz f_{max}, *IEEE Microwave and Guided Wave Letters*, 5, 230, 1995.
111. Akazaki, T., Enoki, T., Arai, K., Umeda, Y., and Ishii, Y., High frequency performance for sub 0.1 μm Gate InAs-inserted-channel InAlAs/InGaAs HEMT, *Electronics Letters*, 28, 1230, 1992.
112. Enoki, T., Arai, K., Kohzen, A., and Ishii, Y., InGaAs/InP Double channel HEMT on InP, *Proceedings of International Conference on InP and Related Materials*, 14, 1992.
113. Matloubian, M., Liu, T., Jelloian, L. M., Thompson, M. A., and Rhodes, R. A., K-Band GaInAs/InP channel power HEMTs, *Electronics Letters*, 31, 761, 1995.
114. Shealy, J. B., Matloubian, M., Liu, T. Y., Thompson, M. A., Hashemi, M. M., DenBaars, S. P., and Mishra, U. K., High-performance submicrometer gatelength GaInAs/InP composite channel HEMT's with regrown contacts, *IEEE Electron Device Letters*, 16, 540, 1996.
115. Shealy, J. B., Matloubian, M., Liu, T. Y., Lam, W., and Ngo, C., 0.9 W/mm, 76% PAE (7 GHz) GaInAs/InP composite channel HEMTs, *Proceedings of International Conference on InP and Related Materials*, 20, 1997.
116. Chevalier, P., Wallart, X., Bonte, B., and Fauquembergue, R., V-band high-power/low-voltage InGaAs/InP composite channel HEMTs, *Electronics Letters*, 34, 409, 1998.

9

RF Power Transistors from Wide Bandgap Materials

Karen E. Moore
University of Florida

9.1 Introduction

Wide bandgap materials such as silicon carbide (SiC) and gallium nitride (GaN) are known as such because their bandgaps are much larger than those of more conventional semiconductors such as silicon, germanium, or gallium arsenide. 4H-SiC, for example, has a bandgap of 3.2 eV, and GaN has a bandgap of 3.4 eV, as compared to 1.11 eV for silicon. These materials have been studied in theory for over 30 years; however, it has only been in the past decade that device development from wide bandgap semiconductors has occurred at any level, with significant breakthroughs occurring in SiC substrate and epi technology in the late 1980s, and in GaN epi technology in the mid-1990s. The first commercial applications for wide bandgap materials were blue LEDs fabricated from SiC, later followed by blue LEDs with greatly increased brightness, from GaN-related materials. While these applications are not RF applications, they have helped drive development of these very new, experimental materials.

In recent years, wide bandgap semiconductors have received a great deal of attention as a nearly ideal material for the fabrication of high speed, high power transistors [1–5], particularly for cellular base

0-8493-1562-X/03/$0.00+$1.50

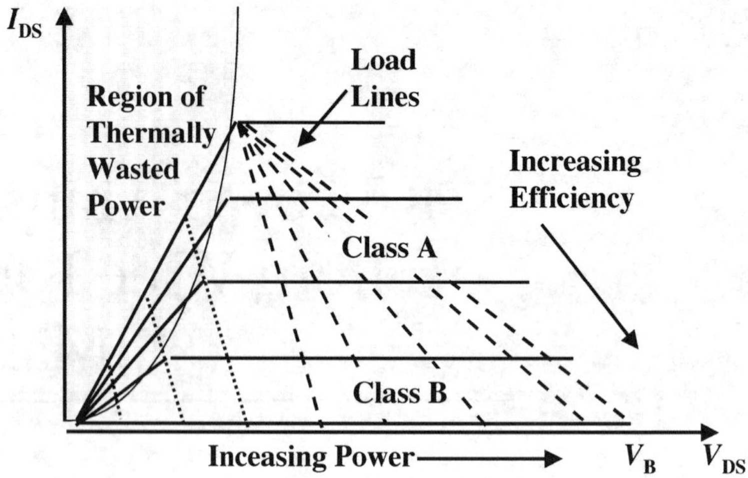

FIGURE 9.1 I-V curves and load lines for a power transistor.

station, broadcast, and high frequency radar applications. The large bandgaps should allow SiC- and GaN-based transistors to have stable DC and RF operation at very high temperatures [6]. Several other material properties of wide bandgap materials also make them attractive for fabrication of high voltage, high power, and high frequency transistors. These include a high electric breakdown field of 3 to 4 × 10^6 V/cm, high saturated electron drift velocities of 1.5 to 2.2 × 10^7 cm/sec, and high thermal conductivity (for SiC substrates) of 4.9 W/cm-K.

This article will present a basic description of figures of merit for microwave devices fabricated from wide bandgap materials, a very basic description of the operation of some of the commonly studied wide bandgap RF power transistors, a discussion of the material properties needed for RF power generation and how those properties translate into improved performance of microwave systems, and a summary of state-of-the-art wide bandgap high frequency device performance. The article will focus primarily on AlGaN/GaN HEMTs, SiC MESFETs, and SiC SITs, since they are currently the most mature and most widely studied wide bandgap RF device technologies.

9.2 Figures of Merit for RF Power Transistors

Virtually all RF systems require active circuit elements for use as oscillators, amplifiers, etc. These elements permit conversion of energy from DC bias sources to RF bands where the energy can be used to provide useful gain at specified frequencies. The ideal RF power transistor has high current, high breakdown voltage, and a low "knee" voltage (the voltage at which the transistor current saturates), as illustrated in Fig. 9.1. The device is given DC bias at one-half its maximum operating voltage and one-half or less of its maximum operating current, and any RF signal superimposed over the device is amplified over the I-V curve as shown in Fig. 9.1. The maximum possible RF output power of a transistor is [7]

$$P_{out,max} = (1/8)(V_{DS} - V_{knee})^2 / R_L \tag{9.1}$$

where P_{out} is the RF output power, V_{DS} is the drain bias, V_{Knee} is the knee voltage, and R_L is the load resistance, which is determined by the bias current and voltage in the device. The Power Added Efficency (PAE), η, of the device is another important figure of merit, and quantifies the amount of DC bias that is converted to RF power:

$$\eta = (P_{out} - P_{in}) / P_{dc} \tag{9.2}$$

The maximum efficiency for an RF transistor is 50% under class A operation (transistor is biased at 50% of its open channel current), or 78.5% under class B operation (transistor is biased at pinch off or I_{dq} = 0A); however, the maximum possible output power for the device remains unchanged with bias as long as the device is operated at 50% or less of its open channel current. Qualitatively, any RF drive in the region below the knee voltage will result only in resistive loss, which leads to lower output power, decreased gain, and lower efficiency. The increased resistive loss also leads to device heating that will further degrade device performance. Thus, the ideal RF transistor would have a knee voltage of 0V.

Two other important figures of merit for a microwave transistor are gain and intermodulation distortion (IMD). Gain is generally quantified under both large-signal and small-signal conditions, and is the ratio of output power to input power for a device [8]. Intermodulation distortion, in its simplest sense, is a measure of how constant gain is in a device over a wide range of instantaneous drive conditions as induced by a large RF signal. Any variation of gain from its linear, or constant value will cause generation of new signals at harmonic frequencies. These harmonics can cause system level problems. For example, a harmonic signal could be in a channel that is adjacent to its carrier, and be mistaken for a signal in that adjacent channel. IMD becomes increasingly more important, particularly in cellular systems, as more and more carriers are used in a given bandwith with increasing technology capabilities.

The unity current gain cutoff frequency of a device, f_t, and the maximum frequency of oscillation, f_{max}, are both very important parameters. These two figures of merit will determine the highest frequency at which a device is useful as an amplifier.

Finally, the power density of an RF transistor is very important for several reasons. Power density is most often expressed in units of watts/mm of gate periphery or watts/square cm of die area. For a given power rating, a high power density results in a smaller device, which will mean higher output impedance, easier matching, and possibly more power per die and/or fewer combining networks.

9.3 Common RF Power Devices from Wide Bandgap Materials

The three most commonly studied wide bandgap RF power devices are the SiC MESFET (MEtal-Semiconductor Field Effect Transistor), the SiC SIT (Static Induction Transistor), and the AlGaN/GaN HEMT (High Electron Mobility Transistor), sometimes called an AlGaN/GaN MODFET (Modulation Doped Field Effect Transistor). A brief description of operation of these three devices follows.

9.3.1 SiC MESFETs

The MESFET was first proposed by Mead in 1966 [9], and fabricated in GaAs by Hooper and Lehrer in 1967 [10]. A basic schematic of a MESFET is shown in Fig. 9.2.a [11]. The MESFET is a planar device fabricated by growth of a thin, doped epitaxial layer located on either a semi-insulating substrate or a low-doped layer of conductivity type opposite to that of the channel material. MESFETs in SiC can be fabricated on substrates of the same conductivity as the device channel (typically, n-type), but with a 1 to 5 micron buffer layer of opposite conductivity between the channel and the substrate. High resistivity substrates are most desirable for high frequency devices and result in improved DC and RF performance for the transistor by better confining electrons to the conducting channel and reducing microwave losses. This will be described in more detail in the next section.

In the MESFET, current is passed through the conducting channel by means of two ohmic contacts (the source and drain), which are typically separated by a distance of 3 to 10 μm. The actual dimensions are dependent upon the operating frequency, and smaller dimensions are used as operating frequency is increased. A rectifying Schottky contact (the gate) is located between the two ohmic contacts, and is typically 0.1 to 2 μm in length for modern microwave devices.

Fig. 9.2b shows the DC I-V curves for a MESFET [11]. During operation, the drain contact is biased at a specified potential (positive drain potential for an n-channel device) and the source is grounded. The flow of current through the conducting channel is controlled by a combination of a negative DC (for n-channel devices) gate bias with a superimposed RF signal also at the gate. The DC bias sets the

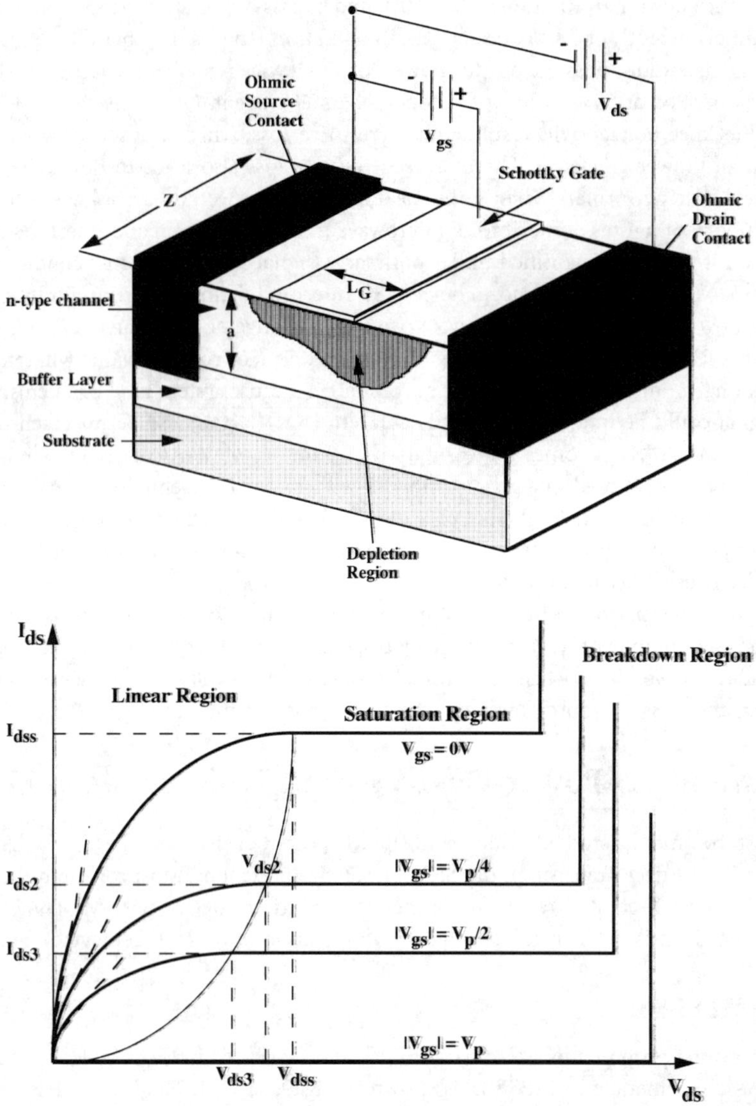

FIGURE 9.2 (a) Cross-section of a typical SiC metal-semiconductor field effect transistor (MESFET); (b) I-V curves of a typical MESFET.

quiescent operating current for the device, while the RF signal modulates the channel current, thereby providing RF gain. The operation of the transistor is determined by the ability of the gate signal to effectively modulate and control the current in the conducting channel. For this reason, any electrons that leak into the substrate or through the gate electrode will lead to performance degradation. The availability of both high resistivity, low leakage substrates and high quality, low leakage Schottky barrier gates in SiC makes high performance SiC MESFETs possible for high voltage RF power applications.

9.3.2 SiC SITs

Static Induction Transistors (SITs) are useful as high power RF sources and amplifiers at UHF and microwave frequencies. This device was originally proposed by Nishizawa [12] as a power amplifier for applications such as audio amplifiers. A cross-section of the transistor is shown in Fig. 9.3a. The device has a structure very similar to a vacuum triode with source and drain contacts separated by a certain

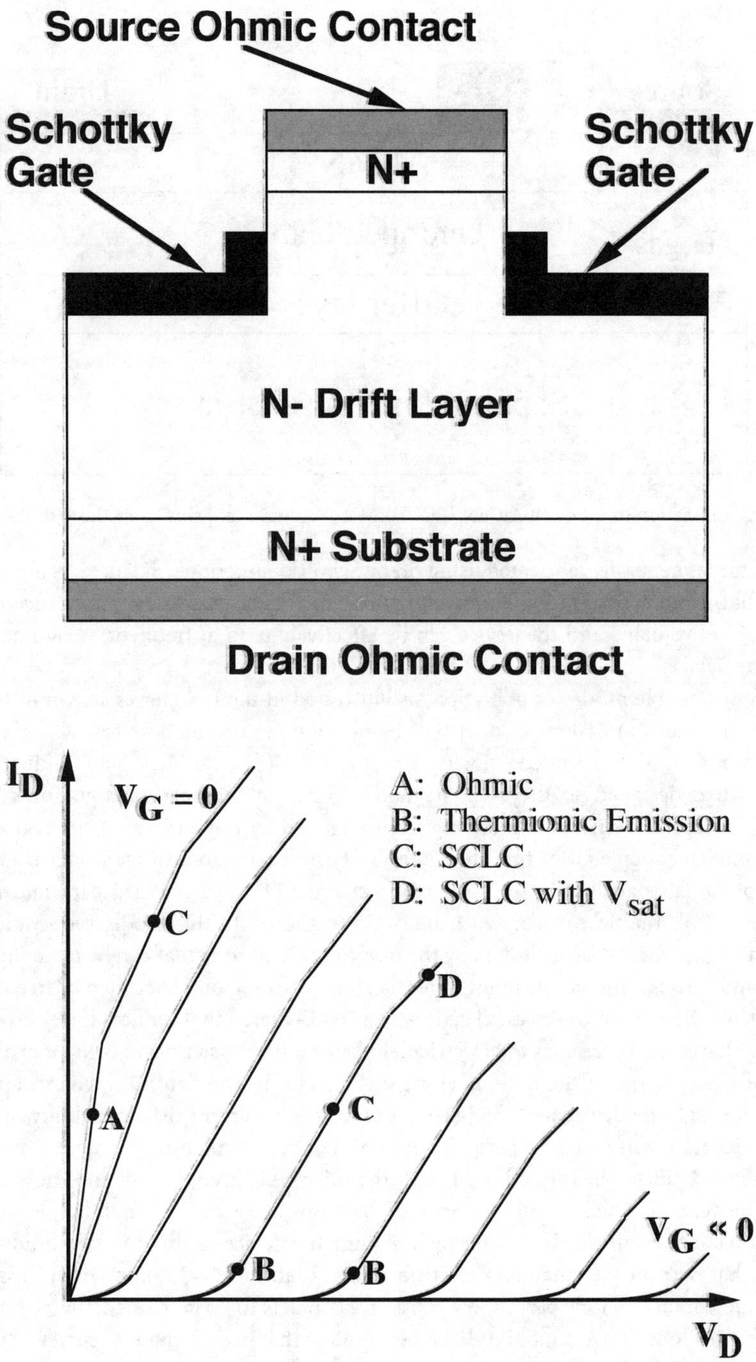

FIGURE 9.3 (a) Cross-section of a single unit cell of a typical SiC static induction transistor (SIT); (b) I-V curves of a typical SIT.

distance. Electrons are emitted from the source, which is generally at ground potential, and are accelerated to the drain, which is biased at a positive potential, where they are collected. A grid structure is located in the space between the source and drain electrodes so the charge carriers can be externally modulated. The RF gain of the device is determined by the efficiency with which the modulation is accomplished.

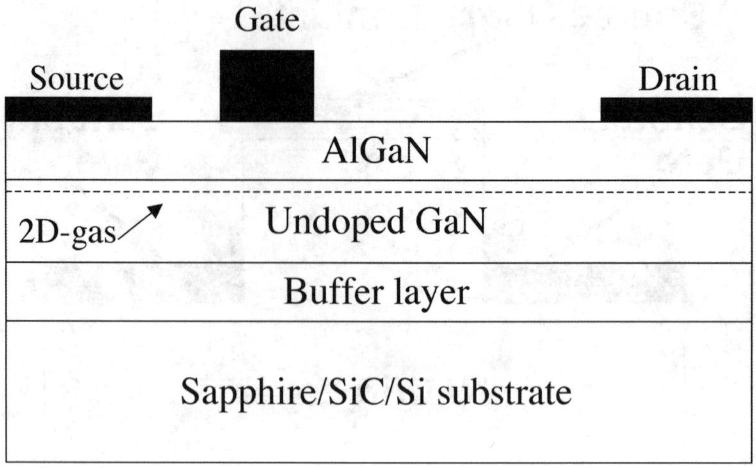

FIGURE 9.4 Cross-section of a typical AlGaN/GaN HEMT. I-V curves are similar to those of the SiC MESFET.

The grid structure is generally fabricated using pn or Schottky junctions. SIT devices are capable of high voltage gain, but limited current gain. However, since the device has good impedance characteristics, high power gain is available and the device can be effectively used at frequencies significantly in excess of the f_t for the device.

SITs have four possible modes of operation, as illustrated in the I-V curves shown in Fig. 9.3b. These modes are ohmic, thermionic emission, space-charge limited current flow (SCLC), and space-charge limited current flow under saturated velocity conditions (SCLC with V_{sat}). Static induction transistors perform best when designed so they operate under either of the space-charge-limited current flow conditions. This occurs when the conducting channel is lightly doped so that the region between the grid bars is completely depleted of free charge under typical bias conditions. Under these conditions a saddle point potential is created in the conducting channel between the grid bars, and modulation of this potential controls the flow of current from drain to source. In this mode of operation the current that can flow through the device is limited by the number of electrons that can be forced to flow between the grid bars and through the saddle point potential. Since this region is depleted of free charge carriers, the injected charge flowing in the channel generates a space-charge that reduces the electric field in front of the moving charge, and increases the electric field behind it. The net result is that current flow is self-limited by the space charge of the moving electrons. This limits the density of charge and RF currents that can flow through the device and establishes a limit to the current drive capability of the device.

The use of SiC offers improved RF performance for SIT transistors since the current density that flows through the semiconductor is a function of both the injected charge density and the velocity at which the electrons move. Since the saturated velocity of electrons in SiC is very high (i.e., $v_s = 2.2 \times 10^7$ cm/sec [13]), a high channel current is possible by designing the device so that the magnitude of the electric field is above that necessary to maintain electron velocity saturation. Also, the use of highly doped n+ source and drain contact regions permit low resistance contacts to be fabricated, thereby minimizing the degrading effects of low electron mobility. The net result is that higher channel current can be achieved, and this results in good RF performance.

9.3.3 AlGaN/GaN HEMTs

AlGaN/GaN HEMTs (High Electron Mobility Transistors) are somewhat similar in structure and operation to SiC MESFETs, with the notable difference that they are heterojunction devices. A cross-section of an AlGaN/GaN HEMT is shown in Fig. 9.4. In this device, all of the conduction is in a channel formed by a sheet of charge immediately under the AlGaN/GaN heterojunction, with the channel usually being about 50 to 250Å total thickness. Typical sheet charge densities in the AlGaN/GaN materials system are

1 to 1.5×10^{13} cm^{-2}, which is 3 to 10 times that normally seen in GaAs-based HEMTs. The charge in this device can either come from doping in the AlGaN layer, which spills over into the lower energy GaN layer or can be induced by the piezoelectric field found in strained GaN-based materials grown on heteroexpitaxial substrates like sapphire or SiC [14]. The thick GaN layer and the buffer layer, most often AlN, provide isolation from the substrate and carrier confinement. These devices have been fabricated most frequently on sapphire substrates, but have also been extensively studied on SiC substrates [15–16], and have recently been fabricated on p-type ⟨111⟩ silicon wafers [17]. To date, the best AlGaN/GaN HEMT performance has been from devices utilizing semi-insulating SiC substrates.

Like in the MESFET, current flows in the AlGaN/GaN HEMT from source to drain, induced by a bias at the drain, and the gate modulates charge through the channel through the same combination of DC bias and RF signal. The high sheet charge density in these structures creates a very high current density to complement the high breakdown voltage already inherent in the GaN channel material. The high sheet charge will also allow very low on-resistance, which will improve the microwave performance of the devices, and the combination of high current density with high voltage operation makes the AlGaN/GaN HEMT the transistor with the highest power density of the wide bandgap devices.

9.3.4 Other Wide Bandgap Devices

There are several other wide bandgap power devices that are beginning to be studied. These include AlGaN/GaN HBTs (Heterojunction Bipolar Transistors), which should have excellent efficiency and breakdown characteristics [18], MOS-HFETs (Metal-Oxide-Semiconductor Heterojunction Field Effect Transistors), which are expected to have improved linearity as compared to AlGaN/GaN HEMTs [19], JFETs (Junction Field Effect Transistors), which should have very good high temperature performance [20], and IMPATT diodes, which are expected to generate power at very high frequencies with much greater efficiency and lower access resistances than their GaAs and InP-based counterparts. All of these devices are in their technological infancy, and have yet to demonstrate the remarkable RF results of the more conventional wide bandgap devices. Nevertheless, the potential for these devices to become important in the future should not be overlooked. In the meantime, the rest of this article will focus only on results from SiC MESFETs, SiC SITs, and AlGaN/GaN HEMTs.

9.4 Desirable Material Properties for RF Power Transistors

There are a number of semiconductor material properties that affect the performance of a high speed, high power transistor [1–2]. These include the bandgap, critical (breakdown) electric field, and thermal conductivity. Electron and hole transport properties, the saturated electron velocity, and the electric field at which electron velocity saturates will also strongly influence the DC and RF characteristics of a high frequency power transistor. Several of these properties are summarized in Table 9.1, which compares the material properties of Si, GaAs, 4H-SiC, and GaN. In addition to the properties summarized in Table 9.1, the electrical conductivity of the substrate material can strongly affect RF losses in a transistor. Finally, the ability to make both low resistance ohmic contacts and good rectifying Schottky contacts are critical to transistor fabrication and performance. Figure 9.5 attempts to capture some of the many relationships

TABLE 9.1 Material Properties of Common Semiconductors

Property	Units	Silicon	GaAs	4H-SiC	GaN
Bandgap	EV	1.11	1.43	3.2	3.4
Breakdown field	V/cm	7×10^5	7×10^5	35×10^5	35×10^5
Saturation velocity	cm/sec	1×10^7	1×10^7	2×10^7	1.5×10^7
Saturation field	V/cm	8×10^3	3×10^3	25×10^3	15×10^3
Thermal conductivity	W/cm-K	1.5	0.46	4.9	1.7/substrate
Electron mobility	Cm^2/V-sec	1350	6000	800	1000
Hole mobility	Cm^2/V-sec	450	330	120	300

between material properties and power device and system performance. The system level advantage is particularly important as a metric for evaluating devices from wide bandgap materials, as that is where their real importance and impact for a customer or consumer will be evident.

9.4.1 Critical or Breakdown Field

SiC and GaN have critical fields of 3.5 MV/cm, five times that of Si or GaAs. This critical, or breakdown field of a material is possibly the most important material parameter for design of a high power density device, as it determines the highest operating voltage of a transistor for a given device design and channel doping, and thus limits the RF power swing in the device. Analagously, for a given voltage requirement, a higher breakdown field will allow a device designer to use a higher doping level in the device than for a lower breakdown field material, and, with the higher doping, tighter device dimensions. Higher operating voltage, as shown in Eq. (9.1), results in both higher power and higher power density in the device. Higher doping and reduced dimensions in an FET will also enable a device with increased transconductance, lower parasitic resistances, increased power gain, higher f_t and f_{max}, and improved efficiency due to the decreased access resistances.

Furthermore, large output power levels can be achieved through either high current or high voltage operation, but a device that obtains its power from high voltage rather than high current will be much smaller due to the fact that voltage scaling in a device only involves design changes in a device channel on the order of 1 to 2 μm, whereas high current devices require complete scaling of the device periphery. This becomes even more important for a very high power device as increased current in the linear region of device operation introduces more resistive heating, and further degrades device performance. In addition, a smaller geometry, high voltage part will have output impedance levels that are much larger than for a larger geometry, high current device of the same output power level, making the high voltage FET easier to incorporate into a circuit design. At a system level, the higher power density of a wide bandgap transistor will lead to more power per die, and thus a smaller die count per system, and greater bandwidth due to improved output impedance characteristics. Higher efficiency will translate into lower total energy usage for the microwave system in question, and smaller die sizes will use smaller and therefore cheaper packages.

9.4.2 Thermal Conductivity

The thermal conductivity of a material determines the ease with which heat generated from unconverted DC power can be removed from the device. Any temperature rise from undissipated heat will further degrade device performance by causing a drop in the mobility and saturated electron velocity in the transistor, which in turn causes the device to become progressively less efficient and to generate more heat. Thermal conductivity can also influence whether or not special packaging and/or system cooling become necessary for successful device operation. The thermal conductivity of SiC of 4.9 W/cm-K is three times that of Si and ten times that of GaAs or sapphire and, as such, is a tremendous advantage for SiC-based devices. Since GaN-based devices are grown on both sapphire and SiC substrates at the present time, the limiting thermal conductivity for these devices depends on the choice of substrate.

9.4.3 Wide Bandgap

A large bandgap is considered desirable for high voltage power devices for two reasons. First, the bandgap determines the upper temperature limit of device operation. This is particularly important when transistor scaling and power density are examined, as higher temperature operation makes it possible to design a smaller, denser device that will withstand the heat it generates under bias. Second, wide bandgap materials have been shown to be more resistant to radiation, such as α particles, which are known to be very destructive in Si MOS devices. The large bandgaps of 3.4 eV for GaN and 3.2 eV for 4H-SiC give these materials a high temperature performance advantage over Si and GaAs. In the system limit, a device that can withstand higher channel and ambient temperatures may be suitable for a cheaper package and relaxed system cooling requirements.

9.4.4 Saturated Electron Velocity

The saturated electron velocity of a material is very important for sub-micron gate length devices, which typically operate at very high electric fields. In this regime, the frequency performance (in particular, f_t) of a device is largely determined by electron velocity. SiC has a measured saturated electron velocity of 2.2×10^7 cm/second, twice that of Si or GaAs, and GaN has an electron velocity of 1.5×10^7 cm/second, still significantly higher than that of Si or GaAs. The electric field at which the electron velocity saturates is also important as it determines how quickly the charge carriers can be accelerated to their saturated values. The saturation fields for GaN and SiC are 1.5 to 2.5×10^4 V/cm, or 2 to 8 times higher than the 3 to 8×10^3 V/cm values for GaAs and Si. The high saturation fields combined with the low mobilities of wide bandgap semiconductors result in devices that will have higher knee voltages and therefore will have to be operated at considerably higher supply voltages before they operate in a saturated electron velocity regime. This will move their optimal performance to much higher voltage levels than for more conventional semiconductor technologies. Overall, the high electron velocities, at whatever voltage is necessary, will contribute to a very high speed device and, as a result, as very high system frequency.

9.4.5 Electron and Hole Mobilities

A primary disadvantage of fabricating transistors from wide bandgap semiconductors is the relatively low values for electron and hole mobilities in these materials. Electron and hole transport properties are also critical to successful device performance, and play a dominant role in determining the on-resistance and knee voltage of a device in its low-field region of operation. Low mobility results in increased parasitic resistance, increased losses, and reduced gain. These problems are worsened both as operating frequency is increased, where series resistances play an increasingly stronger role, and at elevated temperatures, where mobility will rapidly decrease.

SiC and GaN have electron mobilities of 800 to 1000 cm^2/V-sec in nominally undoped material. These values are much less than the 1350 cm^2/V-sec electron mobility of silicon, let alone the 6000 cm^2/V-sec electron mobility of GaAs. At the typical 10^{17} cm^{-3} doping levels commonly seen in high voltage MESFETs, the electron mobility in 4H-SiC drops to around $\mu_e \sim 500$ cm^2/V-sec. Fortunately, the electron mobility in a 2D-gas in an AlGaN/GaN heteostructure typically remains close to 1000 cm^2/V-sec as the electrons in the AlGaN/GaN heterojunction are physically separated from their donor atoms and thus are not affected by ionized impurity scattering as in the case of SiC. The hole mobility in both SiC and GaN is very low, on the order of $\mu_p \sim 120$ to 300 cm^2/V-sec. The low value for hole mobility severely limits the use of p-type SiC in RF transistors that are intended for operation above about 1 GHz.

9.4.6 Substrate Conductivity

In addition to the material properties detailed in Table 9.1, there are several other material parameters than can affect RF power device performance. The electrical conductivity of the device substrate is a very important material property for higher frequency operation of RF transistors. The SIT, which is a vertical device, must have a conducting substrate in order to accomodate its backside source contact. The effects of substrate conductivity are much more complicated in MESFET and HEMTs. Substrate conductivity is strongly related to losses at high frequency in microwave MESFETs and HEMTs, since they are lateral devices. At microwave frequencies, the contact pads of the transistor act as lossy transmission lines, and as such create both loss and dispersion in the small-signal characteristics of the device, including small-signal gain, which is the upper limit of power gain in a transistor. Frequency dispersion of various device elements makes device model extrapolation, wideband device operation, and circuit design all very difficult. Development of semi-insulating SiC substrates of both 6H and 4H polytypes has clearly demonstrated the importance of this material improvement. The device gain in MESFETs on semi-insulating

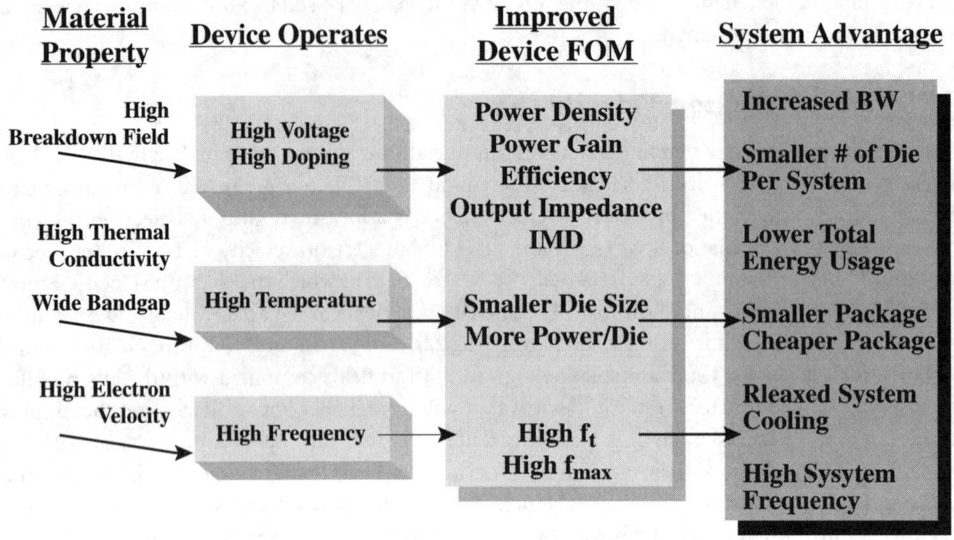

FIGURE 9.5 Relationships between material properties and device and system level performance for transistors and systems using wide bandgap materials.

SiC is higher, the substrate losses are smaller, and the dispersion seen in parasitic device elements is decreased as compared to devices on highly doped SiC substrates. Similarly, AlGaN/GaN HFETs fabricated on sapphire (an insulator) or semi-insulating silicon carbide have much better microwave performance than those fabricated on highly doped silicon substrates. This will be discussed in more detail in the device results in the next section.

9.4.7 Electrical Contacts

Electrical contacts, which are critical to successful device operation, are also strongly influenced by material properties. One of the disadvantages of wide bandgap materials is that they tend to form ohmic contacts with higher contact resistance than ohmic contacts fabricated on smaller bandgap semiconductors. Unfortunately, ohmic contact resitance can strongly influence both DC and RF operation of a device by contributing to the resistive region of the I-V curves shown in Fig. 9.1. As discussed earlier, this increase in the on-resistance adversely affects the extrinsic transconductance, small-signal gain, frequency performance, and power-added efficiency of a device. With optimization, reasonably low contact resistances have been realized in both SiC and GaN n-type materials.

High quality Schottky contacts are also critical to successful transistor performance. The quality of its Schottky contact determines the ability of the gate to completely pinch off the device channel, and controls the gate leakage current in the device under both DC and RF conditions. High gate leakage will decrease RF gain and power-added efficiency, and, in turn, the maximum output power of a device. Since good Schottky contacts are typically easy to fabricate on wide bandgap materials, good Schottky gate operation is usually possible in SiC technology.

9.5 State-of-the-Art Wide Bandgap Microwave Transistor Data

The effects of the material properties of wide bandgap semiconductors are best illustrated through the phenomenal device performance being reported today. Despite the immaturity of wide bandgap technology, the DC, small-signal, and large signal results from wide bandgap transistors are already challenging or exceeding the best results reported in many semiconductor material systems.

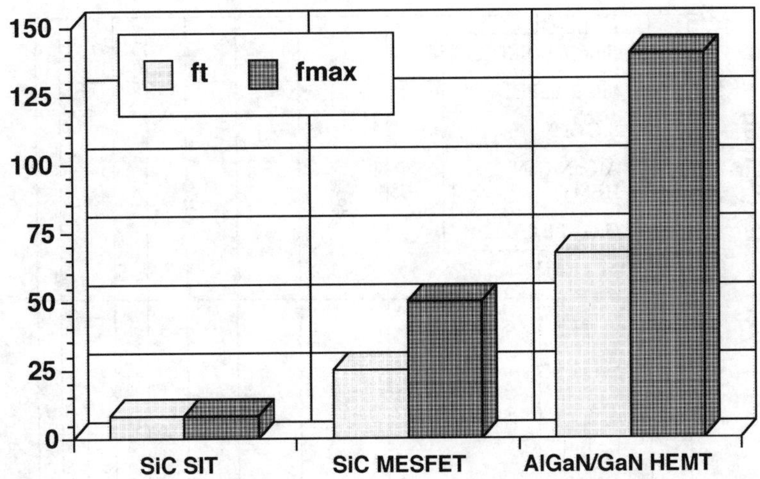

FIGURE 9.6 Comparison of highest reported f_t and f_{max} for SiC SITs, SiC MESFETs, and AlGaN/GaN HEMTs.

9.5.1 High Frequency Performance

f_t and f_{max} are commonly cited for SiC MESFETs, SITs, and AlGaN/GaN HEMTs because they provide an upper limit to the frequencies at which these devices are useful in system applications. Figure 9.6 illustrates both the best performance reported to date for these devices and the performance differences from the device types. The best performance is from the AlGaN/GaN HEMT, with f_t = 67 GHz and f_{max} = 140 GHz [21]. The SiC MESFET is also very fast, with peak f_t = 25 GHz and f_{max} = 50 GHz [22]. The SiC SIT is the weakest performer, with an f_t of only 7 GHz [23] and f_{max} of 8 GHz.

The reasons for the differences in performance can be explained through simple device physics. f_t is linear with transconductance, which is very high (~240 to 250 mS/mm) for the AlGaN/GaN HEMT. Transconductance in the SiC MESFET is typically on the order of 50 mS/mm, leaving the MESFET with much lower f_t. The strong effect of gate capacitance on f_t can also be seen with the SiC MESFET. The MESFETs cited with 25 GHz f_t are fabricated on semi-insulating substrates. When the same device is fabricated on a conducting SiC substrate, which adds a very large parasitic pad capacitance to both the gate and the drain of the device, the best reported f_t drops to about 8 GHz and f_{max} falls to 16 GHz [24]. Similarly, when an AlGaN/GaN HEMT was fabricated on a highly doped p-type silicon substrate, f_t and f_{max} were both 25 GHz [17]. The SiC SIT has a very large parasitic gate-drain capacitance due to its vertical design, and low transconductance due to the lower doping levels used in this device, so even with careful design and processing, the upper limit of f_t for the SIT is expected to be around 20 GHz.

f_{max} is linear with f_t, but also depends strongly on gate, source, drain, and channel resistances, output conductance, and gate-drain capacitance. The large differences in f_{max} among the wide bandgap devices are probably primarily due to the dramatic differences in transconductances in the different devices. This is validated by the fact that the f_{max}/f_t ratio is nearly the same for the AlGaN/GaN HEMT and the SiC MESFET.

9.5.2 Power Density

Power density, shown in Fig. 9.7, is also an important parameter for evaluation as it is instrumental in determining the minimum device, die, and package size of a microwave transistor. The trend for power density is similar to that of frequency performance. AlGaN/GaN HEMTs have the highest reported power density, 9.1 W/mm at Vds = 30 V [25], due to both the high current density and high breakdown voltage of these devices. SiC MESFETs follow with a significant power density of 5.6 W/mm at Vds = 60 V [26]. While the drain voltage of the MESFET is higher, its current carrying capability (drain current density) is significantly lower than that of the AlGAN/GaN HEMT, and so the power density of the SiC MESFET ends up being lower than that of the AlGaN/GaN HEMT. SiC SITs have lower power density at higher

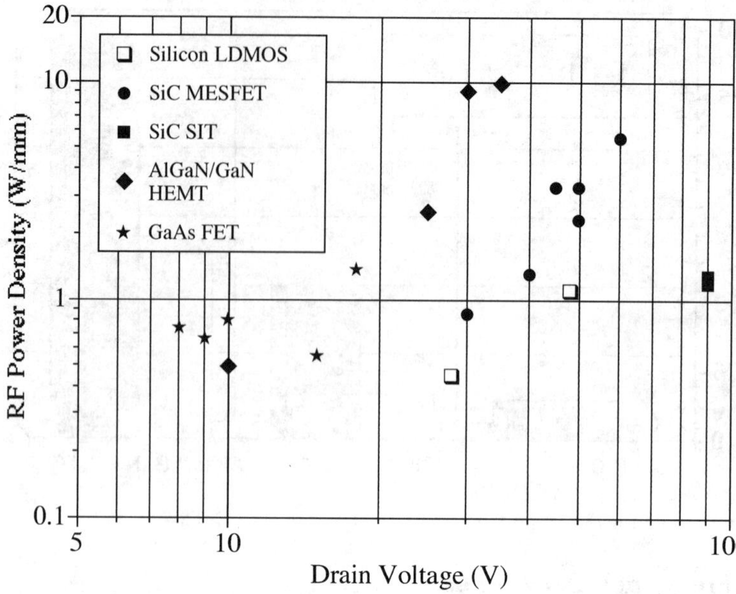

FIGURE 9.7 RF power density vs. drain voltage for different RF power technologies.

drain voltage, due to their lower current carrying capability. The peak power density for a SiC SIT is 1.3 W/mm at Vds = 90 V [27]. Overall, as seen in Fig. 9.7, the wide bandgap power devices significantly outperform silicon LDMOS and GaAs devices at higher supply voltages. Also worth noting is the fact that the highest power density for an AlGaN/GaN HEMT on a sapphire substrate is only 3.1 W/mm, much less than the 9.1 W/mm reported for a very similar device design on a SiC substrate. This further underscores the importance of substrate thermal conductivity.

9.5.3 Total Power

Total RF output power, shown in Fig. 9.8, is very important. In the final analysis, it will not matter what power density a transistor achieves if it cannot be scaled up into a total power that is useful for a system-level application. In the case of wide bandgap devices, the power requirements for their intended uses (basestations, radar, broadcast, etc.) range from approximately 30 to 1500 W. For total power, the SiC SIT is the device with superior performance. Some of the best reported SIT power data includes a 34.5 cm gate periphery module with 470 W total output power at 600 MHz, and a higher frequency, 3 GHz, 3 cm gate periphery module with 36 W output power and 42% power-added efficiency [27]. The SIT has also been used to make a UHF high power amplifier module designed for HDTV transmission [28]. This module, designed for operation in the 470 to 806 MHz frequency range, had 200 W average output power and 1 kW peak output power. SiC MESFETs have been fabricated with total CW output power of 80 W, and total pulsed output power of 120 W, both at 3.1 GHz [26]. The AlGaN/GaN HEMTs are most limited in total output power at the present time, with the highest reported power of 9.86 W at 8.2 GHz [25]. Also represented in Fig. 9.8 are currently available Si LDMOS parts with output powers as high as 120 W, and GaAs HFETs with 200 W total output power at 2.16 GHz [29] and 15.8 W at 12 GHz [30].

9.6 Challenges to Production

While there is little doubt that wide bandgap materials have tremendous potential in the high power RF arena, there are still several challenges to production that have yet to be answered. The challenges include the size, defectivity, and choice (SiC/sapphire/silicon) of the device substrate along with the quality and

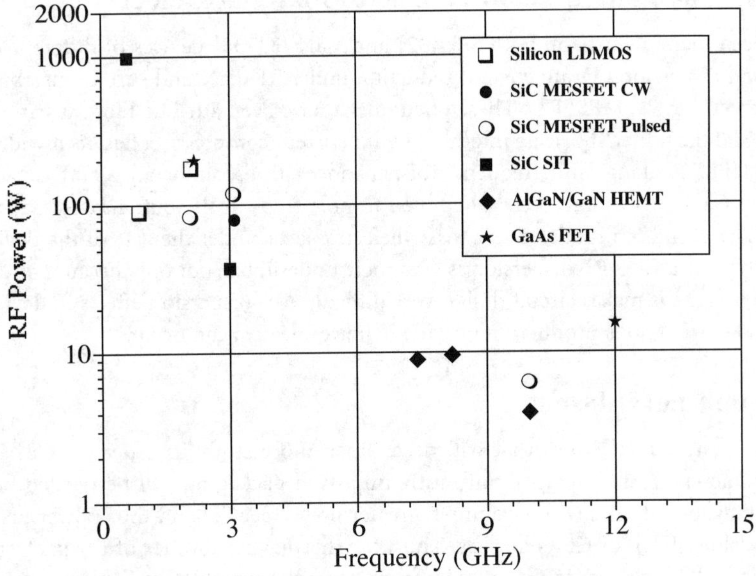

FIGURE 9.8 Total output power vs. frequency for different RF power technologies.

defect density of the epilayers, elimination of dispersion phenomena currently seen in many of these devices, and improved understanding of system packaging and thermal limitations. In addition, market drivers will be needed to provide the impetus for continuing material and device development. It will be very difficult to implement any type of large-scale production of wide bandgap power transistors until these issues are resolved.

9.6.1 SiC Substrates

SiC substrates have come a long way since they were first introduced over a decade ago. However, they still have a long way to go. The most common "killer" defect in SiC wafers, the micropipe (a screw dislocation that can run the length of a boule of material), has been reduced in density from over 100/cm^2 in 1993 to less than 1/cm^2 in 1998 [31]. Those numbers, however, are best research results, and commercially available wafers will have higher micropipe densities, are significantly more expensive than silicon wafers, and are only available in 2″ to 3″ diameters, depending on polytype and doping. The high defect densities in SiC substrates make it difficult to yield large, high voltage, high power devices, while the small wafer sizes increase cost through decreased economy of scale, and can actually be a limiting factor in that it is now difficult to even obtain used semiconductor process equipment for 2″ to 3″ wafers.

9.6.2 Substrates for GaN Devices

As mentioned earlier, GaN devices have been grown on a number of different substrates. This is because larger diameter (2″) GaN substrates will become available, in prototype quantities only, for the first time in 2000. In the absence of a homoepitaxial GaN wafer, substrate choice for GaN is something of a conundrum. Sapphire is relatively cheap, is offered in large diameter (4″ to 6″) wafers, and provides an excellent low-loss microwave substrate. However, the thermal conductivity of sapphire is extremely poor (0.46 W/cmK) and will severely limit the power density and total power performance of devices fabricated on it. On the other hand, semi-insulating SiC is also an excellent microwave substrate, but has the cost, size, and defectivity issues described above. Silicon is yet another substrate possibility, but there is very limited data as to its performance at this time.

9.6.3 Dispersion and Instability Due to Material Defects

There have been many reports on both SiC [32] and GaN [33–35] devices of different types of device instabilities and dispersions. Drain current reduction under RF drive and surface instabilities have both been documented for SiC MESFETs. These phenomena have been attribued to traps at both the surface of the device and the buffer/substrate interface. Drain current compression has been widely reported for AlGaN/GaN HEMTs, along with frequency dispersion of transconductance and capacitances. These phenomena in GaN devices have been attributed to hole traps in the buffer layer, traps in the AlGaN barrier, and have even been modeled as a lossy dielectric layer under the gate of the device. In the final analysis, any type of DC-to-RF dispersion is extremely undesirable, not only because of device degradation, but also because is makes circuit design very difficult. Any dispersion effects in devices will have to be eliminated before serious production of wide bandgap devices can occur.

9.6.4 System Level Issues

There are two system level issues that will need much more study in order to make wide bandgap devices a commercial reality. First, significantly improved packaging will be needed because devices from wide bandgap materials will have much greater power per die and thus power per package than currently available silicon or GaAs devices. Thus, the thermal capabilities of the package for the wide bandgap device will have to be significantly improved over the currently available packaging technology used for silicon power devices. Second, the overall temperature handling and cooling capabilities of the systems using RF power devices will have to be improved. Put simply, there cannot be any advancement in technology if a very high temperature tolerant device operating at a very high temperature and power level causes thermal runaway in the non wide bandgap, temperature-sensitive devices that surround it.

9.6.5 Market Drivers

One bit of good news for wide bandgap technology is that it appears to be in high demand despite its immature status. Cree Inc. has already announced release of a SiC MESFET product. The U.S. government is providing very large amounts of funding to drive development of wide bandgap technology for military applications [36]. Most importantly, development and sales of blue LEDs from GaN are growing very fast in both demand and level of technology development [37]. The large LED market will drive nitride-based technology to improved materials, which will in turn feed back into microwave device technology, which should in turn mitigate many of the issues described above.

9.7 Conclusions

This article has discussed the properties of wide bandgap, high frequency, high power transistors, and how the materials from which they are fabricated affect device performance. SiC and GaN are shown to be outstanding materials for use in microwave device applications. This is clearly exemplified by the outstanding device performance seen from these relatively immature technologies. Device results including f_{max} = 139 GHz and power densities of 9.1 W/mm for AlGaN/GaN HEMTs, power densities of 5.6 W/mm and power-added efficiency of 65.7% for SiC MESFETs, device modules of over 470 W at power densities of 175 kW/in^2, and the design and execution of an HDTV transmitter module in SiC SIT technology all show the tremendous potential of wide bangap materials for use in microwave transistors. With the improvements of larger diameter, low defect density substrates, more perfect epitaxial films, and improved package and system thermal tolerances, wide bandgap materials should prove an excellent high power, high temperature technology for microwave applications.

References

1. R. Trew, J. Yan, and P. Mock, The Potential of Diamond and SiC Electronic Devices for Microwave and Millimeter-Wave Power Applications, *Proceedings of the IEEE*, 79, 5, May 1991, 598–620.
2. R. Davis, Thin Films and Devices of Diamond, Silicon Carbide, and Gallium Nitride, *Physica B*, 185, 1993, 1–15.
3. C.E. Weitzel et al., Silicon Carbide High-Power Devices, *IEEE Transactions on Electron Devices*, 43, 10, 1996, 1732–1741.
4. C.E. Weitzel and K.E. Moore, Performance Comparison of Wide Bandgap Semiconductor RF Power Devices, *Journal of Electronic Materials*, 27, 4, 1998, 365–369.
5. J. Zolper, Wide Bandgap Semiconductor Microwave Technologies: From Promise to Practice, IEEE International Electron Devices Meeting, 1999.
6. R.J. Trew, Wide Bandgap Semiconductor Transistors for Microwave Power Amplifiers, *Microwave*, March 2000, 46–54.
7. I. Bahl and P. Bhartia, eds., *Microwave Solid State Circuit Design*, John Wiley & Sons, Inc., New York, 1988, 483–536.
8. S. Liao, *Microwave Circuit Analysis and Amplifier Design*, Prentice-Hall, Englewood Cliffs, NJ, 1987, 78–122 and 236–274.
9. C.A. Mead, Schottky Barrier Gate Field-Effect Transistor, *Proceedings of the IEEE*, 54, 1966, 307.
10. W. W. Hooper and W. I. Lehrer, An Epitaxial GaAs Field-Effect Transistor, *Proceedings of the IEEE*, 55, 1967, 1237.
11. S. Sze, *Physics of Semiconductor Devices*, John Wiley & Sons, New York, 1981, 312–361.
12. J. Nishizawa, T. Terasaki, and J. Shbata, Field-Effect Transistor versus Analog Transistor (Static Induction Transistor), *IEEE Transactions on Electron Devices*, ED-22, 1975, 185–197.
13. I. Khan and J. Cooper, Measurement of High-Field Electron Transport in Silicon Carbide, *IEEE Transactions on Electron Devices*, ED-47, 2, 2000, 269–273.
14. M.S. Shur, A.D. Bykhovski, and R. Gaska, Pyroelectric and Piezoelectric Propeties of GaN-Based Materials, *MRS Internet Journal of Nitride Semiconductor Research*, Res. 4S1, G1.6, 1999.
15. M.A. Khan et al., GaN Based Heterostructure for High Power Devices, *Solid State Electronics*, 41, 10, 1997, 1555–1559.
16. U.K. Mishra, Y.-F. Wu, B.P. Keller, S. Keller, and S. Denbaars, GaN Microwave Electronics, *IEEE Transactions on Electron Devices*, ED-46, 6, 1998, 756–761.
17. E.M. Chumbes, A.T. Schremer, J.A. Smart, D. Hogue, J. Komiak, and J. Shealy, Microwave Performance of AlGaN/GaN High Electron Mobility Transistors on Si(111) Substrates, *IEEE IEDM Digest*, 1999, 397–400.
18. L.S. McCarthy, P. Kozodoy, M. Rodwell, S. Denbaars, and U.K. Mishra, AlGaN/GaN Heterojunction Bipolar Transistor, *IEEE Electron Device Letters*, 20, 6, 1999, 277–279.
19. M.A. Khan, X. Hu, G. Sumin, A. Lunev, J. Yang, R. Gaska, and M. Shur, AlGaN/GaN Metal Oxide Semiconductor Heterostructure Field Effect Transistor, *IEEE Electron Device Letters*, 21, 2, 2000, 63–65.
20. L. Zhang et al., Epitaxially-Grown GaN Junction Field Effect Transistors, *IEEE Transactions on Electron Devices*, ED-47, 3, 2000, 507–511.
21. L. Eastman, K. Chu, J. Smart, and R. Shealy, GaN Materials for High Power Microwave Amplifiers, *Proceedings of the Materials Research Society Symposium*, 512, 1998, 3–7.
22. S. Allen, R. Sadler, T. Alcorn, J. Palmour, and C. Carter, Jr., Silicon Carbide MESFETs for High-Power S-Band Applications, *1997 MTT-S Digest*, 57–60.
23. J. Henning, A. Przadka, M. Melloch, and J. Cooper, Design and Demonstration of C-Band Static Induction Transistors in Silicon Carbide, *1999 57th Annual Device Research Conference Digest*, 48–49, 1999.

24. C. Weitzel, J. Palmour, C. Carter Jr., K. Nordquist, K. Moore and S. Allen, SiC Microwave Power MESFETs and JFETs, *Compound Semiconductors 1994,* U. Mishra and H. Goronkin, eds., 389–392, 1994.
25. Y.-F. Wu, D. Kapolnek, J. Ibbetson, N.-Q. Zhang, P. Parikh, B. Keller, and U. Mishra, High Al-Content AlGaN/GaN HEMTs on SiC Substrates with Very High Power Performance, *IEEE IEDM Digest*, 1999, 925–927.
26. J. Palmour, S. Allen, S. Sheppard, W. Pribble, R. Sadler, T. Alcorn, Z. Ring, and C. Carter, Jr., Progress in SiC and GaN Microwave Devices Fabricated on Semi-Insulating 4H-SiC Substrates, *1999 57th Annual Device Research Conference Digest*, 38–41, 1999.
27. R. Siergiej, R. Clarke, A. Agarwal, C. Brandt, A. Burke, A. Morse, and P. Orphanoa, High Power 4H-SiC Static Induction Transistors, *IEEE IEDM Digest,* 1995, 353–356.
28. A First for Silicon Carbide, *Compound Semiconductor,* July/August 1996, 4.
29. H. Ishida, T. Yokoyama, H. Furukawa, T. Tanaka, M. Maeda, S. Morimoto, Y. Ota, D. Ueda, and C. Hamaguchi, 200W GaAs-Based MODFET Power Amplifier for W-CDMA Base Stations, *IEEE IEDM Digest*, 1999, 393–396.
30. K. Matsunaga, Y. Okamoto, I. Miura, and M. Kuzuhara, Ku-Band 15W Single-Ship HJFET Power Amplifier, *1996 MTT-S Digest*, 697–700.
31. B. Foutz, Industry Shows Support for Wide Bandgap Semiconductors at the Spring MRS Meeting, *Compound Semiconductor,* June 1999, 21–28.
32. K. Hilton, M. Uren, P. Wilding, H. Johnson, J. Guest, and B. Smith, Surface Induced Instabilities in 4H-SiC Microwave MESFETs, International Conference on Silicon Carbide and Related Materials, 1999, paper 416.
33. C. Nguyen, N. Nguyen, and D. Grider, Drain Current Compression in GaN MODFET's Under Large-Signal Modulation at Microwave Frequencies, *Electronics Letters*, 35, 16, 1380–1382.
34. E. Kohn, I. Daumiller, P. Schmid, N. Nguyen, and C. Nguyen, Large Signal Frequency Dispersion of AlGaN/GaN Heterostructure Field Effect Transistors, *Electronics Letters*, 35, 12, 1022–1024.
35. S. Trassaert, B. Boudart, C. Gaquiere, D. Theron, Y. Crosnier, F. Huet, and M. Poisson, Trap Effects Studies in GaN MESFETs by Pulsed Measurements, *Electronics Letters*, 35, 16, 1386–1388.
36. J. Zolper, C. Wood, and M. Yoder, Nitride Semiconductor Transistors Poised to Revolutionize DOD Systems, *Compound Semiconductor,* June 1999, 29–30.
37. R. Dixon, Who's Who in Blue and Green LEDs, *Compound Semiconductor,* June 1999, 15–20.

10

Monolithic Microwave IC Technology

Lawrence P. Dunleavy
University of South Florida

10.1 Overview

10.1.1 MMIC Definition and Concepts

Pucel gives an excellent review of Monolithic Microwave Integrated Circuit (MMIC) technology in a 1981 paper.[1] Pucel went on to assemble a collection of papers on the subject in which he states in the introduction:[2]

> the monolithic approach is an approach wherein all active and passive circuit elements and interconnections are formed, in situ on or within a semi-insulating semi-conductor substrate by a combination of deposition schemes such as epitaxy, ion implantation, sputtering, and evaporation.

Figure 10.1 is a conceptual MMIC chip illustrating most of the major components. These include field effect transistor (FET) active devices, metal-insulator-metal (MIM) capacitors, thin film resistors, spiral strip inductors, via hole grounding, and air bridges.

As implied by the above quote and Fig. 10.1, in a MMIC all of the circuit components, including transistors, resistors, capacitors, and interconnecting transmission lines are integrated onto a single semi-insulating/semiconducting (usually GaAs) substrate. Use of a mask set and a corresponding series of processing steps achieves the integrated circuit fabrication. The mask set can be thought of as a mold. Once the mold has been cast, the process can be repeated in a "turn-the-crank" fashion to batch process tens, hundreds, or thousands of essentially identical circuits on each wafer.

0-8493-1562-X/03/$0.00+$1.50

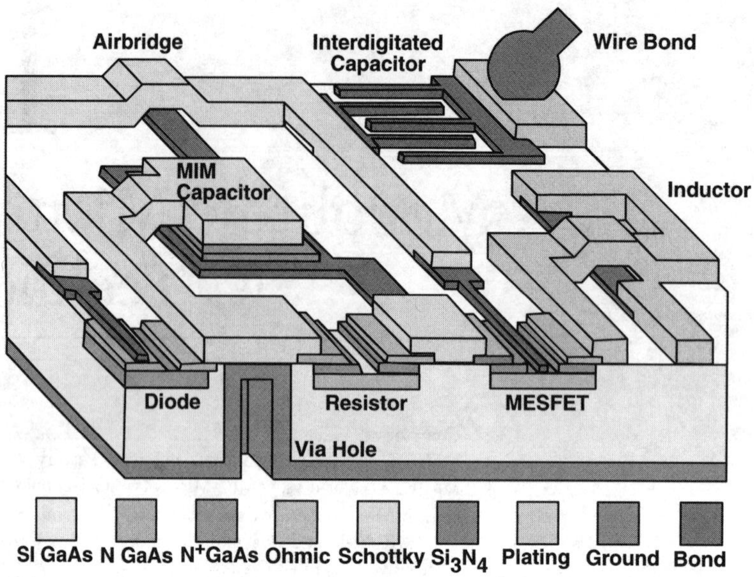

FIGURE 10.1 Three-dimensional conceptual illustration of MMIC technology (from Ladbrooke[15]).

10.1.2 A Brief History of GaAs MMICs

As noted by Pucel,[3] the origin of MMICs may be traced to a 1964 government program at Texas Instruments. A few key milestones are summarized in the following:

- 1964 — U.S. government funded a research program at TI based on silicon integrated circuit technology:[4] The objective was a transmit/receive module for a phased array radar antenna. The results were disappointing due to the poor semi-insulating properties of silicon.
- 1968 — Mehal and Wacker[5] used semi-insulating gallium arsenide (GaAs) as the substrate with Schottky diodes and Gunn devices as active devices to fabricate an integrated circuit comprising a 94 GHz receiver "front-end."
- 1976 — Pengelly and Turner[6] used MESFET devices on GaAs to fabricate an X-band (~10 GHz) amplifier and sparked an intense activity in GaAs MMICs.
- 1988 (approximately) — U.S. government's Defense Advanced Research Projects Agency (DARPA, today called ARPA) launched a massive research and development program called the MIMIC program (included Phase I, Phase II, and Phase III efforts) that involved most of the major MMIC manufacturing companies.

In the early 1980s a good deal of excitement was generated and several optimistic projections were made predicting the rapid adoption of GaAs MMIC technology by microwave system designers, with correspondingly large profits for MMIC manufacturers. The reality is that there was a much slower rate of progress to widespread use of MMIC technology, with the majority of the early thrust being provided by the government for defense applications. Still, steady progress was made through the 1980s and the government's MIMIC program was very successful in allowing companies to develop lower cost design and fabrication techniques to make commercial application of the technology viable. The 1990s have seen good progress toward commercial use with applications ranging from direct broadcast satellite (DBS) TV receivers, to automotive collision avoidance radar, and the many wireless communication applications (cell phones, WLANs, etc.).

TABLE 10.1 Features of Hybrid and Monolithic Approaches

Feature	Hybrid	Monolithic
Type of substrate	Insulator	Semiconductor
Passive components	Discrete/Deposited	Deposited
Active components	Discrete	Deposited
Interconnects	Deposited and wire-bonded	Deposited
Batch processed?	No	Yes
Labor intensive/chip	Yes	No

[a] After Pucel[7] with permission from IEEE.

10.1.3 Hybrid vs. MMIC Microwave IC Technologies

The conventional approach to microwave circuit design that MMIC technology competes with, or is used in combination with, is called "hybrid microwave integrated circuit," "discrete microwave integrated circuit" technology, or simply MIC technology. In a hybrid MIC, the circuit pattern is formed using photolithography. Discrete components are then assembled onto the substrate (e.g., using solder or silver epoxy) and connected using bondwires. In contrast to the batch processing afforded the MMIC approach, MICs have to be assembled with discrete components attached using relatively labor-intensive manufacturing methods. Table 10.1 summarizes some of the contrasting features of hybrid and monolithic approaches.

The choice of MMICs vs. the hybrid approach is mainly a matter of volume requirements. The batch processing of MMICs gives this approach advantages for high volume applications. Significant cost savings can be reaped in reduced assembly labor, however, for MMIC the initial design and mask preparation costs are considerable. The cost of maintaining a MMIC manufacturing facility is also extremely high and this has forced several companies out of the business. A couple of examples are Harris, which sold its GaAs operation to Samsung and put its resources into silicon. Another is AT&T, which is also relying on silicon for is anticipated microwave IC needs.

The high cost of maintaining a facility can only be offset by high volume production of MMICs. Still this does not prevent companies without MMIC foundries from using the technology, as there are several "commercial foundries" who offset the costs of maintaining their facilities by manufacturing MMIC chip products for third party companies through a foundry design working relationship.

A key advantage of MMICs is small size. To give an example, a hybrid MIC the size of a business card can easily be reduced to a small chip one or two millimeters on a side. An associated advantage is the ease of integration that allows several functions to be integrated onto a single chip. For example, Anadigics and Raytheon have both manufactured DBS-related MMICs, wherein the functions of amplification (LNA and IF amplifiers), signal generation (VCO), and signal conversion (mixer) and filtering are all accomplished on a 1 mm × 2 mm, or smaller, chip.

FIGURE 10.2 An example of combined MMIC and hybrid MIC technologies, this radar module includes several MMIC chips interconnected using microstrip lines. The hybrid substrate (white areas) is an alumina insulating substrate; the traces on the hybrid substrate are microstrip lines (courtesy Raytheon Systems Company).

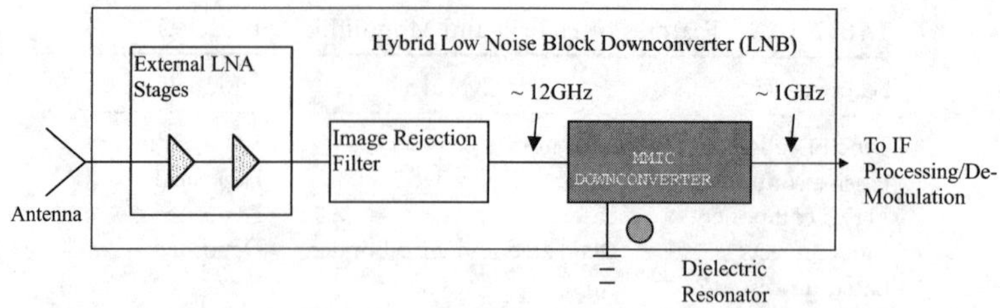

FIGURE 10.3 Application of MMICs in Ku-Band Direct Broadcast Satellite downconverters. The antenna is typically a small parabolic dish.

In contrast to MMIC, MIC lithography is quite inexpensive and a much smaller scale investment is required to maintain a MIC manufacturing capability. There are also some performance advantages of the hybrid approach. For example, it is much easier to tune or repair a hybrid circuit after fabrication than it is for a MMIC. For this reason, for applications where the lowest noise figure is required, such as in a satellite TV receiver, an individually tuned hybrid LNA may be preferred as the first stage.

Ultimately, there is no such thing as a truly all MMIC system. Monolithic technology can be used to integrate single functions, or several system functions, but cannot sustain a system function in isolation. Usually, a MMIC is packaged along with other circuitry to make a practically useful component, or system. Figure 10.2, a radar module made by Raytheon, exemplifies how the advantages of both MMIC and hybrid approaches are realized in a hybrid connection of MMIC chips. Another example of combined hybrid/MMIC technology is shown in Fig. 10.3, in the form of a low noise block downconverter for Direct Broadcast Satellite applications. An example of combined MMIC and hybrid MIC technologies, this radar module includes several MMIC chips interconnected using microstrip lines. The hybrid substrates (white areas) is an alumina insulating substrate; the traces on the hybrid substrate are microstrip lines (courtesy Raytheon Systems Company).

10.1.4 GaAs MMICs in Comparison to Silicon VLSI Computer Chips

Everyone is familiar with silicon digital IC chips, or at least the enormous impact silicon-based Very Large Scale Integrated (VLSI) circuits have had on the computer industry. Silicon computer chips are digital circuits that contain hundreds to thousands of transistors on each chip. In a digital circuit the transistors are used as switches that are in one of two possible states depending on the "logic" voltage across a pair of terminals. The information processed by a digital circuit consists of a sequence of "1s" and "0s," which translate into logic voltages as the signal passes through the digital IC. Noise distorts the logic waveform in much the same way that it distorts a sinusoidal signal, however, as long as the signal distortion due to noise is not severe, the digital circuitry can assign the correct (discrete) logic levels to the signal as it is processed. Signal interpretation errors that occur due to noise are measured in terms of a bit error rate (BER).

The speed of the digital processing is related to how fast the transistors can switch between one state and another, among other factors. Because of certain material factors, such as electron mobility, digital circuits made on GaAs have been demonstrated to have speed advantages over silicon digital ICs, however, the speed advantages have not been considered by the majority of companies to outweigh the significant economic advantages of well-established, lower cost, silicon processing technology.

Because of the large volume of silicon chips that have been produced over the last twenty years, silicon processing techniques are significantly more established and in many cases standardized as compared to GaAs processing techniques, which still vary widely from foundry to foundry. The digital nature of the signals and operation modes of transistors in digital ICs makes uniformity between digital ICs, and even similar ICs made by different manufacturers, much easier to achieve than achieving uniformity with analog GaAs MMICs.

In contrast, GaAs MMICs are analog circuits that usually contain less than 10 transistors on a typical chip. The analog signals processed, which can take on any value between certain limits, may generally be thought of as combinations of noisy sinusoidal signals. Bias voltages are applied to the transistors in such a way that each transistor will respond in one of several predetermined ways to an applied input signal. One common use for microwave transistors is amplification, whereby the result of a signal passing through the transistor is for it to be boosted by an amount determined by the gain of the transistor.

A complication that arises is that no two transistors are identical in the analog sense. Taking gain for example, while there will be a statistical distribution of gain for a set of amplifier chips measured on the same GaAs MMIC wafer, a different (wider) set of statistics applies to variations in gain from wafer-to-wafer for the same design. These variations are caused primarily by variations in transistors, but also by variations in other components that make up the MMIC, including MIM capacitors, spiral inductors, film resistors, and transmission line interconnects. Successful foundries are able to control the variations within acceptable limits in order to achieve a satisfactory yield of chips meeting a customer's requirements. However, translating a MMIC design mask set to a different manufacturing foundry is a different story altogether.

This is not to say that foundry translation of MMIC designs cannot be accomplished. Under the federally funded MIMIC program, mentioned earlier, several pairs of foundries were tasked to translate designs from one to the other to demonstrate a "second-sourcing" capability. These efforts met with varying degrees of success, but not without considerable effort on the part of the participating GaAs MMIC foundries. Two pairs of companies involved in this second-sourcing demonstration effort for the MIMIC program are Raytheon and Texas Instruments, and Hughes Aircraft Company (GaAs foundry since bought and closed by Raytheon Company) and General Electric Company (now part of Lockheed-Martin).

10.1.5 MMIC Yield and Cost Considerations

Yield is an important concept for MMICs and refers to the percentage of circuits on a given wafer with acceptable performance relative to the total number of circuits fabricated. Since yield may be defined at several points in the MMIC process, it must be interpreted carefully.

- DC yield is the number of circuits whose voltages and currents measured at DC are within acceptable limits.
- RF yield is generally defined as the number of circuits that have acceptable RF/microwave performance when measured "on-wafer," before circuit dicing.
- Packaged RF yield is the final determination of the number of acceptable MMIC products that have been assembled using the fabricated MMIC chips.

If measured in terms of the total number of circuits fabricated, each of these yields will be successively lower numbers. For typical foundries, DC yields exceed 90%, while packaged yields may be around 50%. Final packaged RF yield depends heavily on the difficulty of the RF specifications, the uniformity of the process (achieved by statistical process control), as well as how sensitive the RF performance of the circuit design is to fabrication variations.

The costs involved with MMICs include:

- Material
- Design
- Mask set preparation
- Wafer processing
- Capital equipment
- Testing
- Packaging
- Inspection

A typical wafer run may cost \$20,000 to \$50,000, with \$5,000 to \$10,000 attributable to the mask set alone. These figures do not include design costs. Per-chip MMIC costs are determined by:

- Difficulty of design specifications
- Yield
- Material (wafer size and quality)
- Production volume
- Degree of automation

Some 1989–90 example prices for MMIC chips are as follows:[8]

1 to 5 GHz Wideband Amplifier — \$30.00
2 to 8 GHz Wideband Amplifier — \$45.00
6 to 18 GHz Wideband Amplifier — \$100.00
DC to 12 GHz Attenuator — \$60.00
DBS downconverter chip — \$10.00.

In comparison, example prices for 1999 MMIC chips are as follows:[9]

DBS Downconverter chip — <\$2.50
DC to 8 GHz HBT MMIC amp. — ~\$3
Packaged power FET MMIC — \$1 for ~2 GHz
4 to 7 GHz and 8 to 11 GHz high power/high efficiency power amps. — \$85 (2 Watt) to \$330 (12 Watt)

Prices for low volume specials at high frequency have not moved much since 1989.

10.1.6 Si vs. GaAs for Microwave Integrated Circuits

The subject of Silicon versus GaAs has been a hotly debated subject since the beginning of the MMIC concept in around 1965 (see "A brief history of GaAs MMICs" above). Two of the main discriminating issues between the technologies are microwave transistor performance and the loss of the semiconductor when used as a semi-insulating substrate for passive components.

A comparison of relevant physical parameters for silicon and GaAs materials is given in Table 10.2. The dielectric constants of the materials mainly affects the velocity of propagation down transmission line interconnects, and for the materials compared, this parameter is on the same order. For the other factors considered, significant differences are observed. The thermal conductivity, a measure of how efficiently the substrate conducts heat (generated by DC currents) away from the transistors, is best for silicon and is one key advantage of the silicon approach. This advantage is offset by silicon's lower mobility and lower resistivity. Mobility in a semiconductor is a measure of how easily electrons can move through the "doped" region of the semiconductor (see discussion of FET operation in the following section). The

TABLE 10.2 Properties of GaAs, Silicon, and Common Insulating Substrates

Property	GaAs	Silicon	Semi-Insulating GaAs	Semi-Insulating Silicon	Sapphire	Alumina
Dielectric constant	12.9	11.7	12.9	11.7	11.6	9.7
Thermal cond. (watts/cm-K)	0.46 (fair)	1.45 (good)	0.46 (fair)	1.45 (good)	0.46 (fair)	0.37 (fair)
Resistivity (ohm-cm)	—	—	10^7–10^9 (fair)	10^3–10^5 (poor)	$>10^{14}$ (good)	10^{11}–10^{14} (good)
Elec. mobility	4300[a] (best)	700[a] (good)	—	—	—	—
Sat. elec. velocity	1.3×10^7 (best)	9×10^6 (good)	—	—	—	—

Adapted from Pucel,[13] with permission from IEEE.
[a] At a doping concentration of $10^{17}/cm^3$.

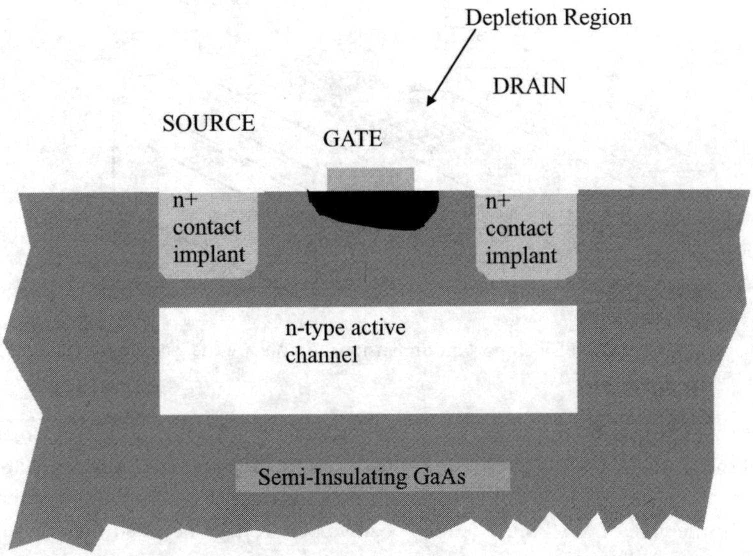

FIGURE 10.4 Cross-section of a MESFET transistor. In operation, current I_{ds} flows between the gate and drain terminals through the doped n-type active channel. An AC voltage applied to the gate modulates the size of the depletion region causing the I_{ds} current to be modulated as well. Notice that in a MMIC the doped n-type region is restricted to the transistor region leaving semi-insulating GaAs outside to serve as a passive device substrate.

mobility, as well as the saturated velocity, have a strong influence on the maximum frequency at which a microwave transistor can have useful gain.

Turning our attention to passive component operation, GaAs has much better properties for lower loss passive circuit realization. With the exception of a resistor, the ideal passive component is a transmission line, inductor, or capacitor that causes no signal loss. Resistivity is a measure of how "resistant" the substrate is to leakage currents that could flow, for example, from the top conductor of a microstrip line and the ground plane below. Looking first at the properties of the insulators sapphire and alumina, the resistivity is seen to be quite high. Semi-insulating GaAs is almost as high, and silicon has the lowest resistivity (highest leakage currents for a given voltage).

These considerations have led many companies to invest heavily in GaAs technology for microwave applications over the last several years. However, silicon remains a strong contender. In fact there has been a very strong renewal in development of silicon MMICs[10] with the advent of numerous markets for microwave "wireless" products. The front lines of the battlefield between silicon and GaAs are at frequencies below 6 GHz, where potential commercial opportunities are numerous. Silicon-based microwave ICs are also beginning to appear in higher frequency applications, such as Ku-band DBS satellite receivers.[11] Silicon-germanium heterojunction bipolar transistor technology is paving the way for increasing the applicability of silicon technology to even higher frequencies.[12]

10.2 Basic Principles of GaAs MESFETs and HEMTs

10.2.1 Basic MESFET Structure

The primary active device in a GaAs MMIC is the metal electrode semiconductor field-effect transistor, or MESFET. The basic construction of a MESFET is shown in Figs. 10.4 and 10.5. An "active layer" is first formed on top of a semi-insulating GaAs substrate by intentionally introducing an n-type impurity onto the surface of the GaAs, and isolating specific channel regions. These channel regions are semiconducting in that they contain free electrons that are available for current flow. When a metal is placed in

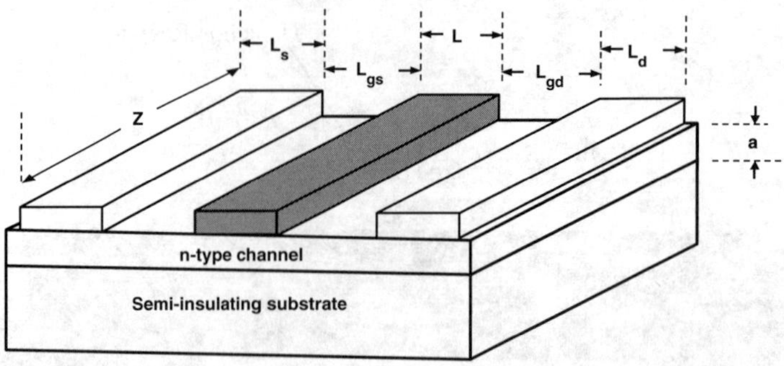

FIGURE 10.5 Aspect view of MESFET. Important dimensions to note are the gate length (Lg) and the gate width (Z). (From Golio.[16])

direct contact with a semiconductor, as in the case of the gate, a "Schottky diode" is formed. One of the consequences of this is that a natural "depletion region," a region depleted of available electrons is formed under the gate. A diode allows current flow easily in one direction, while impeding current flow in the other direction. In the case of a MESFET gate, a positive bias voltage between the gate and the source "turns on" the diode and allows current to flow between the gate and the source through the substrate. A negative bias between the gate and the drain "turns off" the diode and blocks current flow, it also increases the depth of the depletion region under the gate.

In contrast to the gate contact, the drain and source contacts are made using what are called "ohmic contacts." In an ohmic contact, current can flow freely in both directions. Whether an ohmic contact or Schottky diode is formed at the metal-semiconductor interface is determined by the composition of the metal placed on the interface and the doping of the semiconductor region directly under the metal. The introduction of "pocket n+ implants" help form the ohmic contacts in the FET structure illustrated in Fig. 10.4. In the absence of the gate, the structure formed by the active channel in combination with the drain and source contacts essentially behaves as a resistor obeying ohms law. In fact this is exactly how one type of GaAs-based resistor commonly used in MMICs is made.

10.2.2 FETs in Microwave Applications

The most common way to operate a MESFET, for example in an amplifier application, is to ground the source (also called "common source" mode), introduce a positive bias voltage between the drain and source, and a negative bias voltage between the gate and source. The positive voltage between the drain and source V_{ds} causes current I_{ds} to flow in the channel.

As negative bias is applied between the gate and source V_{gs} the current I_{ds} is reduced as the depletion region extends farther and farther into the channel region. The value of current that flows with zero gate-to-source voltage is called the saturation current I_{dss}. Eventually, at a sufficiently large negative voltage, the channel is completely depleted of free electrons and the current I_{ds} will be reduced to essentially zero. This condition is called "pinch off." In most amplifier applications, the negative gate voltage is set to a "bias condition" between 0 volts and the pinch-off voltage V_{po}.

Figure 10.6 gives a simplified view of a FET configured in an amplifier application. An input sinusoidal signal $V_{gs}(t)$ is shown offset by a negative DC bias voltage. The sinusoidal variation in V_{gs} causes a likewise sinusoidal variation in depth of the depletion region that in turn creates a sinusoidal variation (or modulation) in the output current. Amplification occurs because small variations in the V_{gs} voltage cause relatively large variations in the output current. By passing the output current through a resistance RL the voltage waveform V_{ds} (t) is formed. The V_{ds} waveform is shown to have higher amplitude than V_{gs} to illustrate the amplification process.

Other common uses, which involve different configurations and biasing arrangements, include use of FETs as the basis for mixers and oscillators (or VCOs).

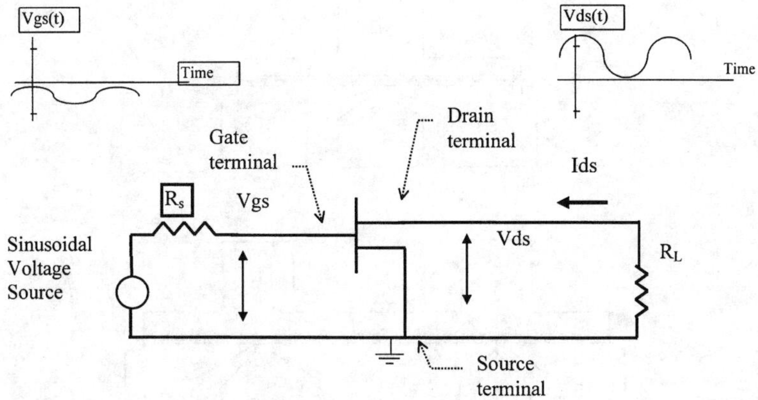

FIGURE 10.6 Simplified schematic of a FET in an amplifier application. Not shown are matching networks needed to match between the source resistance R_s and the FET input impedance, and the load resistance R_L and the FET output impedance. Also omitted are the networks needed to properly apply the DC bias voltages to the device and provide isolation between the RF and DC networks.

10.2.3 FET Fabrication Variations and Layout Approaches

Figure 10.7 illustrates a MESFET fabricated with a recessed gate, along with a related type of FET device called a high electron mobility transistor (HEMT). A recessed gate is used for a number of reasons. First in processing it can aid in assuring that the gate stripe is placed in the proper position between the drain and source, and it can also result in better control and uniformity in I_{dss} and V_{po}. A HEMT is a variation of the MESFET structure that generally produces a higher performing device. This translates, for example, into a higher gain and a lower noise figure at a given frequency.

In light of the above cursory understanding of microwave FET structures, some qualitative comments can be made about some of the main factors that cause intended and unintended variations in FET performance. The first factor is the doping profile in the active layer. The doping profile refers to the density of the charge carriers (i.e., electrons) as a function of depth into the substrate. For the simplest

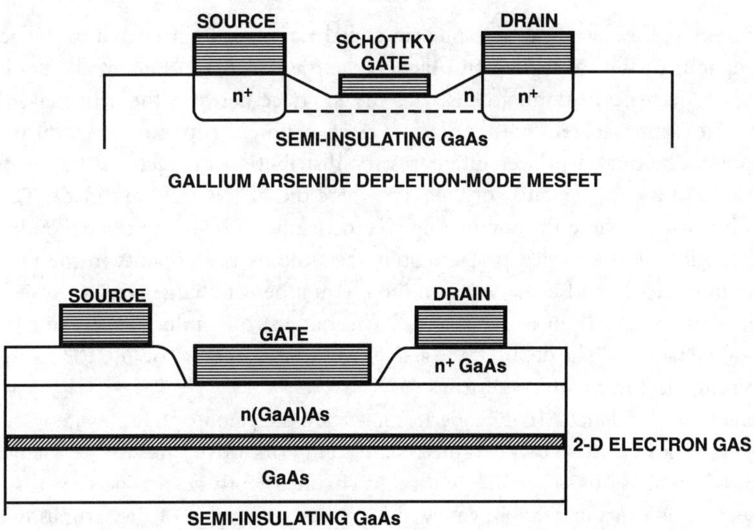

FIGURE 10.7 MESFET (top) and HEMT (bottom) structures showing "gate recess" structure whose advantages include better control of drain-to-source saturation current I_{dss} and pinch-off voltage V_{po}. (From Goyal.[17])

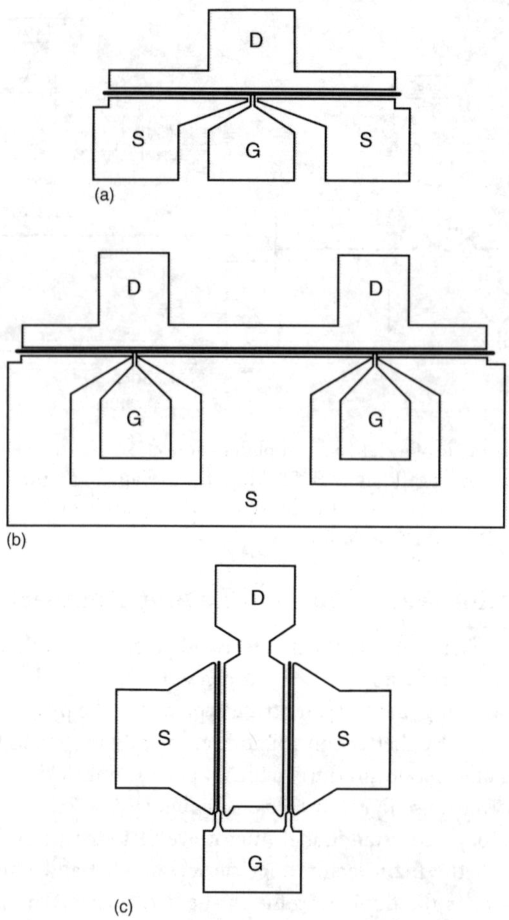

FIGURE 10.8 Three common layout approaches for MESFETs and HEMTs. (From Ladbrooke.[18])

type, uniform doping, the density of dopants (intended impurities introduced in the active region) is the same throughout the active region. In practice, there is a natural "tail" of the doping profile that refers to a gradual decrease in doping density as the interface between the active layer and the semi-insulating substrate is approached. One approach to create a more abrupt junction is the so-called "buried P-layer" technique. The buried p-layer influences the distribution of electrons versus depth from the surface of the chip in the "active area" of the chip where the FET devices are made. The idea is create better definition between where the conducting channel stops and where the nonconducting substrate begins. (More specifically the buried p-layer counteracts the n-type dopants in the tail of the doping profile.) The doping profile and density determine the number of charge carriers available for current flow in a given cross-section of the active channel. This has a strong influence on the saturation current I_{dss} and pinch-off voltage V_{po}. The depth of the active layer (dimension "a" of Fig. 10.5) also plays a critical role in determining the current characteristics.

Other variables that influence MESFET performance are the gate length and gate width ("L" and "Z" of Fig. 10.5). The names for these two parameters are counterintuitive since the gate length refers to the shorter of the two dimensions. The gate width and channel depth determine the cross-sectional area available for current flow. An increase in gate width increases the value of the saturation current, which translates into the ability to operate the device at higher RF power levels (or AC voltage amplitudes). Typical values for gate widths are in the range of 100 microns for low noise devices to over 10 millimeters for high power devices. The gate length is usually the minimum feature size of a device and is the most significant factor in determining the maximum frequency where useful gain can be obtained from a FET;

generally, the smaller the gate length the higher the gain for a given frequency. However, the fabrication difficulty increases, and processing yield decreases, as the gate length is reduced. The difficulty arises from the intricacy in controlling the exact position and length (small dimension) of the gate.

The geometrical layout of the FET also influences performance. Figure 10.8 shows common FET layouts. The layout affects what are called "external parasitics," which are undesired effects that can be modeled as a combination of capacitors, inductors, and resistors added to the basic FET electrical model.

In MMIC fabrication, variations in the most of the above-mentioned parameters are a natural consequence of a real process. These variations cause variations in observed FET performance even for identical microwave FETs made using the same layout geometry on the same wafer. Certainly there are many more subtle factors that influence performance, but the factors considered here should give some intuitive understanding of how unavoidable variations in the physical structure of fabricated FETs cause variations in microwave performance. As previously mentioned, successful GaAs MMIC foundries use statistical process control methods to produce FET devices within acceptable limits of uniformity between devices.

10.3 MMIC Lumped Elements: Resistors, Capacitors, and Inductors

10.3.1 MMIC Resistors

Figure 10.9 shows three common resistor types used in GaAs MMICs. For MMIC resistors the type of resistor material, and the length and width of the resistor determine the value of the resistance. In practice, there are also unwanted "parasitic" effects associated with MMIC resistors that can be modeled generally as a combination of series inductance, and capacitance to ground in addition to the basic resistance of the component.

10.3.2 MMIC Capacitors

The most commonly used type of MMIC capacitor is the metal-insulator-metal capacitor shown in Fig. 10.10. In a MIM capacitor the value of capacitance is determined from the area of the overlapping metal (smaller dimension of two overlapping plates), the dielectric constant ε_r of the insulator material, typically silicon nitride, and the thickness of the insulator. For values less than about 0.2 pF, series connected MIM capacitors can be used.

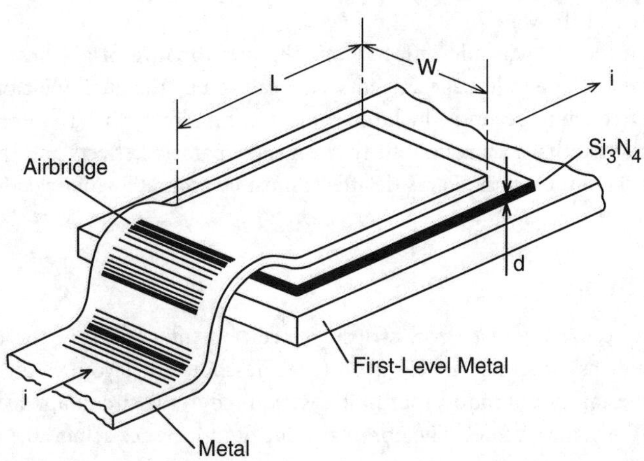

FIGURE 10.9 Common MMIC resistor fabrication approaches: (a) implanted GaAs, (b) mesa-etched/epitaxialy grown GaAs, (c) thin film (e.g., TaN). (From Goyal.[19])

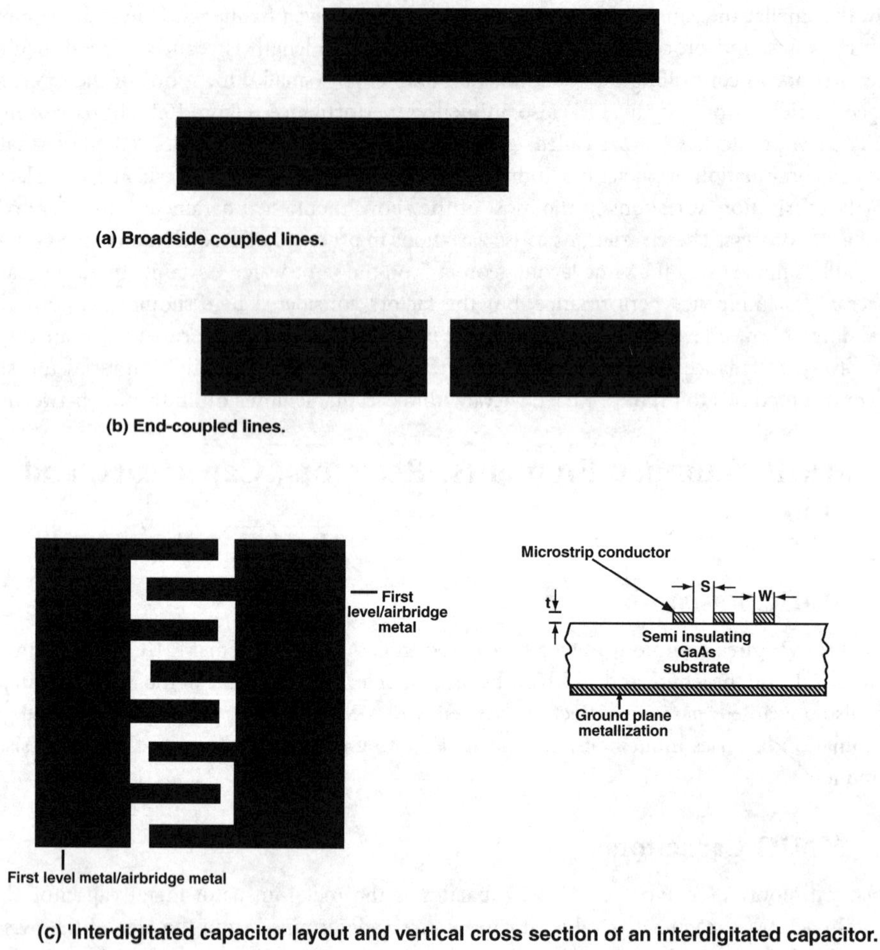

FIGURE 10.10 Metal-insulator-metal conceptual diagram. (From Ladbrooke.[20])

Smaller values of capacitance can be achieved with one of the various arrangements of coupled lines illustrated in Fig. 10.11. For these capacitors, the capacitance is determined from the width and spacing of strips on the surface of the wafer.

At microwave frequencies, "parasitic" effects limit the performance of all these capacitors. The two main effects are signal loss due to leakage currents, as measured by the quality factor Q of the capacitor, and a self-resonance frequency, beyond which the component no longer behaves as a capacitor. The final wafer or chip thickness can have a strong influence on these parasitic effects and the associated performance of the capacitors in the circuit. Parasitic effects must be accurately modeled for successful MMIC design usage.

10.3.3 MMIC Inductors

MMIC inductors are realized with narrow strips of metal on the surface of the chip. Various layout geometries are possible as illustrated in Fig. 10.12(a). The choice of layout is dictated mainly by the available space and the amount of inductance L that is required in the circuit application, with the spiral inductors providing the highest values. The nominal value of inductance achievable from strip inductors is determined from the total length, for the simpler layouts, and by the number of turns, spacing, and line width for the spiral inductors.

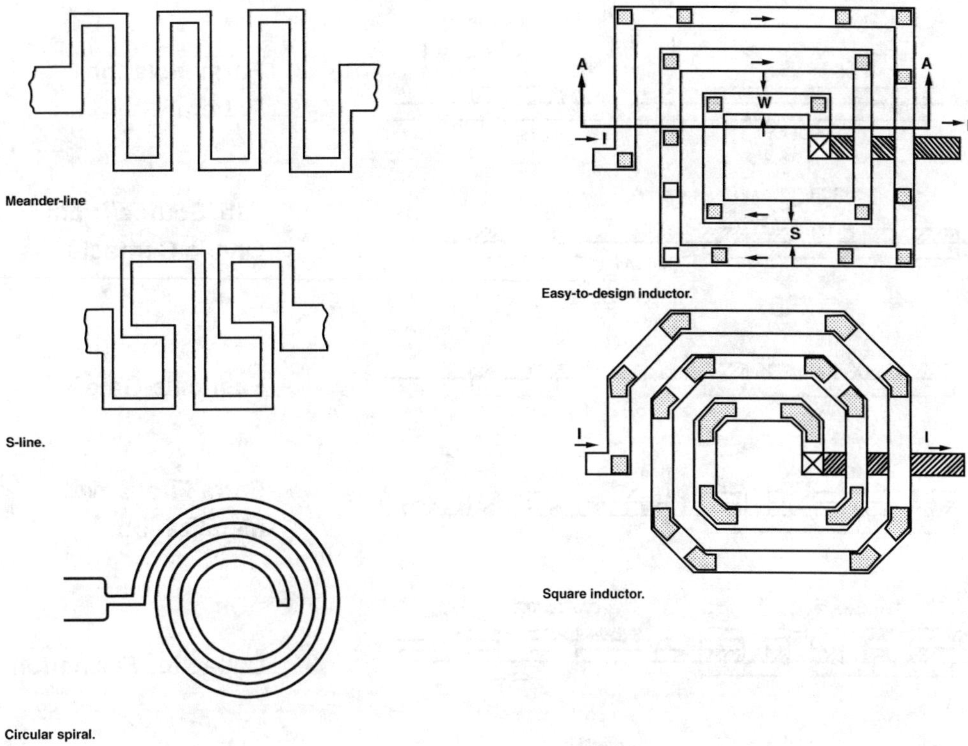

FIGURE 10.11 MMIC approaches for small valued capacitors. (From Goyal.[21])

At microwave frequencies, "parasitic" effects limit the performance of these inductors. The two main effects are signal loss due to leakage currents, as measured by the quality factor Q of the capacitor, and a self-resonance frequency, beyond which the component no longer behaves as an inductor. The final thickness of the substrate influences not only the nominal value of inductance, but also the quality factor and self-resonance frequency. Inductor parasitic effects must be accurately modeled for successful MMIC design usage.

10.3.4 Air Bridge Spiral Inductors

Air bridge spiral inductors are distinguished from conventional spiral inductors by having the metal traces that make up the inductor suspended from the top of the substrate using MMIC air bridge technology. MMIC air bridges are generally used to allow crossing lines to jump over one another without touching and are almost invariably used in conventional spiral inductors to allow the center of the spiral to be brought through the turns of the spiral inductor for connection to the circuit outside of the spiral. In an air bridge spiral inductor, all of the turns are suspended off the substrate using a series of air bridges supported by metalized posts. The reason for doing this is to improve inductor performance by reducing loss as well as the effective dielectric constant of the lines that make up the spiral. The latter can have the effect of reducing inter-turn capacitance and increasing the resonant frequency of the inductor. Whether or not air bridge inductors are "worth the effort" is a debatable subject as the air bridge process is an important yield-limiting factor. This means circuit failure due to collapsed air bridges, for example, occur at an increasing rate, the more air bridges that are used.

10.3.5 Typical Values for MMIC Lumped Elements

Each MMIC fabrication foundry sets its limits on the geometrical dimensions and range of materials available to the designer in constructing the MMIC lumped elements.

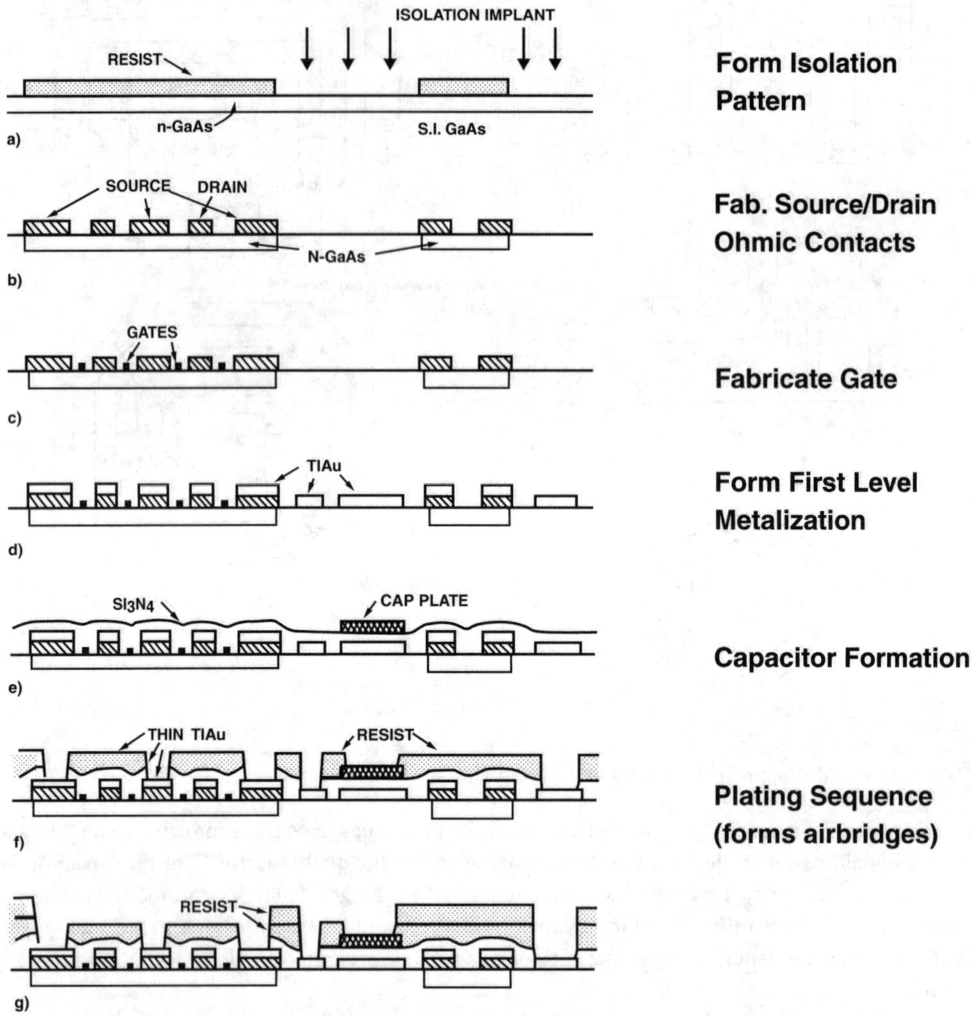

FIGURE 10.12 (a) Various MMIC inductor layouts. (From Goyal.[22])

Accordingly, the range of different resistor, capacitor, and inductor values available for design will vary from foundry to foundry. With this understanding, a "typical" set of element values associated with MMIC lumped elements are presented in Table 10.3.

10.4 MMIC Processing and Mask Sets

The most common MMIC process approach in the industry can be characterized as having 0.5 micron gate length MESFETs fabricated on a GaAs wafer whose final thickness is 100 microns, or 4 mils. The back side of the wafer has plated gold; via holes are used to connect from the back side of the wafer to the topside of the wafer. Although specific procedures and steps vary from foundry to foundry, Fig. 10.12b illustrates a typical process.

Defining Terms

Active load: A transistor connected in a way as to replace a function that would conventionally be performed by a passive component such as a resistor, capacitor, or inductor.

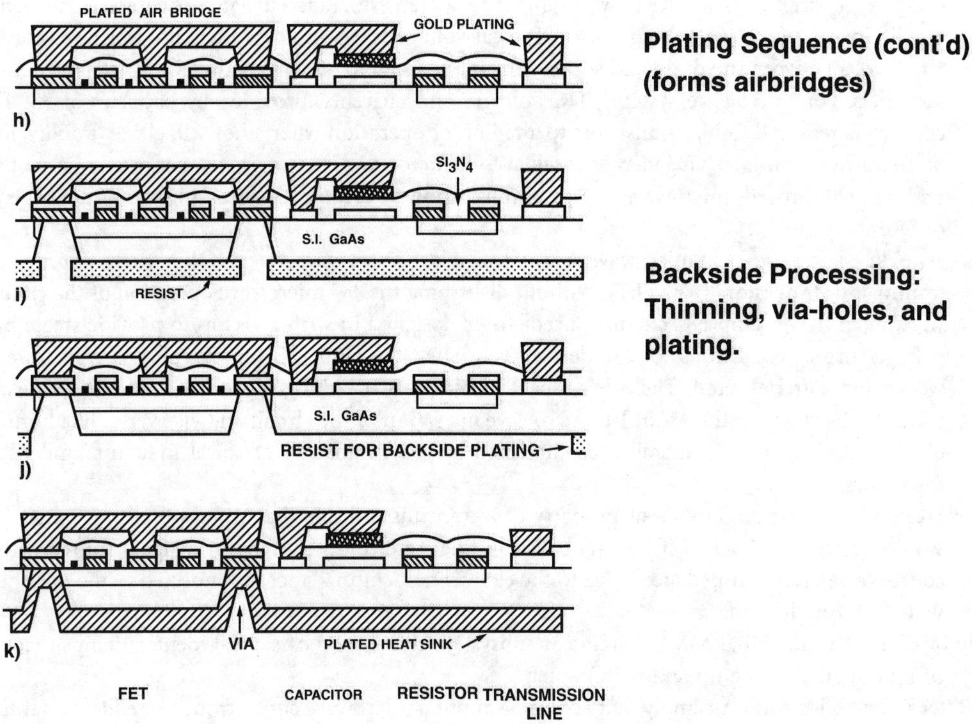

FIGURE 10.12 (b) Conceptual diagrams illustrating flow for typical MMIC process. (From Williams.[23,25])

TABLE 10.3 Ranges of MMIC Lumped Element Values Available to the Designer for a "Typical" Foundry Process

Type	Value	Dielectric or Metal	Application
Inductor: Single loop, meander line, etc.	0.01–0.5 nH	Plated gold	Matching
Inductor: Spiral	0.5–10 nH	Plated gold	Matching, DC power (bias) supply choke
Capacitor: MIM	0.1–20 pF	Plated and unplated gold, silicon nitride	Matching, RF/DC signal separation
Capacitor: Coupled lines and interdigital	0.001–0.1 pF	Plated or unplated gold	Matching, RF/DC signal separation
Resistor: Thin film	5 Ω–1 kΩ	NiCr, TaN	DC bias ckts., feedback, matching, stabilization
Resistor: GaAs Monolithic	10 Ω–10 kΩ	Implanted or epitaxial GaAs	DC biasing, feedback, matching, stabilization

Adapted from Goyal.[14]

Air bridge: A bridge made of metal strip suspended in air that can connect components on an integrated circuit in such a way as to cross over another strip. Air bridges are also used to suspend metallization in spiral inductors off of the semiconducting substrate, in a way that can lead to improved performance in some cases.

Attenuation: Ratio of output signal to input signal, usually expressed in dB (see below), for a device that reduces the signal level.

Bias voltage or current: The DC power applied to a transistor allowing it to operate as an active amplifying or signal-generating device. Typical voltage levels in GaAs FETs used in receivers are 1 to 7 volts between the drain and source terminals, and 0 to –5 volts between the gate and source terminals. For microwave systems, DC voltages and currents, provided by batteries or AC/DC converters required to bias transistors to a region of operation where they will either amplify, mix or frequency translate, or generate (oscillators) microwave energy. Since energy can be neither created or destroyed, microwave energy amplification or creation is accomplished at the expense of DC energy.

Bias network: A key aspect of microwave circuit design is to apply the proper DC bias to the appropriate terminals of transistors (e.g., FETs) without disturbing the AC microwave operation of the circuit. In some cases, on-chip DC circuitry needs to be designed in such a way as to provide stable bias voltage/current conditions for the device even when the chip DC supply voltages vary (due to weakening batteries, etc.). The other aspect of bias network design is to isolate the DC network form interfering with the AC or RF/microwave operation of the circuit and vice versa. In a lumped element design, this is generally accomplished by a combination of spiral inductors and MIM capacitors.

Characteristic impedance: Inherent property of a transmission line that defines the impedance that would be seen by a signal if the transmission line were infinitely long. If a signal source with a source or reference impedance equal to the characteristic impedance is connected to the line there will be zero reflections.

Chip or die: An individual MMIC circuit or subsystem that is one of several identical chips that are produced after dicing up an MMIC wafer.

DBS Receiver: Electronic assembly that accepts as input a microwave signal from a satellite containing transmitted TV signals modulated onto the signal. The receiver first amplifies the low level signal, then processes the signal by first converting it to a lower IF frequency and then demodulating the signal to separate the TV signals from the microwave carrier signal. A basic way of looking at the relationship of the microwave carrier to the TV signal is to think of the carrier signal as an envelope with a message inside. The message is the TV signal. Demodulation is the process of carefully removing the message from the envelope (carrier). The noise figure of receiver is a measure of the amount of noise that will be added to the signal (carrier and TV signal) by the receiver. If the receiver adds too much noise, the result will be a snowy picture on the TV screen.

DBS: Direct Broadcast Satellite: Refers to TV signal transmission and distribution from a base station up to a satellite, and then down to consumers who have suitable satellite receiving antennas and downconverter receivers.

$$F(\text{dB}) = 10\text{Log}_{10}\left(\text{SNR}_{\text{in}}/\text{SNR}_{\text{out}}\right)$$

dB: Decibel: A unit of measure that describes the ratio between two quantities in terms of a base 10 logarithm. For example, the ratio between the power level at the input and output of an amplifier is called the power gain and may be expressed in decibels as follows:

$$G(\text{dB}) = 10\text{Log}_{10}\left(P_{\text{out}}/P_{\text{in}}\right)$$

dBm: Decibel referenced to milliwatts: A common unit of power in decibels referenced to 1 milliwatt.

$$P(\text{dBm}) = 10\text{Log}_{10}\left(P\text{ in mW}/1\text{ mW}\right)$$

DC: Direct current/voltage: Constant voltage or current with no variation over time. This can be considered in general terms as an alternating current/voltage (AC) with a frequency of variation of zero, or a zero frequency signal. For microwave systems, DC voltages and currents, provided by batteries or AC/DC converters required to bias transistors to a region of operation where they will either amplify, mix or frequency translate, or generate (oscillators) microwave energy.

DRO: Dielectric resonator (stabled) oscillator: A dielectric resonator is a cylindrically shaped piece of material, or "puck," that has low loss resonant frequencies that are determined primarily by the size of the cylinder. Placing a dielectric resonator near a microstrip line can form a resonant circuit that will frequency stabilize a voltage controlled oscillator.

Dielectric constant (e_r): The dielectric constant is an electric property of an insulator or semiconducting material that describes how differently electric fields will behave inside the material as compared to air. As an example, $e_r = 12.9$ for GaAs as compared to $e_r = 1$ for air. In integrated circuits, an effective dielectric constant (e_{eff}) is used since the electric fields supported by the signals traveling through the conductors on the circuit flow through both air and the insulator or semiconductor simultaneously.

FET: Field Effect Transistor: The MESFET (Metal-Electrode-Semiconductor-Field-Effect-Transistor) is a specific type of FET that is the dominant active (amplifying) device in GaAs MMICs. A FET is composed of three terminals called the gate, drain, and source, and a conducting "channel." In an amplifier application, the source is connected to ground, and DC bias is applied between the drain and source causing a current to flow in the channel. The current flow is controlled and modulated by the AC or DC voltage applied to the gate.

Filter: A network, usually composed of inductors and capacitors (for lumped circuit), or transmission lines of varying length and characteristic impedance (for distributed circuit), that passes AC signals over a certain frequency range while blocking signals at other frequencies. A bandpass filter passes signals over a specified range (f_{low} to f_{hi}), and rejects frequencies outside this range. For example, for a DBS receiver that is to receive satellite transmitted microwave signals in a frequency range of 11 GHz to 12 GHz, a bandpass filter (BPF) would allow signals in this frequency range to pass through with minimum signal loss, while blocking all other frequencies. A low pass filter (LPF) would allow signals to pass with minimum signal loss as long as their frequency was less than a certain cutoff frequency above which significant signal blocking occurs.

Frequency: The repetition rate of a periodic signal used to represent or process a communication signal. Frequency is expressed in units of Hertz (Hz). One Hz represents one cycle per second, 1 MHz represents one million cycles per second, and 1 GHz represents one billion cycles per second.

Gain: Ratio of the output signal over the input signal of a component. See example given with dB definition. For an amplifying device the gain will be greater than 1 when expressed as a ratio, and greater than 0 dB when expressed in decibels.

Image rejection filter: A filter usually placed before a mixer to preselect the frequency that when mixed with the local oscillator frequency, will downconvert to the desired intermediate frequency. In the absence of an image rejection filter, there are two RF frequencies which, when mixed with the local oscillator frequency f_{lo}, will down convert to the intermediate frequency f_{if}. Specifically these are given by $f_{lo} + f_{if}$ and $f_{lo} - f_{if}$. For example, if the local oscillator frequency is 10 GHz and the local oscillator frequency is 1 GHz, then signals at both 11 GHz and 9 GHz will produce an intermediate frequency signal of 1 GHz without an image rejection filter.

Impedance (Z): Electrical property of a network that measures its ability to conduct electrical AC current for a given AC voltage. Impedance is defined as the ratio of the AC Voltage divided by the AC current at a given point in the network. In general, impedance has two parts, a real (resistive) part, and an imaginary (inductive or capacitive reactive) part. Unless the circuit is purely resistive (made up of resistors only), the value of impedance will change with frequency.

Impedance matching: One of the main design activities in microwave circuit design. An impedance matching network is made up of a combination of lumped elements (resistors, capacitors, and inductors), or distributed elements (transmission lines of varying characteristic impedance and length). Impedance matching networks transform network impedance from one value to another. For example, on the input to a low noise transistor, the impedance of an incoming 75 ohm transmission line would be transformed by the input matching network to the impedance Z_{opt}, required to achieve the minimum noise figure of the transistor. The Smith Chart is a tool commonly used by microwave engineers to aid with impedance matching.

Insulator: A nonconducting material, also called a dielectric. An example is Teflon. Insulators are often used as substrates onto which hybrid electrical circuits are constructed.

Ku-Band: Frequency band of approximately 11 to 12 GHz.

L-Band: Frequency band of approximately 1 to 2 GHz.

LNA: Low noise amplifier: Boosts low level radio/microwave signal received without adding substantial distortions to the signal.

Loss: See **attenuation.**

Mask or mask set: A mask defines the geometrical pattern to be used for a single step in the fabrication process of a MMIC. A mask set consists of the dozen or so (varies with process and company) individual masks that are required to complete a MMIC wafer fabrication from start to finish. Examples of masks or mask levels are first level metal (defines all the primary metal structure on the circuit), capacitor top plate (defines the pattern for the metal used to form the top plate of MIM capacitors), and dielectric etch (defines areas where dielectric (insulator) material will be removed after coating the entire wafer with it).

Microstrip: A transmission line commonly used in MMICs, a microstrip consists of a conducting strip suspended over a ground plane on a slab of insulating or semiconducting material. The characteristic impedance of a microstrip line is determined by the width of the line and the height or thickness of the insulating or semiconducting slab. Conceptually, microstrip can be visualized as a transmission line as follows: envision a section of coaxial cable that is sliced lengthwise down to the center conductor. Now, uncurl the line so that the outer shield is laying flat and the center conductor is suspended over the shield by the insulator (now a flat slab). If we now flatten out the center conductor into a strip, a microstrip will be the result. Microstrip is a convenient way to route signals between components in a MMIC.

Microwave: Term used to refer to a radio signal at a very high frequency. One broad definition gives the microwave frequency range as that from 300 MHz to 300 GHz.

MIM Capacitor: Metal-Insulator-Metal Capacitor: (also called a Thin Film Capacitor) Integrated circuit implementation of a common electrical element that stores electric energy (a car battery can be thought of as a big capacitor). Two extreme behaviors of a capacitor are that it will act as an open circuit to low frequencies or DC (zero frequency), and as a short frequency at sufficiently high frequency (how high is determined by the capacitor value).

Mixer: A mixer is a nonlinear device containing either diodes or transistors, the function of which is to combine signals of two different frequencies in such a way as to produce energy at other frequencies. In a typical downconverter application, a mixer has two inputs and one output. One of the inputs is the modulated carrier RF or microwave signal at a frequency f_{rf}, the other is a well-controlled signal from a local oscillator or VCO at a frequency f_{lo}. The result of downconversion is a signal at the difference frequency $f_{rf} - f_{lo}$, which is also called the intermediate frequency f_{if}. A filter is usually connected to the output of the mixer to allow only the desired IF frequency signal to be passed on for further processing. For example, for an RF frequency of 10.95 GHz (=10,950 MHz) and an LO frequency of 10 GHz (=10,000 MHz), the IF frequency would be 950 MHz.

MMIC: Monolithic Microwave Integrated Circuit: The word monolith refers to a single block of stone that does not (in general) permit individual variations. MMICs are made of gallium arsenide (GaAs), silicon, or other semiconducting materials. In a MMIC, all of the components needed to make a circuit (resistors, inductors, capacitors, transistors, diodes, transmission lines) are formed onto a single wafer of material using a series of process steps. Attractive features of MMICs over competing hybrid (combination of two or more technologies) circuits are that a multitude of nearly identical circuits can be processed simultaneously with no assembly (soldering) using batch processing manufacturing techniques. A disadvantage is that circuit adjustment after manufacture is difficult or impossible. As a consequence, significantly more effort is required to use accurate computer-aided design (CAD) techniques to design MMICs that will perform as desired without adjustment. Of course, eventually assembly and packaging of MMICs is performed in order to connect them into a system such as a DBS receiver. MMICs are only cost effective for very high volume applications because the cost of the initial design is very high, as is the cost of wafer manufacture. These costs can only be recovered through high volume manufacture.

Noise: Random perturbations/distortions in signal voltage, current, or power.

Noise figure: Property of a microwave component that describes the amount of noise added to a signal passing through the component. Technically defined as the signal-to-noise ratio at the component input to the signal-to-noise ratio at the component output. For a transistor, the noise figure is highly dependent on the impedance the transistor sees when it looks back at the input matching network from its input (gate for FET) terminal. The minimum noise figure F_{min} is the lowest noise figure that a FET can exhibit under optimum input impedance matching conditions (Z_{opt} or in terms of reflection coefficient G_{opt}). Noise figure is usually specified in decibels.

Package: In MMIC technology, die or chips have to ultimately be packaged to be useful. An example of a package is the T07 "can." The MMIC chip is connected within the can with bond wires connecting from pads on the chip to lead pins on the package. The package protects the chip from the environment and allows easy connection of the chip with other components needed to assemble an entire system, such as a DBS TV receiver.

P1dB: 1 dB Compression power: Like TOI, this gives a measure of the maximum signal power level that can be processed without causing significant signal distortion or saturation effects. Technically, this refers to the power level at the input or the output of a component or system at which the saturation of active devices like transistors causes the gain to be compressed by 1dB from the linear gain.

Reflection coefficient (G): Another way of expressing the impedance. The reflection coefficient is defined as how much signal energy would be reflected at a given frequency. Like impedance, the reflection coefficient will vary with frequency if inductors or capacitors are in the circuit. The reflection coefficient is always defined with respect to a reference or characteristic impedance $(=(Z - Z0)/(Z + Z0))$. For example, the characteristic impedance of one typical TV transmission line is 75 ohms, whereas another type of TV transmission line has a characteristic impedance of 300 ohms. Hooking up a 75 ohm transmission line to a 300 ohm transmission line will result in a reflection coefficient of value $(300 - 75)/(300 + 75) = 0.6$, which means 60% of the energy received from the antenna.

RF: Radio frequency: A general term used to refer to radio signals in the general frequency range from thousands of cycles per second (kHz) to millions of cycles per second (MHz). It is also is often used generically and interchangeably with the term microwave to distinguish the high frequency AC portion of a circuit or signal from the DC bias signal or the downconverted intermediate frequency (IF) signal.

Self bias: A technique employed whereby a transistor only needs a single bias supply voltage between the drain terminal and ground. This is commonly accomplished by placing a parallel combination of a resistor and capacitor between the source terminal and ground.

Semiconductor (or semi-insulator): A material that is partially conducting (can support electrical current flow), but also has properties of an insulator. Common examples are silicon and GaAs. The amount of current conduction that can be supported can be varied by doping the material with appropriate materials that result in the increased presence of free electrons for current flow.

Smith chart: A complicated looking two-dimensional chart used by RF/microwave engineers that allows for impedance to be plotted.

Spiral inductor: Integrated circuit implementation of a common electrical element that stores magnetic energy. Two extreme behaviors of an inductor are that it will act as a short circuit to low frequency or DC energy, and as an open circuit to energy at a sufficiently high frequency (how high is determined by the inductor value). In a MMIC, a spiral inductor is realized by a rectangular or circular spiral layout of a narrow strip of metal. The value of the inductance increases as the number of turns and total length of the spiral is increased. Large spiral inductors are very commonly used as bias chokes to isolate the DC input connection from the RF circuit. Since a large-valued inductor essentially looks like an open circuit to high frequency RF/microwave energy, negligible RF/microwave energy will leak through and interact with the DC bias circuitry.

Third Order Intercept (TOI) Point: This gives a measure of the power level where significant undesired nonlinear distortion of a communication signal will occur. It is related to the maximum signal that can be processed without causing significant problems to the accurate reproduction of the desired information (e.g., TV signal). Technically, the TOI is the hypothetical power in dBm at which the power of the third order intermodulation nonlinear distortion product between two signals input to a component would be equal to the linear extrapolation of the fundamental power.

VCO: Voltage Controlled Oscillator: A device that produces microwave energy at a frequency that is adjustable over a certain range depending on an input DC voltage. An oscillator contains an active device, such as a FET that is connected in such a way as to be susceptible to breaking into oscillation at a frequency that is controlled by a resonant circuit. A voltage-controlled oscillator typically contains a diode that allows the resonant frequency of the resonator to be varied according to the voltage placed across its terminals. Without external stabilization with a high quality factor (low loss) resonant circuit, for example a dielectric resonator, the frequency of a VCO will not be very steady or stable. This will cause unacceptable noise and instability in the received signal.

Via holes: Holes chemically etched from the back of a MMIC wafer and filled with metal in such a way as to allow an electrical connection between the back side of a wafer and the topside of the wafer.

VSWR: Voltage Standing Wave Ratio: Another way of expressing impedance mismatch resulting in signal reflection. With respect to reflection coefficient G (see **Reflection Coefficient**) the VSWR may be expressed mathematically as:

$$\mathrm{VSWR} = \left(1 + /G/\right) / \left(1 - /G/\right)$$

Yield: Percentage of acceptably good chips to the total chips considered at a certain level of a MMIC process. High yield is one of the most important parameters of a cost-efficient process. DC yield refers to the percentage of chips that behave appropriately to the application of DC biasing voltages and currents (see **Bias Voltage or Current**). RF yield refers the percentage of chips that properly process RF/microwave signals.

References

1. Pucel, R. A., Design considerations for monolithic microwave circuits, *IEEE Trans. Microwave Theory and Techniques*, MTT-29, 513–534, June 1981.
2. Pucel, R. A., ed., *Monolithic Microwave Integrated Circuits*, IEEE Press, 1985, 1.
3. Pucel, R.A., ed., *Monolithic Microwave Integrated Circuits*, IEEE Press, 1985, 1–2.
4. Hyltin, T.M., Microstrip transmission on semiconductor substrates, *IEEE Trans. Microwave Theory and Techniques*, MTT-13, 777–781, Nov. 1965.
5. Mehal, E., and Wacker, R., GaAs integrated microwave circuits, *IEEE Trans. Microwave Theory and Tech.*, MTT-16, 451–454, July 1968.
6. Pengelly, R. S., and Turner, J.A., Monolithic broadband GaAs FET amplifiers, *Electron Lett.*, 12, 251–252, May 13, 1976.
7. Pucel, R.A., ed., *Monolithic Microwave Integrated Circuits*, IEEE Press, 1985, 1.
8. Goyal, R., *Monolithic Microwave Integrated Circuits: Technology and Design*, Artech House, Norwood, MA, 1989, 17.
9. Pengelly R., Cree, and Schmitz, Bert, M/A-Com personal communication, August 1999.
10. GaAs fights back against Silicon assault, *Military and Aerospace Electronics*, 27–30, October 18, 1993.

11. Mike Frank, Hewlett Packard Company, personal communication, December 22, 1995.
12. Cressler, J.D. Re-Engineering Silicion: Si-Ge heterojunction bipolar technology. *IEEE Spectrum,* 49–55, March 1995.
13. Pucel, R., Ed., *Monolithic Microwave Integrated Circuits,* IEEE Press, 1985, 2.
14. Reprinted with permission from *Monolithic Microwave Integrated Circuits: Technology and Design,* by R. Goyal, Artech House, Norwood, MA, (www.artech-house.com), 1989, 320.
15. Reprinted with permission from, *MMIC Design GaAs FETs and HEMTs*, by P. Ladbrooke Artech House, Norwood, MA, (www.artech-house.com), 1989, 29.
16. Reprinted with permission from, *Microwave MESFETs and HEMTs*, Edited by J. M. Golio, Ed., Artech House, Norwood, MA, (www.artech-house.com), 1991, 28.
17. Reprinted with permission from *Monolithic Microwave Integrated Circuits: Technology and Design,* by R. Goyal, Artech House, Norwood, MA, (www.artech-house.com), 1989, 113.
18. Reprinted with permission from, *MMIC Design GaAs FETs and HEMTs*, by P. Ladbrooke Artech House, Norwood, MA, (www.artech-house.com), 1989, 92.
19. Reprinted with permission from *Monolithic Microwave Integrated Circuits: Technology and Design,* by R. Goyal, Artech House, Norwood, MA (www.artech-house.com), 1989, 342.
20. Reprinted with permission from, *MMIC Design GaAs FETs and HEMTs*, by P. Ladbrooke Artech House, Norwood, MA, (www.artech-house.com), 1989.
21. Reprinted with permission from *Monolithic Microwave Integrated Circuits: Technology and Design,* by R. Goyal, Artech House, Norwood, MA, (www.artech-house.com), 1989, 331.
22. Reprinted with permission from *Monolithic Microwave Integrated Circuits: Technology and Design,* by R. Goyal, Artech House, Norwood, MA, (www.artech-house.com), 1989, 320–325.
23. Reprinted with permission from *Modern GaAs Processing Methods,* by R. Williams, Artech House, Norwood, MA, (www.artech-house.com), 11–15.

11

Semiconductors

Mike Harris
Georgia Tech Research Institute

11.1 Introduction

Semiconductor is a class of materials that can generally be defined as having an electrical resistivity in the range of 10^{-2} to 10^9 ohm-cm.[1] Addition of a very small amount of impurity atoms can make a large change in the conductivity of the semiconductor material. This unique materials property makes all semiconductor devices and circuits possible. The amount of free charge in the semiconductor and the transport characteristics of the charge within the crystalline lattice determine the conductivity. Device operation is governed by the ability to generate, move, and remove free charge in a controlled manner. Material characteristics vary widely in semiconductors and only certain materials are suitable for use in the fabrication of microwave and RF devices.

Bulk semiconductors, for microwave and RF applications, include germanium (Ge), silicon (Si), silicon carbide (SiC), gallium arsenide (GaAs), and indium phosphide (InP). Electronic properties of these materials determine the appropriate frequency range for a particular material. Epitaxial layers of other important materials are grown on these host substrates to produce higher performance devices that overcome basic materials limitations of homogeneous semiconductors. These specialized semiconductors include silicon germanium, gallium nitride, aluminum gallium arsenide, and indium gallium arsenide among others. Many of the advanced devices described in this book are made possible by the optimized properties of these materials. Through the use of "bandgap engineering," many of the material compromises that limit electronic performance can be overcome with these hetero-structures.

Electron transport properties determine, to a large extent, the frequency at which the various semiconductors are used. On the basis of maturity and cost, silicon will dominate all applications in which it can satisfy the performance requirements. Figure 11.1 is a plot showing the general range of frequencies over which semiconductor materials are being used for integrated circuit applications. It should be noted that the boundary tends to shift to the right with time as new device and fabrication technologies emerge. Discrete devices may be used outside these ranges for specific applications.

This section provides information on important materials properties of semiconductors used for microwave and RF applications. Basic information about each semiconductor is presented followed by

0-8493-1562-X/03/$0.00+$1.50

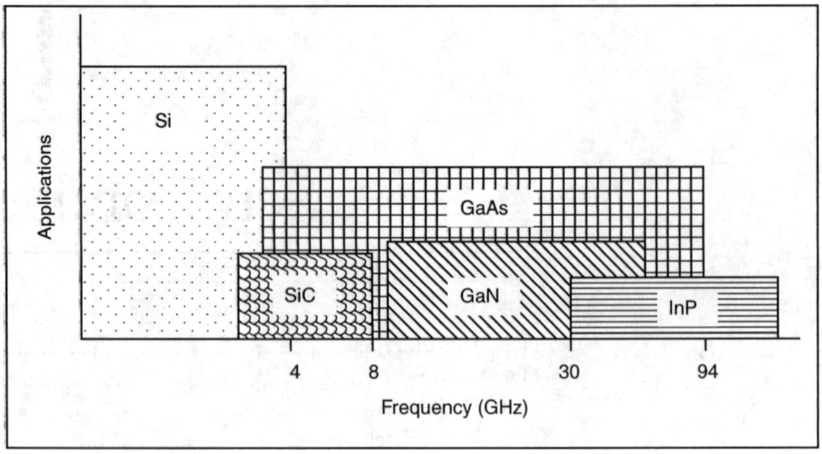

FIGURE 11.1 Frequency range for semiconductor materials.

tables of electronic properties, thermal properties, and mechanical properties. In order to use these materials in microwave and RF applications, devices and circuits must be fabricated. Device fabrication requires etching and deposition of metal contacts and this section provides a partial list of etchants for these semiconductors along with a list of metallization systems that have been used to successfully produce components.

11.2 Silicon

Silicon is an elemental semiconductor that is by far the best known and most mature. A measure of the maturity of silicon is the fact that it is available in 300-mm (12″) diameter wafers. Single crystal silicon wafers are made from electronic grade silicon, which is one of the most refined materials in the world, having an impurity level of no more than one part per billion. Silicon and germanium have the same crystal structure as diamond. In this structure, each atom is surrounded by four nearest neighbor atoms forming a tetrahedron as shown in Figure 11.2. All of the atoms in the diamond lattice are silicon. Silicon is a "workhorse" material at lower frequencies; however, its electron transport properties and low bulk resistivity limit its application in integrated circuit form to frequencies typically below 4 GHz. Silicon-based discrete devices such as PIN diodes find application at higher frequencies.

Si wafers are available with dopant atoms that make them conductive. Wafers having phosphorus impurities are n-type containing excess electrons. Boron-doped wafers are p type and have an excess of holes. Flats are cut on the wafers to distinguish between conductivity types and crystal orientation as shown in Fig. 11.3.[2]

Silicon is an indirect band gap material meaning that when an electron and hole recombine, the energy produced is dissipated as a lattice vibration. This should be compared to material like gallium arsenide that is direct gap. When an electron and hole recombine in GaAs, energy is released in the form of light.

11.3 Gallium Arsenide

Silicon is the most widely used semiconductor to make electronic devices, however, there are compounds that perform functions beyond the physical limits of the electronic properties of silicon. There are many different kinds of compound semiconductor materials, but the most common material combinations used in microwave and RF applications come from the group III and group V elements.

Gallium arsenide has a zincblende crystalline structure and is one of the most important compound semiconductors. Zincblende consists of two interpenetrating, face-centered cubic (fcc) sublattices as seen

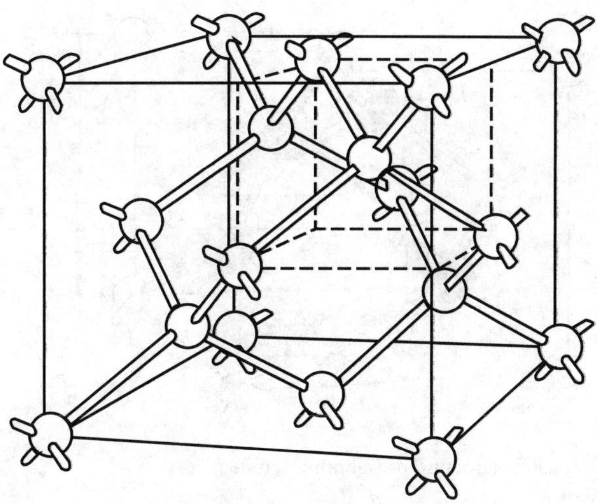

FIGURE 11.2 Cyrstalline structure of Si.

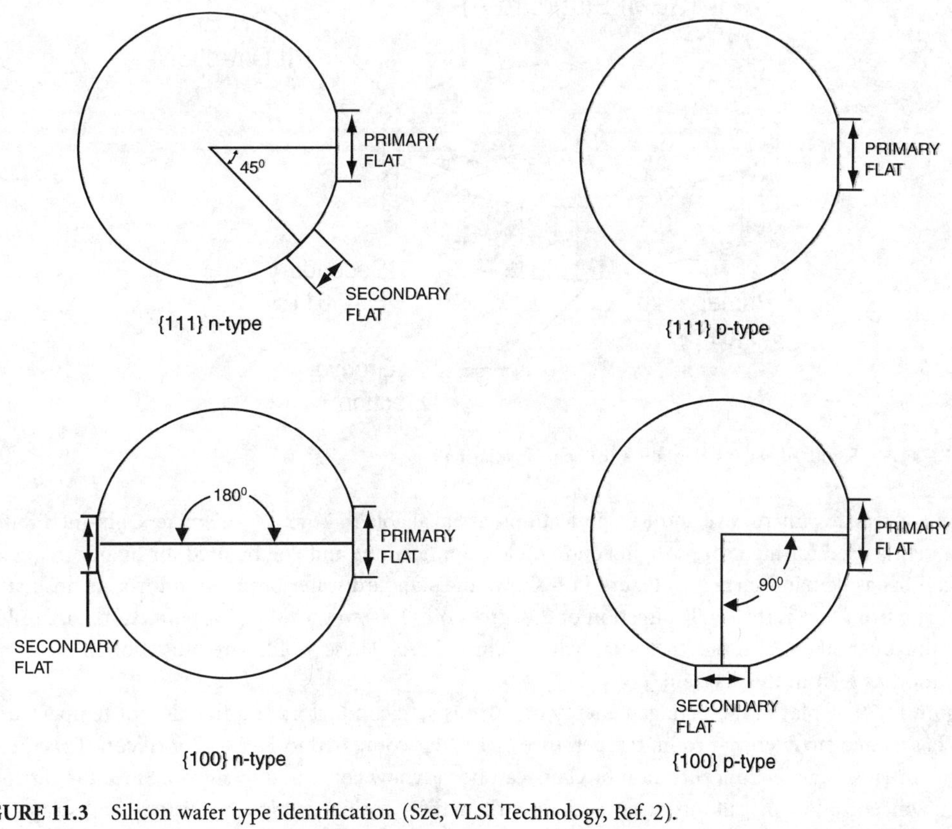

FIGURE 11.3 Silicon wafer type identification (Sze, VLSI Technology, Ref. 2).

in Fig. 11.4. One sublattice is displaced by 1/4 of a lattice parameter in each direction from the other sublattice, so that each site of one sublattice is tetrahedrally coordinated with sites from the other sublattice. That is, each atom is at the center of a regular tetrahedron formed by four atoms of the opposite type. When the two sublattices have the same type of atom, the zincblende lattice becomes the diamond lattice as shown above for silicon. Other examples of compound semiconductors with the zincblende lattice include indium phosphide and silicon carbide.

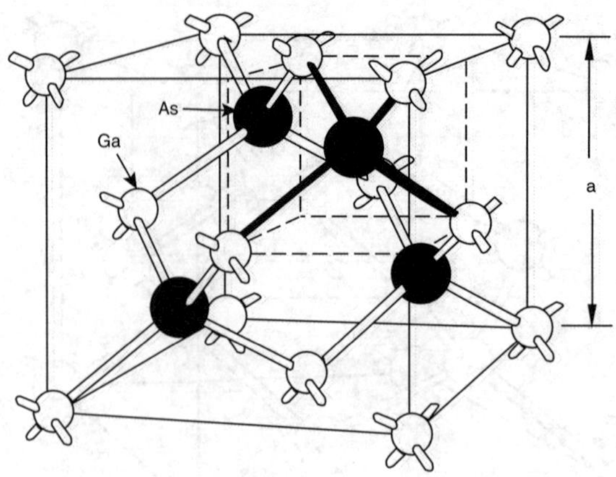

FIGURE 11.4 Zinc blende crystalline structure of gallium arsenide (GaAs).

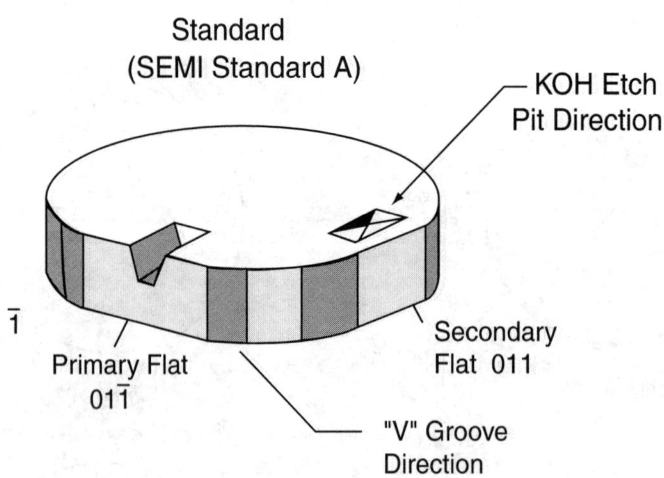

FIGURE 11.5 Standard wafer orientation for semi-insulating GaAs.

GaAs wafers, for microwave and RF applications, are available in 4″ and 6″ diameters. Six-inch wafers are currently used for high speed digital and wireless applications and will be used for higher frequency applications as demand increases. Figure 11.5 shows the standard wafer orientation for semi-insulating GaAs. The front face is the (100) direction or 2 degrees off (100) toward the [110] direction. Figure 11.6 shows the different edge profiles that occur when etching GaAs. These profiles are a function of the crystal orientation as seen in the diagram.[3]

Figure 11.7 is a plot of the bandgap energy of various semiconductors as a function of temperature.[4] GaAs has a bandgap energy at room temperature of 1.43 eV compared to 1.12 eV for silicon. This means that the intrinsic carrier concentration of GaAs can be very low compared to silicon. Since the intrinsic carrier concentration in gallium arsenide is less than that in silicon, higher resistivity substrates are available in gallium arsenide. High resistivity substrates are desirable since they allow the active region of a device to be isolated using a simple ion implantation or mesa etching. Availability of high resistivity gallium arsenide substrates is one reason that this material has found such widespread use for microwave and wireless applications.

The ability to move charge is determined by the transport characteristics of the material. This information is most often presented in the charge carrier velocity electric field characteristic as shown in Figure 11.8.[5] For low values of electric field, the carrier velocity is linearly related to the electric field

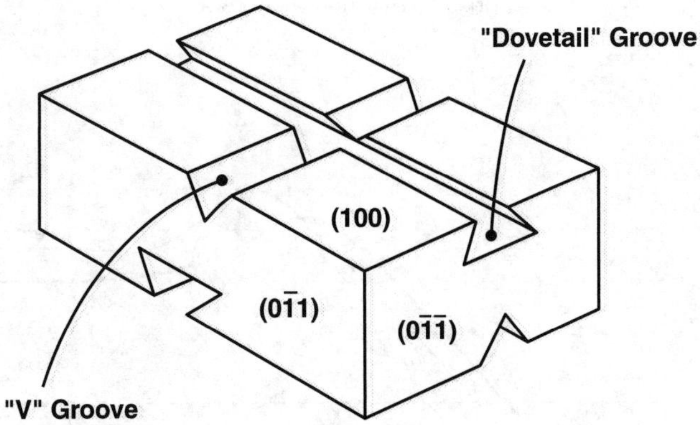

FIGURE 11.6 Orientation-dependent etching profiles of GaAs[3].

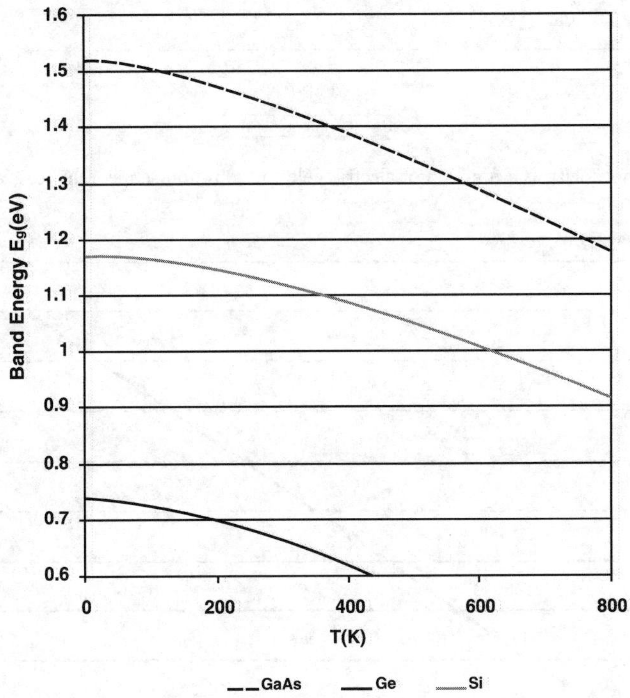

FIGURE 11.7 Energy bandgaps of GaAs, Si, and Ge as a function of temperature (after Sze, Ref. 3).

strength. The proportionality constant is the mobility and this parameter is important in determining the low field operation of a device. Generally a high value of mobility is desired for optimum device performance. Since the mobility of electrons in gallium arsenide is about six times that of silicon, gallium arsenide is a more attractive material for high frequency RF and high-speed digital applications. This mobility advantage along with the availability of high resistivity substrates makes gallium arsenide the preferred and most widely used semiconductor material for these applications.[5]

The relative dielectric constant of GaAs is of critical importance in the design of monolithic microwave integrated circuits (MMICs). This parameter is used to determine the width of transmission lines used to interconnect transistors. The characteristic impedance of the transmission line is a function of the relative dielectric constant, the substrate thickness, and the width of the line. Figure 11.9 shows how the relative dielectric constant behaves with temperature. Microwave and RF designs must accommodate this variation over the expected temperature range of operation.

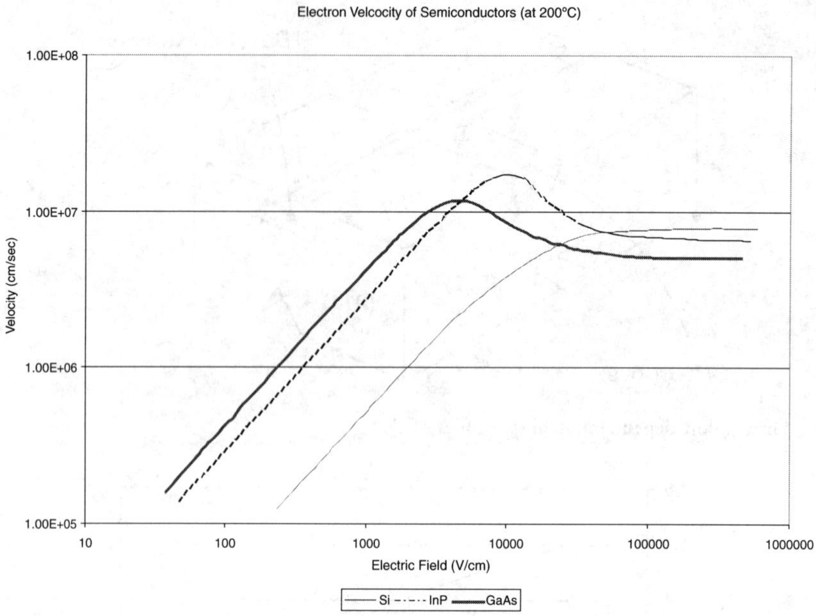

FIGURE 11.8 Electron velocity as a function of electric field for common semiconductors (after Bahl, Ref. 5).

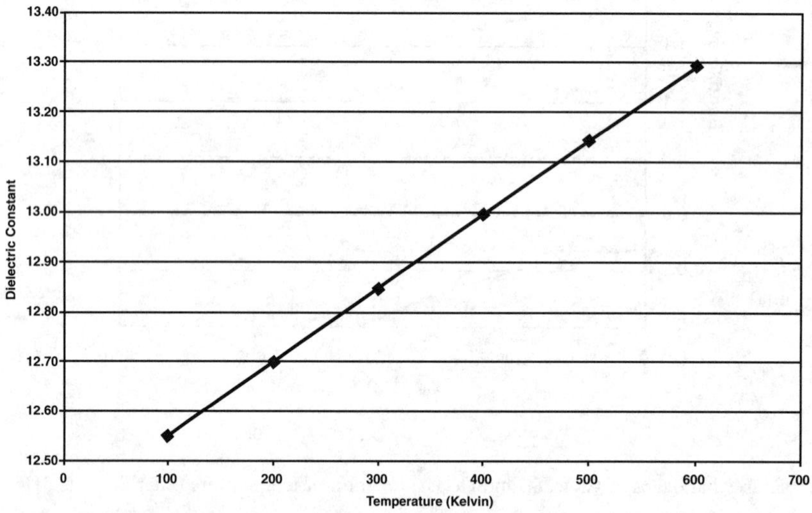

FIGURE 11.9 Relative dielectric constant of GaAs as a function of temperature.

Reliable design and use of GaAs-based MMICs depends on keeping the device channel temperature below a certain absolute maximum level. Channel temperature is defined by the equation,

$$\mathbf{Tch} = \mathbf{Tsink} + \theta \times \mathbf{P}$$

where **Tch** is the channel temperature in (K), and **Tsink** is the heat sink temperature in (K). θ is the thermal resistance defined as L/kA, where L is the thickness of the GaAs substrate, k is the thermal conductivity of GaAs, and A is the area of the channel. **P** is the power dissipated (W). Thermal conductivity is a bulk material property that varies with temperature. Figure 11.10 shows the variation in thermal

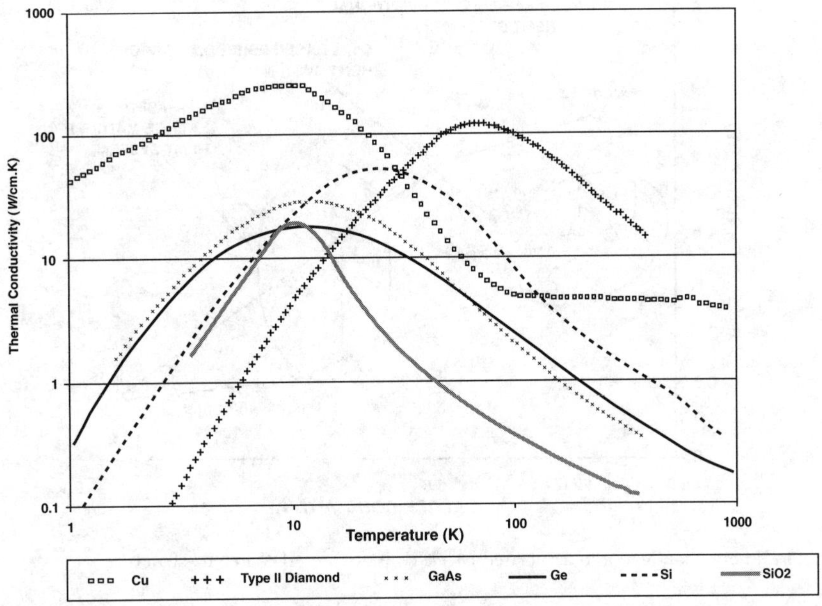

FIGURE 11.10 Thermal conductivity of various materials as a function of temperature (after Sze, Ref. 4).

conductivity as a function of temperature for various semiconductors. At temperatures on the order of 100 K, the thermal conductivity of GaAs approaches that of copper.

11.4 III-V Heterostructures

A new class of high performance materials, based on GaAs and InP, has been developed using advanced epitaxial processes such as molecule beam epitaxy (MBE). These materials have heterojunctions that are formed between semiconductors having different compositions and bandgaps. The bandgap discontinuity can provide significant improvement in the transport properties of the material and allows optimization of device characteristics not possible with homojunctions. In FET devices, the current density can be high while retaining high electron mobility. This is possible because the free electrons are separated from their donor atoms and are confined to a region where the lattice scattering is minimal. Table 11.1 lists the most common hetero-structures for microwave and RF applications. Material compositions vary and are designated by mole fraction using subscripts. In order to match the lattice spacing of GaAs, $Al_{0.25}Ga_{0.75}As$ is routinely used as shown in the diagram of Fig. 11.11. This diagram also shows the compounds that are lattice matched to InP. When a compound is formed with materials that have different lattice spacing, there is strain in the material and under certain conditions, improved performance is possible. This materials system is called pseudomorphic. Figure 11.12 is a cross-section diagram of a double pulsed doped pseudomorphic layer structure used for microwave power transistors.

TABLE 11.1 Common Heterostructures Used for Microwave and RF Applications

$Al_xGa_{1-x}As/GaAs$
$Al_xIn_{x-1}As/InGaAs$
$Al_xGa_{1-x}As/In_yGa_{1-y}As/GaAs$
$InGaAs/InP$
$Al_xIn_{x-1}As/InP$

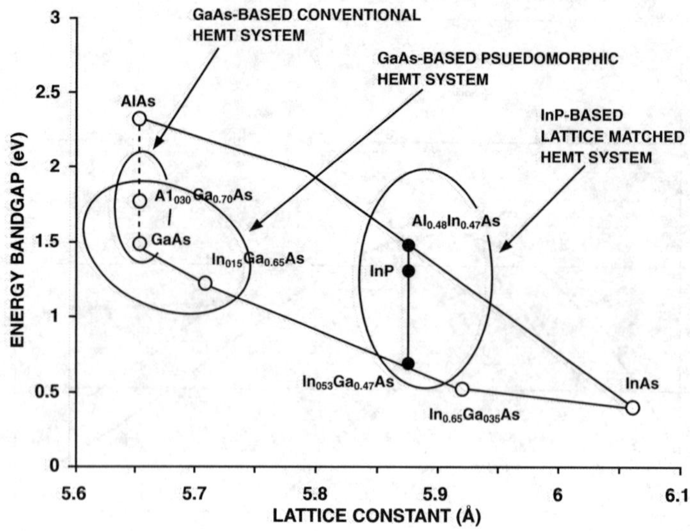

FIGURE 11.11 Energy bandgap and associated lattice constants for II-V hetero-structures.

	Thickness	Dopant	Doping
N^+ GaAs Cap	500 Å	Si	5.0 x 10^{18}cm^{-3}
i AlGaAs Donor	300Å	None	-------
Si Planar Doping	-------	Si	5.0 x 10^{12}cm^{-2}
i AlGaAs Spacer	20 Å	None	--------
i In GaAs Channel	120Å	None	--------
i GaAs Spacer			
Si Planar Doping	-------	Si	1.2 x 10^{12}cm^{-2}
S/L Buffer	1000Å	None	--------
i GaAs Buffer	5000Å	None	--------
GaAs Substrate			

FIGURE 11.12 Double pulsed doped pseudomorphic HEMT layer structure. (From Quantum Epitaxial Designs. With permission.)

11.5 Indium Phosphide

Indium phosphide (InP) is an important compound semiconductor for microwave and RF devices due to its physical and electronic properties. Some of the most important properties of InP are high peak electron velocity, high electric field breakdown, and relatively high thermal conductivity. Three-inch diameter bulk, semi-insulating InP wafers are available and four-inch diameter wafers are being validated. Indium phosphide has a zincblende crystalline structure like gallium arsenide and its lattice constant is 5.8687 angstroms compared to 5.6532 angstroms for GaAs. This materials property is important in the growth of hetero-structures as discussed below. Bulk InP is used in the fabrication of opto-electronic devices but is not used directly for microwave and RF applications. In microwave and RF devices, InP is used as a substrate for epitaxial growth to support GaInAs/AlInAs pseudomorphic high electron mobility transistors. This material system has proved to be a more desirable choice for microwave power amplifiers and millimeter-wave low noise amplifiers.

A fundamental design goal for microwave and RF device engineers, is to achieve the best electron transport properties possible within the reliability, breakdown, and leakage requirements of the application.

Transition from GaAs/AlGaAs high electron mobility transistors to pseudomorphic GaAs/InGaAs/AlGaAs HEMTs resulted in significant improvements in device capability for both low noise and power applications.[6] This was due primarily to the increased electron velocity associated with the smaller electron effective mass in InGaAs compared to GaAs. However, the InAs lattice parameter of about 0.606 nm is considerably larger than the GaAs lattice constant of 0. 565 nm, and due to strain effects the compositional limit of psuedomorphic InGaAs on GaAs substrates is limited to about x = 0.30. One method to achieve higher InAs content in the InGaAs channel is to use an indium phosphide substrate with a lattice constant of 0.587 nm. This newer generation of HEMTs devices uses $In_{0.53}Ga_{0.47}As$ channel lattice matched to InP substrates. InP-based pseudomorphic HEMTs with $In_xGa_{1-x}As$ channel compositions of up to about x = 0.80 have achieved improvements in performance capability for both low noise and power amplifier applications compared to the best GaAs-based devices.

InP-based MMICs require semi-insulating substrates with resistivities from 10^{-6} to 10^{-8} ohm cm. To achieve such resistivities, in nominally undoped crystals would require the residual donor concentration to be reduced by a factor of at least 10^6 from current values. This is not practical and an alternate method is required that employs acceptor doping to compensate the residual donors. In principal, any acceptor can compensate the donors. However, because bulk indium phosphide crystals commonly have a short range variation of at least 5% in donor and acceptor concentration, the maximum resistivity that can be obtained by compensation with shallow acceptors in the absence of p-n junction formation, is only about 15 ohm cm. Resistivities in the semi-insulating range are usually obtained by doping with a deep acceptor such as iron (Fe). Fe is by far the most widely used deep acceptor to produce semi-insulating substrates. As a substitutional impurity, Fe is generally stable under normal device operating temperatures. At high temperatures there is concern of possible diffusion of Fe into the epitaxial layer leading to adverse effects on the devices. Diffusion studies of Fe in InP at temperature of 650°C for four hours indicated virtually no change from the control sample.[7]

Use of InP materials is sometimes restricted by its cost. InP substrates are significantly more costly than those made from GaAs, and even more so when compared to silicon. In addition, the technology of indium phosphide substrate manufacturing is much more difficult than for GaAs or silicon. This situation is not simply the result of lower market demand but is linked fundamentally to the high vapor pressure of phosphorus that creates an obstacle to the synthesis of single crystal boules of larger diameters. While 8-inch silicon substrates and 6-inch GaAs substrates are the rule in commercial fabrication, indium phosphide substrates are still primarily 2 inch. Three-inch diameter wafers are becoming available and 4-inch wafers are being validated. New concepts of single crystal growth may provide larger diameter wafers at low cost leading to wider acceptance of this compound semiconductor for microwave and RF applications.

11.6 Silicon Carbide

Silicon carbide possesses many intrinsic material properties that make it ideal for a wide variety of high power, high temperature, and high frequency electronic device applications. In comparison with silicon and gallium arsenide, SiC has greater than 2.5x larger bandgap, 3x greater thermal conductivity, and a factor of 10 larger breakdown electric field. These characteristics enable SiC devices to operate at higher temperatures and power levels, with lower on-resistances, and in harsh environments inaccessible to other semiconductors. However, electron transport properties may limit SiC applications to frequencies less than 10 GHz. While excellent prototype SiC devices have been demonstrated, SiC devices will not be widely available until material quality improves and material cost drops.[8]

SiC belongs to a class of semiconductors commonly known as "wide bandgap," which makes it, among other things, less sensitive to increased temperatures. Properly designed and fabricated SiC devices should operate at 500°C or higher, a realm that Si does not even approach. Furthermore, the thermal conductivity of SiC exceeds even that of copper; any heat produced by a device is therefore quickly dissipated. The inertness of SiC to chemical reaction implies that devices have the potential to operate even in the most

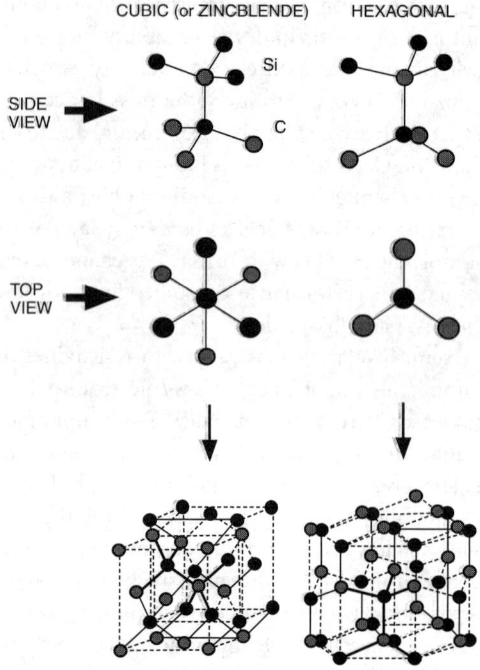

FIGURE 11.13 Difference between the cubic and hexagonal polytypes of SiC. (From Virgil B. Shields, 1994. With permission.)

caustic of environments. SiC is extremely hard and is best known as the grit coating on sandpaper. This hardness again implies that SiC devices can operate under conditions of extreme pressure. SiC is extremely radiation hard and can be used close to reactors or for electronic hardware in space. Properties of particular importance to the microwave and RF device design engineer are high electric field strength and relatively high saturation drift velocity.

SiC exists not as a single crystal type, but as a whole family of crystals known as polytypes. Each crystal structure has its own unique electrical and optical properties.[9,10] Polytypes differ not in the relative numbers of Si and C atoms, but in the arrangement of these atoms in layers as illustrated in Fig. 11.13. The polytypes are named according to the periodicity of these layers. For example, one of the most common polytypes is called 6H, which means a hexagonal type lattice with an arrangement of 6 different Si+C layers before the pattern repeats itself. In total, more than 200 different polytypes of SiC have been shown to exist, some with patterns that do not repeat for hundreds of layers. The exact physical properties of SiC depend on the crystal structure adopted. Some of the most common structures used are 6H, 4H, and 3C, the last being the one cubic form of SiC.

Silicon carbide (SiC) is also emerging as a substrate material that may meet the challenging requirements for GaN growth.[11,12] Type 6H-SiC is lattice matched to within 3.5% of GaN compared to 16% for sapphire.[13] SiC has a thermal conductivity that is over 10 times higher than sapphire.

Figure 11.14 compares the energy band and lattice constants of the conventional III-V semiconductor material systems shown in the circle and the wide gap semiconductors. As shown in the figure, SiC is reasonably lattice matched to GaN and makes a good candidate for a substrate material.

11.7 Gallium Nitride

Gallium nitride in the wurtzite form (2H polytype) has a bandgap of 3.45 eV (near UV region) at room temperature. It also forms a continuous range of solid solutions with AlN (6.28 eV) and a discontinuous range of solid solutions with InN (1.95 eV). The wide bandgap, the heterojunction capability, and the strong atomic bonding of these materials make them good candidates for RF and microwave devices.

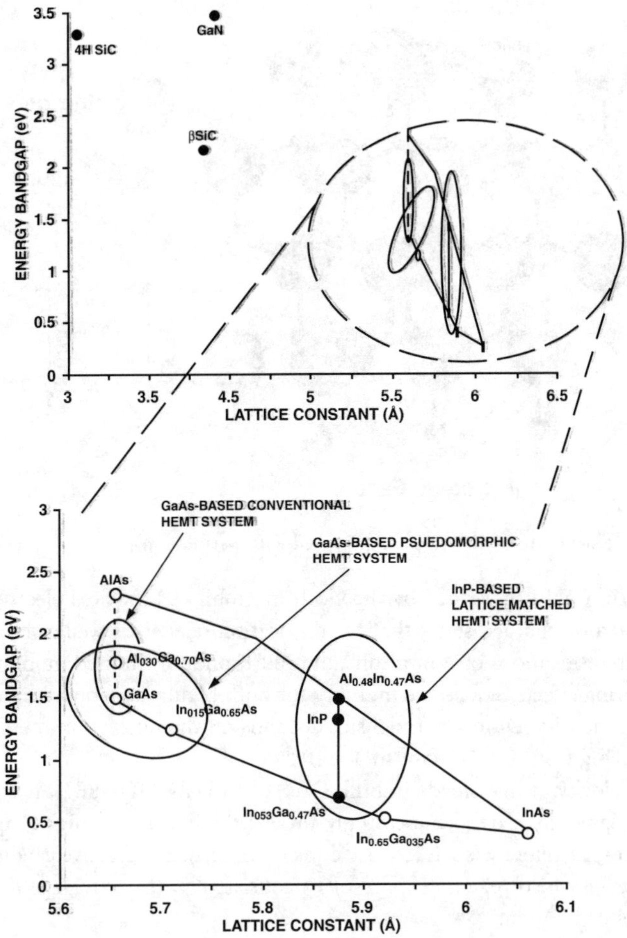

FIGURE 11.14 Comparison of conventional and wide bandgap materials.

Other pertinent device-related parameters include a good thermal conductivity of 1.5 W/cm-K, a type I heterojunction with AlN and AlGaN alloys, large band discontinuities with resultant large interface carrier concentrations, and a large breakdown voltage.

Intrinsic material properties of gallium nitride combined with hetero-structure designs are producing revolutionary improvement in output power for microwave devices. GaN-based FET devices have demonstrated power densities of near 10 W/mm of gate width compared to 1 W/mm for the best GaAs-based devices.[14] GaN has a bandgap of 3.45 eV compared to 1.43 eV for GaAs. This property leads to orders of magnitude lower thermal leakage. GaN has a thermal conductivity almost 3 times higher than GaAs, a parameter that is critically important for power amplifier applications. The dielectric breakdown strength of GaN is a factor of 2 greater than GaAs, further supporting the consideration of this material for microwave power amplifier applications. A key limitation in the development of GaN devices is the fact that there is no native GaN substrate. Currently, GaN must be deposited on host substrates such as sapphire or SiC.

GaN, AlN, and InGaN have a polar wurtzite structure as shown in Fig. 11.15 and epitaxial films of these materials typically grow along the polar axis. Although the polarity of these nitrides has been studied by a number of techniques, many results in the literature are in conflict.[15] The wurtzite lattice can be considered as two interpenetrating hexagonal close-packed lattices. The wurtzite structure has a tetrahedral arrangement of four equidistant nearest neighbors, similar to the GaAs zincblende structure.

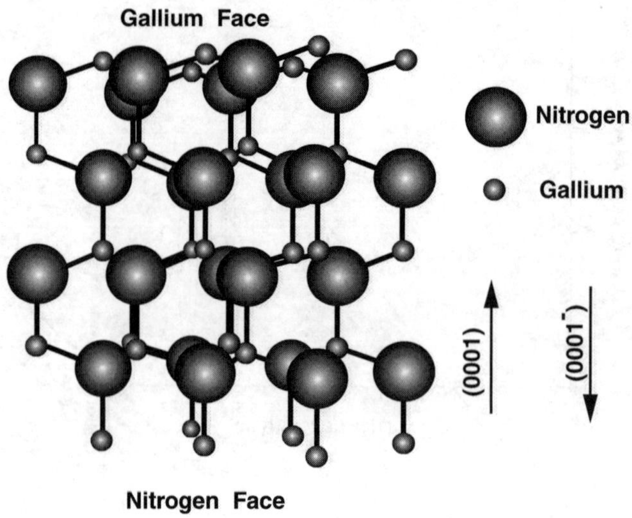

FIGURE 11.15 Wurtzite structure of GaN (Hellman[15] used with permission).

Electron transport in GaN is governed by the electron mobility. Low field electron mobility, in bulk wurtzite GaN, is limited by charged states that are dislocation related, as well as by isolated donor ions and phonons. With no dislocations or donor ions, the 300°K phonon-limited mobility is near 2000 cm^2/V-s. Combined scattering effects reduce the mobility for both highly-doped and lightly-doped material, for a given dislocation density. Dislocation density depends on the nucleation methods used at growth initiation, usually ranging from $3 \times 10^9/cm^2$ to $3 \times 10^{10}/cm^2$.[16]

The average drift velocity at high fields in bulk GaN, or in GaN MODFETs, is expected to be above 2×10^7 cm/s. These high fields extend increasingly toward the drain with higher applied drain-source voltage. The average transit velocity is $>1.25 \times 10^7$ cm/s when only the effective gate length is considered, without the extended high-field region. This result is consistent with an average drift velocity of $\sim 2 \times 10^7$ cm/s over an extended region.

Growth of GaN and its related compounds is dominated by metal-organic vapor phase epitaxy (MOVPE) or MOCVD metal-organic chemical vapor deposition, techniques with obvious advantages in fields where high throughput is required. Molecular beam epitaxy (MBE) is also used to grow GaN films. The III-nitride community grows films of GaN and related nitride materials using hetero-epitaxial growth routes because of the lack of bulk substrates of these materials. This results in films containing dislocations because of the mismatches in the lattice parameters and the coefficients of thermal expansion between the buffer layer and the film and/or the buffer layer and the substrate. These high concentrations of dislocations may also limit the performance of devices.

Advanced growth procedures, including selective area growth (SAG) and lateral epitaxial overgrowth (LEO) techniques for GaN deposition, are being used specifically to significantly reduce the dislocation density. The latter technique involves the initial vertical growth of a configuration and material composition through windows etched in an SiO_2 or Si_3N_4 mask previously deposited on an underlying GaN seed layer and the subsequent lateral and vertical growth of a film over and adjacent to the mask. Reduction in the number of defects in the material and growth of GaN on larger diameter wafers will lead to the production of GaN-based devices and circuits for microwave and higher frequency applications.

11.8 Selected Material Properties

Table 11.2 contains the most common material properties on semiconductors used for microwave and RF applications. Sources of information in this table include textbooks, articles, and World Wide Web sites. Web sites are a convenient way to find materials data and several of the sites used to prepare

TABLE 11.2 Selected Material Properties of Semiconductors for Microwave and RF Applications

Property	Si	SiC	InP	GaAs	GaN
Atoms/cm^3	5.0×10^{22}	3.95×10^{22}	4.43×10^{22}	4.96×10^{22}	
Atomic weight	28.09	40.1	72.90	72.32	41.87
Breakdown Field (V/cm)	3×10^5 [32]	20×10^4 3C-SiC[27] 30×10^5 4H-SiC[27]	5×10^5 [32]	6×10^5	$>10 \times 10^5$
Crystal structure	Diamond	Zincblende	Zincblende	Zincblende	Wurtzite
Density (g/cm^3)	2.3283[23]	4.787[23]	5.316[33]	6.1[23]	
Dielectric constant	11.8[23]	9.75[17] 9.66[18]	12.4[23]	12.5[23]	9[35]
Effective mass m^*/m_0 Electron	1.1	0.37 3C-SiC[19] 0.45 6H-SiC[20]	0.067[23]	0.068[23]	0.22[35,36]
Electron Affinity, eV	4.05[31]	—	4.38[31]	4.07[31]	3.4[34]
Energy Gap (eV) at 300 K	1.107[23]	2.403 3C-SiC[23] 3.101 6H-SiC[23]	1.29[31]	1.35[31]	3.34[37]
Intrinsic carrier concentration (cm^{-3})	1.45×10^{10} [23]	3×10^6 3C-SiC[21] 10^{15}–10^{16} 6H-SiC[22]	1.6×10^7 [23]	1.8×10^6 [23]	$3\text{-}6 \times 10^9$ [23]
Lattice constant (Angstroms)	5.431[31]	4.3596[27]	5.860[31]	5.651[31]	3.190[38]
Linear Coeff. of thermal expansion (10^{-6} K^{-1})	2.49[23]	5.48[22]	4.6[23]	5.4[23]	5.6[27]
Melting point (K)	1685[23]	3070[23]	1335[31]	1511[31]	—
Electron mobility (cm^2/V-S) μ_m	1900[23]	1000 3C-SiC[24] 600 6H-SiC[24]	4600[23]	8800[23]	1000[39]
Holes mobility μ_p (cm^2/V-S)	500[23]	40 3C-SiC[24] 40 6H-SiC[24]	150[23]	400[23]	30[39]
Optical phonon energy (eV)	.063 eV[31]	—	0.43[31]	0.35[31]	.912[40]
Refractive index	3.42[23]	2.65 3C-SiC[25] 2.72 6H-SiC[26]	3.1[31]	3.66[31]	2.7[41] (at band edge)
Resistivity, intrinsic (Ω-cm)	1000[31]	150 3C-SiC[27] $>10^{12}$ 4H-SiC[27]	8.2×10^7 [31]	3.8×10^8 [31]	$>10^{13}$ [27]
Specific heat (J/kg°K)	702[23]	640[28]	310[31]	325[32]	847.39[42]
Thermal conductivity at 300°K (Watt/cm°K)	1.24[23]	3.2 3C-SiC[29] 4.9 6H-SiC[30]	.77[32]	.56[31]	1.3[43]

Table 11.2 are listed in Table 11.3. Most of these sites include detailed references that will provide the user with more extensive information. When using these material properties in the design of components, it is recommended that the user examine the temperature and frequency behavior of the specific property before finalizing the design.

11.9 Etching Processes for Semiconductors

Table 11.4 lists typical etching processes for the various semiconductors described in this chapter.

TABLE 11.3 World Wide Web Sites for Semiconductor Material Properties

http://www.ioffe.rssi.ru/SVA/NSM/Nano/index.html
http://mems.isi.edu/mems/materials/
http://www.webelements.com/webelements/elements/
http://www.sensors-research.com/senres/materials.htm
http://nsr.mij.mrs.org/
http://nina.ecse.rpi.edu/shur/nitride.htm

TABLE 11.4 Etching Processes for Various Semiconductors

Etchant	Substrate and Conditions	Applications	Etch Rate	Ref.
15 HNO_3, 5 CH_3COOH, 2 HF (planar)	Si	For general etching	Variant	44
110 ml CH_3COOH, 100 ml HNO_3, 50 ml HF, 3 g I_2 (iodine etch)	Si	For general etching	Variant	45
KOH solutions; hydrazine (a) KOH (3–50%) (b) 60 hydrazine, 40 H_2O	(100) Si 70–90 C, SiO_2 unmasked 110 C, 10 min, unmasked	For texturing and V grooving solar etch (no texturizing) for diffuse-reflectivity texturizing	Variant	46
28% by weight KOH at 90–95°C	[100] Si, with oxide mask	V-shaped channel etching for optical waveguides	~2 microns/min	47
3CH3OH, 1H3PO4, 1H2O2	[110], [100], Ga [111]GaAs	Preferential structural etching	~2 microns/min, except Ga[111] reduced twofold	48
3 ml HPO_4, 1 ml H_2O_2, 50 ml H_2O	GaAs, 300 K with Shipley 1813 mask	Wet etch	~.1 microns/min	
BCL_3 Gas, reactive ion etch	GaN, BCL_3, 150 W, 10 mTorr	Slow etching, used to obtain a vertical profile for mesa isolation	~.02 microns/min	

TABLE 11.5 Metallization Systems for Ohmic and Schottky Barrier Contacts

Contact	SiC	GaAs	GaN	InP
Ohmic contact mettalization	Ta, Al/Ti, Ni, Ti, Mo, Ta[50]	Au/Ge/Ni, In, Sn, Ag and Ag alloys, Au, Ni[52]	Ti/Al/Ni/Aun-GaN Ni/Aup-GaN[53]	Au and Au alloys, near noble transition metals (Co, Ni, Ti, Pd, Pt)[49]
Schottky gate mettalization	Au and Au alloys[50]	Ti/Pt/Au multilayer structures, Au alloys[51]	Ni/Au/Ti n-GaN Aup-GaN[53]	For AlInP/InP 15% Al superlattice, 20% Al quantum well[52]

11.10 Ohmic and Schottky Contacts

Table 11.5 is a list of metallizations that have been used to successfully fabricate low resistance ohmic contacts for drain and source contacts in FET devices and low leakage Schottky barrier contacts for metal semiconductor diode gates.

References

1. Yu, P.Y. and Cardona M., *Fundamentals of Semiconductors,* New York, Springer-Verlag, 1999, 1.
2. Sze, S.M., Ed., *VLSI Technology,* New York, McGraw-Hill, 1983, 35.
3. Sumitomo Electric III-V Semiconductors Specifications, 32.
4. Sze, S.M., Ed., *Physics of Semiconductors,* New York, John-Wiley & Sons, 1981, 15, 43.
5. Bahl, I. and Bhartia P., *Microwave Solid State Circuit Design*, New York, John Wiley & Sons, 1988, 307.
6. Pearsall, T.P., *Properties, Processing and Applications of Indium Phosphide,* Eisele, H. and Haddad, G.I., Eds., INSPEC, London, 2000, 40.
7. Katz, A., *Induim Phosphide and Related Materials: Processing, Technology and Devices,* Byrne, E.K. and Katz, A., Eds., Boston, Artech House, 1992, 169.
8. Brandes G.R., Growth of SiC Boules and Epitaxial Films for Next Generation Electronic Devices, American Physical Society Centennial Meeting, March 20–26, Atlanta, GA, 1999.

9. von-munch, W., Silicon Carbide, in *Physik der Elemete der* IV. Gruppe und der III-V Verbindungen, K.-H. Hellwege, Ed. Berlin, Heidelberg, Springer-Verlag, 1982, 132–142.
10. Powell, J. A., Pirouz, P., and Choyke, W.J., Growth and Characterization of Silicon Carbide Polytypes for Electronic Applications, in *Semiconductor Interfaces, Microstructures, and Devices: Properties and Applications*, Feng Z.C., Ed. Bristol, United Kingdom, Institute of Physics Publishing, 1993, 257–293.
11. Mohammad, S.N., Salvador A.A and Morkoc Hardel, Emerging Gallium Nitride Based Devices, *Proceedings of the IEEE*, 83, 10, 1306–1355, Oct. 1995.
12. Lin, M.E., Sverdlov, B., Zhou, G.L., and Morkoc, H., A Comparative Study of GaN Epilayers Grown on Sapphire and SiC Substrates by Plasma-assisted Molecular-beam Epitaxy, *Appl. Phys. Lett.*, 62, 26, 3479–3481, 1993.
13. Lin, M.E. et al. Low Resistance Ohmic Contacts on Wide Band-gap GaN, *Appl. Phys. Lett.*, 64, 1003–1005, 1994.
14. Sheppard, S., Doverspike, K., Leonard, M., Pribble, W., Allen, S., and Palmour, J., Improved Operation of GaN/AlGaN HEMTS on Silicon Carbide, International Conference on Silicon Carbide and Related Materials 1999, Paper 349.
15. Hellman, E.S., The Polarity of GaN: a Critical Review. *Materials Research Society Internet Journal*, Vol 3, Res 3, May 19, 1998.
16. Eastman, L., Chu, K., Schaff, W., Murphy, M.,Weimann, M., and Eustis, T., High Frequency AlGaN/GaN MODFET's, *Materials Research Society Internet Journal*, Vol. 2, Res 2, August 12, 1997.
17. Patrick, L., Choyke, W.J., *Physics Review Journal,* 2, 2255, 1977.
18. Choyke, W.J., *NATO ASI Ser. E, Applied Sciences,* 185, and references therin, 1990.
19. Suzuki, A., Ogura A., Furukawa, K., Fujii, Y., Shigeta, M., and Nakajima, S., *Journal of Applied Physics,* 64, 2818, 1988.
20. Wessels, B.W. and Gatos, J.C., *Journal of Physics Chemical Solids*, 15, 83, 1977.
21. Anikin, M.M., Zubrilov, A.S., Lebedev, A.A., Sterlchuck, A.M., and Cherenkov, A.E., *Fitz. Tekh. Polurpovdn.*, 24, 467, 1992.
22. Schackelford, J., Alexander, W., and Parker, J., Eds., *CRC Materials Science and Engineering Handbook*, 2nd ed., Boca Raton, FL, CRC Press, 1994, 304.
23. Lide, D.R., Properties of Semiconductors, *CRC Handbook of Chemistry and Physics*, 12–98, 1999–2000 ed. Boca Raton, FL, CRC Press, 2000.
24. http://vshields.jpl.nasa.gov/windchime.html.
25. Schaffer, P.T.B. and Naum, R.G., *Journal of Optical Society of America*, 59, 1498, 1969.
26. Scaffer, P.T.B., *Applied Optics*, 10, 1034, 1977.
27. Yoder, M.N.,*Wide Bandgap Semiconductor Materials and Devices* , 43, 10, 1634, 1966.
28. Kern, E.L., Hamill, H.W., Deem, H.W., and Sheets, H.D., *Mater Res. Bull.*, 4, 107, 1969.
29. Morelli, D., Hermans, J., Bettz, C., Woo, W.S., Harris, G.L., and Taylor, C., *Inst. Physics Conf. Ser.*, 137, 313–6, 1990.
30. Parafenova, I.I., Tairov, Y.M., and Tsvetkov, V.F., *Sov. Physics-Semiconductors,* 24, 2, 158–61, 1990.
31. http://www.ioffe.rssi.ru/SVA/NSM/Nano/index.html.
32. Garland, C.W. and Parks, K.C., *Journal Applied Physics,* 33, 759, 1962.
33. http://mems.isi.edu/mems/materials/measurements.cgi?MATTAG=galliumarsenidegaas-bulk&PAGE_SIZE=20.
34. Benjamin, M.C., Wang, C., Davis, R.F., and Nemanich, R.J., *Applied Phys. Letter,* 64, 3288, 1994.
35. Mohammad, S.N. and Morkoc, H., *Prog. Quant. Electron.*, 20, 361, 1996.
36. Barker, A.S. and Ilegems, M., *Phys. Rev. B,* 7, 743, 1973.
37. Maruska, H.P. and Tietjen, J.J., *Appl. Phys. Lett.,* 15, 327, 1969.
38. Shur, M.S. and Khan, A.M., *Mat. Res. Bull.*, 22, 2, 44, 1977.
39. Bhapkar, U.V. and Shur, M. S. , *Journal of Applied Physics*, 82, 4, 1649, 1997.
40. Chin, V. W. L., Tansley, T. L., andOsotchan, T., *Journal of Applied Physics*, 75, 7365, 1994.

41. Billeb, A., Grieshhaber, W., Stocker, D., Schubert, E.F., and Karlicek, R.F., *Applied. Physics. Letter,* 69, 2953, 1996.
42. Koshchenko, V.I., Grinberg, Ya, K.H., and Demidienko, A.F., *Inorganic Matter,* 11, 1550–3, 1984.
43. Sichel, E.K. and Pankove, J.I., *J. Phys. Chem. Solids*, 38, 330, 1977.
44. Ruynun, W.R., *Semiconductor Measurements and Instrumentation,* McGraw-Hill, New York, 1975, 129, Table 7.3, Chaps. 1, 2, 7, 9.
45. Integrated Circuit Silicon Device Technology; X-Chemical Metallurgical Properties of Silicon, ASD-TDR-63-316, Vol. X, AD 625, 985. Research Triangle Inst., Research Triangle Park, North Carolina.
46. Baroana, C.R. and Brandhorst, H.W., *IEEE Photovoltaic Spec. Conf. Proc.*, Scottsdale, AZ, 44–48, 1975.
47. http://peta.ee.cornell.edu/~jay/res/vgroove/
48. Merz, J.L. and Logan, R.A., *J. Appl. Physics*, 47, 3503, 1976.
49. Pearshall, T.P., Processing Technologies, in *Properties, Processing, and Applications of Induim Phosphide,* Katz A. and Pearshall, T.P., Eds., Inspec, London, 2000, 246.
50. Harris, G.L., SiC Devices and Ohmic Contacts, in *Properties of SiC,* Harris, G.L, Kelner, G., and Shur, M., Eds., Inspec, London, 1995, 233, 243.
51. Misssous, M., Interfaces and Contacts, in *Properties of Gallium Arsenide*, Morgan, D.V. and Wood, J., Eds., Inspec, London, 1990, 386–387.
52. Lammasniemi, J., Tappura K., and Smekalin K., *Applied Physics Letter,* 65, 20, 2574–5, 1998.
53. Edgar, J. H., Strite, S., Akasaki, I., Amano, H., and Wetzel, C., Specifications, characterisation and applications of GaN based devices, in *Properties, Processing and Applications of Gallium Nitride and Related Semiconductors,* Mohney, S.E., Ed., Inspec, London, 1999, 491–96.

12 Metals

Mike Golio
Motorola

12.1 Introduction

Metals serve several different functions in the realization of RF and microwave products. These functions include:

- The wire or guided wave boundary material for circuits and transmission media.
- The carrier or structural support for dielectric substrates or semiconductor chips.
- The heat sink for devices or circuits that exhibit high power density.
- The reflector element for antennas or screen room applications.

Each of these functions imposes different electrical, thermal, chemical, and mechanical requirements on the metal material selection. Thus the optimum metal for each application will vary. Consideration of a wide range of material properties for each metal is needed to choose an appropriate metal for most applications.

12.2 Resistance, Resistivity, and Conductivity

A first-order consideration in the choice of metals for many electrical applications is the electrical resistance of the metal conductor. DC resistance of a metal rod is given by

$$R = \frac{\rho L}{A} \tag{12.1}$$

where R is the resistance of the rod, ρ is the resistivity of the metal, L is the length of the rod, and A is the cross-sectional area of the rod. The DC electrical properties of metals are also sometimes discussed in terms of conductivity. Conductivity is the inverse of resistivity given by

$$\sigma = \frac{1}{\rho} \tag{12.2}$$

0-8493-1562-X/03/$0.00+$1.50

TABLE 12.1 Electrical Resistivity at Room Temperature of Several Metals in 10^{-8} Ω m

Metal	Electrical Resistivity
Aluminum	2.7
Beryllium	34.0
Chromium	12.6
Copper	1.7
Gold	2.3
Lead	21.1
Magnesium	4.5
Manganese	144
Molybdenum	5.5
Nickel	7.1
Palladium	10.7
Platinum	10.7
Silver	1.6

TABLE 12.2 Temperature Coeffictient of Metal Resistivity at Room Temperature in (1/K)

Metal	Temp. Coefficient of Resistivity
Aluminum	4.38×10^{-3}
Copper	3.92×10^{-3}
Silver	3.71×10^{-3}
Gold	3.61×10^{-3}

where σ is the conductivity of the material. For most applications, high concductivity, or conversely, low resistivity, is desirable. The resistivity of a number of metals is listed in Table 12.1.

The resistivity of metal is also a function of temperature. For small perturbations in temperature, this temperature dependence may be characterized by the equation

$$\alpha = \frac{1}{\rho}\frac{\partial \rho}{\partial T} \tag{12.3}$$

where α is the temperature coefficient of resistivity. The measured temperature dependence of resistivity changes as a function of the nominal temperature and can vary significantly at temperatures near 0 K, or well above room temperature. For most applications, a low value for the temperature coefficient of resistivity is desirable. Table 12.2 presents the temperature coefficient of resistivity at room temperature for several metals that might be chosen for their low resistivity values.

12.3 Skin Depth

An electromagnetic field can penetrate into a conductor only a minute distance at microwave frequencies. The field amplitude decays exponentially from its surface value according to

$$A = e^{-x/\delta_s} \tag{12.4}$$

where x is the normal distance into the conductor measured from the surface, and δ_s is the skin depth. The skin depth or depth of penetration into a metal is defined as the distance the wave must travel in order to decay by an amount equal to e^{-1} = 0.368 or 8.686 dB. The skin depth δ_s is given by

$$\delta_s = \frac{1}{\sqrt{\pi f \mu \sigma}} \tag{12.5}$$

where f is the frequency, σ is the metal conductivity, and μ is the permeability of the metal given as

$$\mu = \mu_o \mu_r \tag{12.6}$$

with μ_o equal to the permeability of free space and μ_r the relative permeability of the metal. For most metals used as conductors for microwave and RF applications, the relative permeability, $\mu_r = 1$. The relative permeability of ferroelectric materials such as iron and steel are typically on the order of several hundred.

Skin depth is closely related to the shielding effectiveness of a metal since the attenuation of electric field strength into a metal can be expressed as in Eq. (12.4). For static or low frequency fields, the only method of shielding a space is by surrounding it with a high-permeability material. For RF and microwave frequencies, however, a thin sheet or screen of metal serves as an effective shield from electric fields.

Skin depth can be an important consideration in the development of guided wave and reflecting structures for high frequency work. For the best conductors, skin depth is on the order of microns for 1 GHz fields. Since electric fields cannot penetrate very deeply into a conductor, all current is concentrated near the surface. As conductivity or frequency approach infinity, skin depth approaches zero and the current is contained in a narrower and narrower region. For this reason, only the properties of the surface metal affect RF or microwave resistance. A poor conductor with a thin layer of high conductivity metal will exhibit the same RF conduction properties as a solid, high conductivity structure.

12.4 Heat Conduction

One of the uses of metal in the development of RF and microwave parts and modules is as a heat spreader. For many applications that involve high power density electronic components, the efficient removal of heat is of great importance in order to preserve component reliability. Chapter 14 discusses heat transfer fundamentals.

The one-dimensional heat flow equation that applies to metals (as well as other media) is given as

$$q = kA\frac{\partial T}{\partial x} \tag{12.7}$$

where q is the heat flow, k is the thermal conductivity of the metal, A is the cross-sectional area for heat flow, and $\frac{\partial T}{\partial x}$ the temperature gradient across the metal. For applications where good heat sinking characteristics are desired, high thermal conductivity, k, is desirable. Table 12.3 lists thermal conductivity values for several metals.

TABLE 12.3 Thermal Conductivities of Typical Metals (W/m K) at Room Temperature

Metal	Thermal Conductivity
Silver	419
Copper	395
Gold	298
Aluminum	156
Brass	101
Lead	32
Kovar	17

TABLE 12.4 Thermal Coefficient of Linear Expansion of Some of the Materials Used in Microwave and RF Packaging Applications (at Room Temperature, in 10^{-6}/K)

Material	Thermal Coefficient of Expansion
	Dielectrics
Aluminum nitride	4
Alumina 96%	6
Beryllia	6.5
Diamond	1
Glass-ceramic	4–8
Quartz (fuzed)	0.54
	Metals
Aluminum	23
Beryllium	12
Copper	16.5
Gold	14.2
Kovar	5.2
Molybdenum	5.2
Nickel	13.3
Platinum	9
Silver	18.9
	Semiconductors
GaAs	5.9
Silicon	2.6
Silicon Carbide	2.2

12.5 Temperature Expansion

Because RF and microwave components often must operate over a wide range of temperatures, consideration of the coefficient of linear expansion of the metals must be included in making a metal selection for many applications. When temperature is increased, the average distance between atoms increases. This leads to an expansion of the whole solid body with increasing temperature. The changes to the linear dimension of the metal can be characterized by

$$\beta = \frac{1}{l}\frac{\Delta l}{\Delta T} \tag{12.8}$$

where β is called the coefficient of linear expansion, l is the linear dimension of the material, ΔT is the change in temperature, and Δl is the change in linear dimension arising from the change in temperature.

Linear expansion properties of metals are important whenever metallic structures are bonded to other materials in an electronic assembly. When two materials with dissimilar thermal expansion characteristics are bonded together, significant stress is experienced during temperature excursions. This stress can result in one or both of the materials breaking or cracking and this can result in degraded electrical performance or catastrophic failure. The best choice of metals to match thermal linear expansion properties is, therefore, determined by the thermal coefficient of linear expansion of the material that is used with the metal. Kovar, for example, is often chosen as the metal material of preference for use as a carrier when alumina dielectric substrates are used to fabricate RF or microwave guided wave elements. Although Kovar is neither a superior electrical conductor nor a superior thermal conductor, its coefficient of linear expansion is a close match to that of the dielectric material, alumina. Table 12.4 presents the coefficients of linear expansion for several metals as well as other materials that are often used for RF and microwave circuits.

12.6 Chemical Properties

The chemical properties of metals can be especially important in the selection of metals to be used for semiconductor device contacts and in integrated circuits. Metal is used extensively in the development of transistors and ICs. Uses include:

- the contact material to establish ohmic and rectifying junctions,
- the interconnect layers, and
- the material used to fabricate passive components such as inductors and transmission line segments.

For these applications, the chemical properties of the metal when exposed to heat and in contact with the semiconductor material play a significant role in the metal selection criteria. The process of fabricating a semiconductor device often involves hundreds of individual process steps and exposure to significant thermal cycling. The temperature ranges associated with device fabrication will far exceed the environment the final device will be exposed to.

For silicon processes, for example, aluminum (or its alloys) is often the metal chosen for contacts and interconnects. Aluminum has high conductivity, but it is also chosen because it adheres well to silicon and silicon dioxide and because it does not interact significantly with silicon during the thermal cycling associated with processing. In contrast, gold also has high conductivity, but its use is typically avoided in silicon fabrication facilities because gold forms deep levels (traps) in silicon that dramatically degrade device performance. Other metals of specific interest in silicon processing include gallium and antiminide, which are often used in the formation of ohmic contacts.

Because interconnects continue to shrink in size as fabricated devices continue to be scaled down, interconnect resistance has begun to pose significant limitations on the levels of integration that can be achieved. One solution to extend these limits is to use copper rather than aluminum for interconnect metal. Copper's higher conductivity translates directly into improved interconnect performance. Rapid progress is being made in this area.

The formation of good ohmic contacts and Schottky barriers is critical to the fabrication of most GaAs devices. Different metals react chemically in distinct ways when exposed to a GaAs surface and high temperature. Metals such as gold, tin, and zinc tend to form ohmic contacts when placed in contact with a GaAs surface. In contrast, aluminum, titanium, and nickel normally form Schottky barriers on GaAs. In order to obtain optimum electrical conductivity and still produce good contacts, sandwiched layers of different metals are sometimes used. For example, a thick layer of gold is often utilized over layers of titanium and platinum to produce Schottky barrier contacts. When this technique is employed, the titanium resting on the surface of the GaAs forms the Schottky barrier, the platinum serves as a diffusion barrier to keep the gold and titanium from diffusing together, and the gold is used to produce a low resistance connection to the contact pads or remaining IC circuitry.

Certain metals can also react with GaAs to produce undesirable effects. Chromium, for example, produces undesirable deep levels in GaAs that can degrade device performance.

12.7 Weight

Over the past several decades a dominant trend in the development of electronic circuits has been the continued reduction of size and weight. Although much of this progress has been made possible by the continued scaling of semiconductor devices and ICs, metal portions of many electronic assemblies still dominate the weight of the system. Metal density can be an important factor in choosing metals for certain applications. Table 12.5 presents the density in g/cm^3 for several metals of interest.

TABLE 12.5 Density of Several Metals in g/cm^3

Metal	Density
Aluminum	2.7
Beryllium	1.85
Copper	8.93
Gold	19.4
Kovar	7.7
Molybdenum	10.2
Nickel	8.9
Platinum	21.45
Silver	10.5

References

Halliday, D. and Resnick, R., *Fundamentals of Physics*, John Wiley & Sons, New York, 1970.
Schroder, D. K., *Semiconductor Material and Device Characterization*, John Wiley & Sons, New York, 1990.
Plonus, M. A., *Applied Electromagnetics*, McGraw-Hill, New York, 1978.
Elliott, D. J., *Integrated Circuit Fabrication Technology*, McGraw-Hill, New York, 1982.
Collin, R. E., *Foundations for Microwave Engineering*, McGraw-Hill, New York, 1966.
Smith, A. A., *Radio Frequency Principles and Applications*, IEEE Press, New York, 1998.

13

RF Package Design and Development

Jeanne S. Pavio
Motorola SPS

13.1 Introduction

Successful RF and microwave package design involves adherence to a rigorous and systematic methodology in package development together with a multi-disciplined and comprehensive approach. This formal planning process and execution of the plan ultimately insures that the package and product will perform as expected, for the predicted lifetime duration in the customer's system, under the prescribed application conditions.

Probably the first concern is having a thorough and in-depth knowledge of the application and the system into which the microwave component or module will be placed. Once these are understood, then package design can begin. Elements that must be considered do not simply include proper electrical performance of the circuit within the proposed package. Mechanical aspects of the package design must be thoroughly analyzed to assure that the package will not come apart under the particular life conditions. Second, the substrates, components, or die within the package must not fracture or lose connection. Third, any solder, epoxy, or wire connections must be able to maintain their integrity throughout the thermal and mechanical excursions expected within the application. Once these elements are thoroughly investigated, the thermal aspects of the package must be simulated and analyzed to appropriately accommodate heat transfer to the system. Thermal management is probably one of the most critical aspects of the package design because it not only contributes to catastrophic circuit overload and failure in out-of-control conditions, but it could also contribute to reduced life of the product and fatigue failures over time. Thermal interactions with the various materials used for the package itself and within the package may augment mechanical stress of the entire package system, ultimately resulting in failure.

Once proper simulation and analysis have been completed from a mechanical and thermal point of view, the actual package design can be finalized. Material and electrical properties and parameters then become the primary concern. Circuit isolation and electromagnetic propagation paths within the package need to be thoroughly understood. In addition, impedance levels must be defined and designed for input and output to and from the package. New developments in package design systems have paved the way for rapid package prototyping through computer integrated manufacturing systems by tieing the design itself to the machining equipment that will form the package. These systems can prototype a part in

0-8493-1562-X/03/$0.00+$1.50

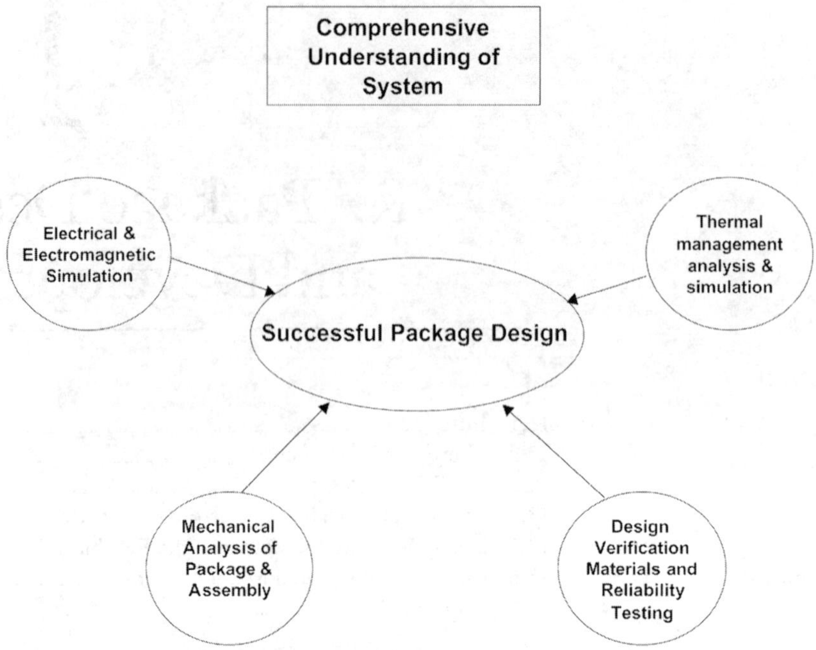

FIGURE 13.1 Elements of successful package design.

plastic for further study or can actually build the prototypes in metal for delivery of prototype samples. Finally, design verification must take place. The verification process typically includes the various long-term reliability tests that gives the designer, as well as his or her customer confidence that the package and its contents will live through the predicted lifetime and application conditions. Other testing may be more specific, such as fracture testing, material properties tests, or precise design tolerance testing. Much of the final testing may also include system-level integration tests. Usually specific power levels are defined and the packages, fully integrated into the system, are tested to these levels at particular environmental conditions.

These are some of the key elements in RF and microwave package development. Although this is not an all-encompassing list, these elements are critical to success in design implementation. These will be explored in the following discussion, hopefully defining a clear path to follow for RF package design and development. Figure 13.1 depicts these key elements leading to successful package design.

13.2 Thermal Management

From an MTBF (mean time before failure) point of view, the thermal aspects of the circuit/package interaction are one of the most important aspects of the package design itself. This can be specifically due to actual heat up of the circuit, reducing lifetime. It may also be due to thermal effects that degrade performance of the materials over time. A third effect may be a materials/heat interaction that causes severe thermal cycling of the materials resulting in stress concentrations and degradation over time.

It is clear that the package designer must have a fully encompassing knowledge of the performance objectives, duty cycles, and environmental conditions that the part will experience in the system environment. The engineer must also understand the thermal material properties within the entire thermal path. This includes the die, the solder or epoxy attachment of that die, the package or carrier base, package system attachment, and material connection to the chassis of the system. There are relatively good databases in the industry that provide the engineer with that information right at his or her fingertips. Among the many are the CINDAS [1] database and the materials' database developed at Georgia Institute of Technology. Other information may be gleaned from supplier datasheets or testing.

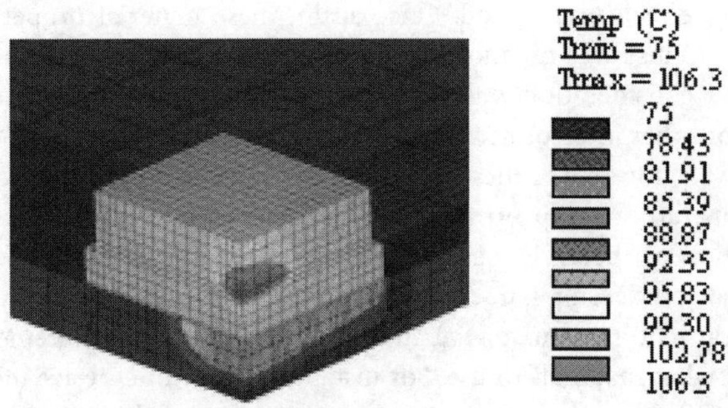

FIGURE 13.2 Ansys output showing thermal gradient across silicon die.

Thermal density within the package, and in particular, at the die level becomes an all-important consideration in the thermal management equation. To insure proper heat transfer and to eliminate any potential thermal failure modes (such as materials breakdown or diffusion and migration), analysis of heat transfer within the die must be completed at the die layout level. This thermal analysis will ultimately be parametrically incorporated into an analysis at the next level up, which may be at the circuit substrate or at the package level itself. The analysis is usually completed with standard finite element simulation techniques present in various software packages available in the industry. Ansys, MSC Nastran, Mechanica, Flowtherm, and Computational Fluid Dynamics (CFD) are some of those available. Material properties that are critical to input into the model would be thermal conductivity and the change in conductivity with temperature. Figure 13.2 shows a typical output of one of these software tools. The FEM simulation uses $^1/_4$ model symmetry. In this particular figure, the analysis demonstrates the thermal gradient across a silicon die, which is an 8 W power amplifier transistor, solder attached to a via structure, with 75°C applied to the bottom of the heat sink. The die junction is at 106.3°C.

It is through such simulated analysis that the entire heat transfer methodology of component to system can be developed. Assuming that there is good correlation between simulated and verified results, the engineer can then gain confidence that the product will have a reasonable lifetime within the specific application. The correlation is typically achieved through the use of infrared microscopy techniques. A number of these infrared microscopes are available in the industry. Usually, the component or module is fixtured on a test station under the infrared camera. The camera is focused on the top surface of the die, which is the heat-generating element. As power is applied to the component or module, the die begins to heat up to a steady-state level. The heat can be measured under RF or DC power conditions. A measurement is done of the die surface temperature. At the same time, a thermocouple impinges on the bottom of the case or package and makes a temperature measurement there. With the maximum die junction temperature (Tjmax in °C), the case temperature (Tc in °C) and the dissipated power in watts, the packaged device junction to case thermal resistance (in °C/W) can be calculated from the following equation:

$$\theta \mathbf{jc} = \left(\mathbf{Tjmax} - \mathbf{Tc}\right) / \mathbf{Pdis}$$

In a fully correlated system, the agreement between simulated and measured results is usually within a few percentage points.

13.3 Mechanical Design

The mechanical design usually occurs concurrently with thermal analysis and heat transfer management. In order to adequately assess the robustness of the package system and its elements, an in-depth understanding of all the material properties must be achieved. In addition to the mechanical properties such as Young's modulus and stress and strain curves for materials, behavior of those materials under thermal loading conditions must be well understood. Once again, these material properties can be found in standard databases in the industry as mentioned above. Typically, the engineer will first insure that the packaging materials under consideration will not cause fracture of the semiconductor devices or of the substrates, solder joints, or other interconnects within the package. This involves knowing the Coefficient of Thermal Expansion (CTE) for each of these materials, and understanding the processing temperatures and the subsequent temperature ramp up that will be experienced under loading conditions in the application. The interaction of the CTEs of various materials may create a mismatched situation and create residual stresses that could result in fracture of any of the elements within the package. The engineer also assesses the structural requirements of the application and weight requirements in order to form appropriate decisions on what materials to use. For instance, a large microwave module may be housed within an iron-nickel (FeNi) package that sufficiently addresses all of the CTE concerns of the internal packaged elements. However, this large, heavy material might be inappropriate for an airborne application where a lightweight material such as AlSiC (aluminum silicon carbide) would be more suitable.

It is not sufficient to treat the packaged component or module as a closed structure without understanding and accounting for how this component or module will be mounted, attached, or enclosed within the actual system application. A number of different scenarios come to mind. For instance, in one situation a packaged component may be soldered onto a printed circuit board (PCB) of a wireless phone. Power levels would not be of concern in this situation, but the mechanical designer must develop confidence through simulation that the packaged RF component can be reliably attached to the PCB. He or she also must insure that the solder joints will not fracture over time due to the expansion coefficient of the PCB compared to the expansion of the leads of the package. Finally, the engineer must comprehend the expected lifetime in years of the product. Cost is obviously a major issue in this commercial application. A solution may be found that is perfectly acceptable from a thermo-mechanical perspective, but it may be cost prohibitive for a phone expected to live for three to five years and then be replaced.

Another scenario on the flip side of the same application is the power device or module that must be mounted into a base station. Here, obviously, the thermal aspects of the packaged device become all important. And great pain must be taken to insure that an acceptable heat transfer path is clearly delineated. With the additional heat from the power device and within the base station itself, heat degradation mechanisms are thoroughly investigated both with simulation techqniques and with rigorous testing. It is common for the RF power chains within base station circuits to dissipate 100 to 200 watts each. Since the expected lifetime of base stations may be over fifteen years, it would be a great temptation for a mechanical engineer to utilize optimum heat transfer materials for the package base, such as diamond for instance, with a thermal conductivity of 40.6 W/in°C. A high power device attached to diamond would operate much cooler than a device attached to FeNi or attached to ceramic. Since, over time, it is the heat degradation mechanisms that eventually cause failure of semiconductor devices, a high power die mounted over a diamond heat sink would be expected to have a much longer lifetime than one mounted over iron nickel or over ceramic. However, once again, the cost implications must enter into the equation. Within the multifunctioned team developing the package and the product, a cost trade-off analysis must be done to examine cost comparisons of materials vs. expected lifetimes. The mechanical package designer may develop several simulations with various materials to input into the cost-reliability matrix. It is necessary that such material substitutions can be done easily and effectively in the parametric model that was initially developed.

The mechanical analysis must encompass attachment of the RF or microwave component or module to the customer system. As we have discussed, in a base station, the thermal path is all important. In order to provide the best heat transfer path, engineers may inadvertently shortcut mechanical stress concerns, which then compromise package integrity. An example was a system mounting condition initially created for the eight-watt power device shown in Fig. 13.2. This semiconductor die was packaged on a copper lead frame to which plastic encapsulation was applied. The lead frame was exposed on the bottom side of the device to insure that there would be a good thermal path to the customer chassis. The copper leads were solder attached (using the typical lead-tin, PbSn, solder) to the printed circuit board. At the same time, the bottom of the device was solder attached to a brass heat sink, as shown in

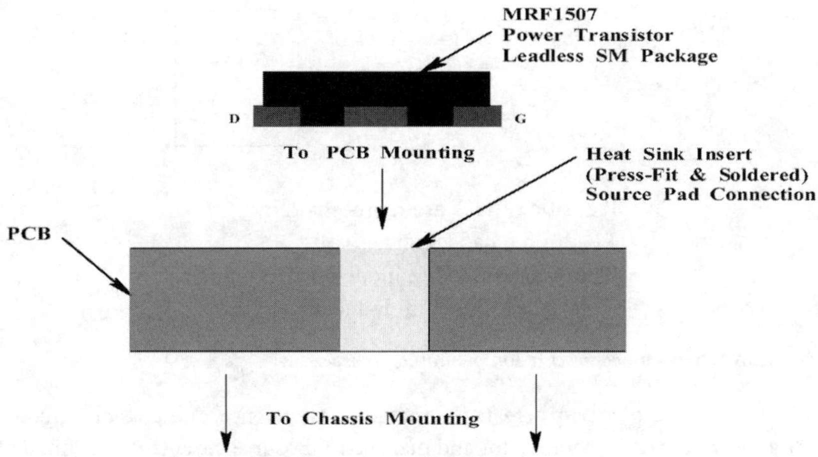

FIGURE 13.3 Eight watt power device attached to brass heat sink.

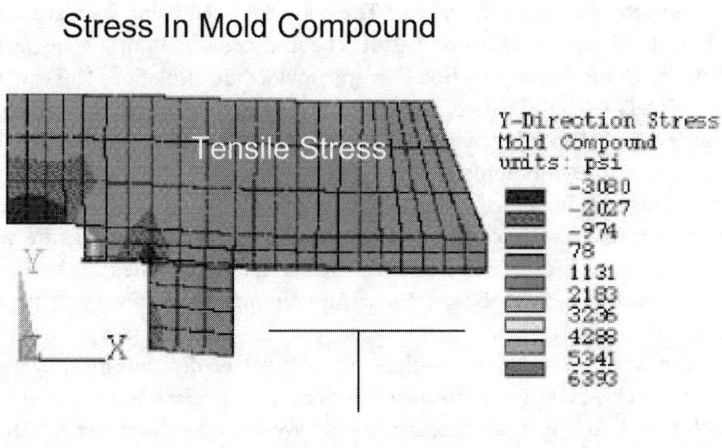

FIGURE 13.4 Modeled stresses in plastic mold compound resulting in failure.

Fig. 13.3, which was then screw mounted to the aluminum chassis to provide thermal transfer to the chassis.

The CTE mismatch of materials resulting in residual stresses during thermal excursions, caused the plastic to rip away from the copper leads. It was the expansion of the aluminum chassis impacting the brass heat sink that created both tensile and shear forces on the leads of the device. The brass heat sink, in effect, became a piston pushing up at the center of the component. The stress levels in the plastic mold compound, which resulted in the failure of the mold/copper interface, can be seen in Fig. 13.4.

Through subsequent simulation, a solution was found that provided the proper heat transfer for the eight-watt device as well as mechanical stability over time and temperature. This was verified through thousands of hours of temperature cycle testing and device power conditioning over temperature excursions.

13.4 Package Electrical and Electromagnetic Modeling

Quite obviously, the electrical design cannot stop at the circuit model for the silicon or GaAs die itself. Particularly at higher frequencies, such as those in the RF or microwave arena, the electromagnetic

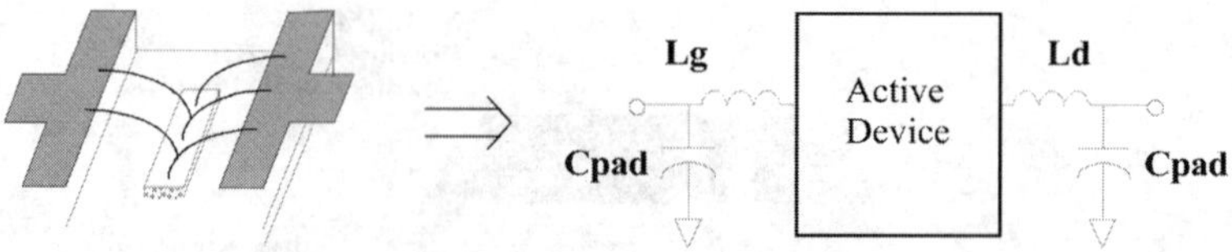

Parasitic effects are represented by:

- Leadframe capacitance, Cpad
- Total gate side wire inductance, Lg
- Total drain side wire inductance, Ld

FIGURE 13.5 Equivalent circuit representation of simple package.

propagation due to all circuit elements create interactions, interference, and possibly circuit oscillations if these electrical effects are not accounted for and managed. Of course the customer's initial requirement will be that a packaged device, component, or module have a specific impedance into and out of their system. Typically, this has been 50 ohms for many microwave systems. It can be achieved through properly dimensioned microstrip input and output leads, through coaxial feeds, or through stripline to microstrip connections that feed into the customer system. These are modeled using standard industry software such as that provided by Hewlett Packard or Ansoft. The next consideration for the package designer is that all of the circuit functions that require isolation are provided that isolation. This can be accomplished through the use of actual metal wall structures within the package. It can also be done by burying those circuit elements in cavities surrounded by ground planes or through the use of solid vias all around the functional elements. These are only some of the predictive means of providing isolation. The need for isolating circuit elements and functions is ascertained by using full wave electromagnetic solvers such as HFSS, Sonnet, or other full wave tools. The EM analysis of the packaged structure will output an S parameter block. From this block, an electrical equivalent circuit can then be extracted with circuit optimization software such as Libra, MDS, ADS, etc. An example of an equivalent circuit representation can be seen in Fig. 13.5.

After proper circuit isolation is achieved within the package, the designer must insure that there will not be inductive or capacitive effects due to such things as wire bonds, leads, or cavities. Wire bonds, if not controlled with respect to length in particular, could have serious inductive effects that result in poor RF performance with respect to things such as gain, efficiency, and intermodulation distortion, etc. In the worst case, uncontrolled wire bonds could result in circuit oscillation. In the same way, RF and microwave performance could be severely compromised if the capacitive effects of the leads and other capacitive elements are not accounted for. These are modeled with standard RF and microwave software tools, and then the materials or processes are controlled to maintain product performance within specifications. Most software tools have some type of "Monte Carlo" analysis capability in which one can alter the material or process conditions and predict the resulting circuit performance. This is especially useful if the processes have been fully characterized and process windows are fully defined and understood. The Monte Carlo analysis then can develop expected RF performance parameters for the characterized process within the defined process windows.

13.5 Design Verification, Materials, and Reliability Testing

After all of the required simulation and package design has been completed, the time has come to begin to build the first prototypes to verify the integrity of the design. During the simulation phase, various material property studies may have been undertaken in order to insure that the correct properties are input into the various models. These may be studies of dielectric constant or loss on a new material, fracture studies to determine when fracture will occur on a uniquely manufactured die or on a substrate, or thermal studies, such as laser flash, to determine the precise thermal conductivity of a material.

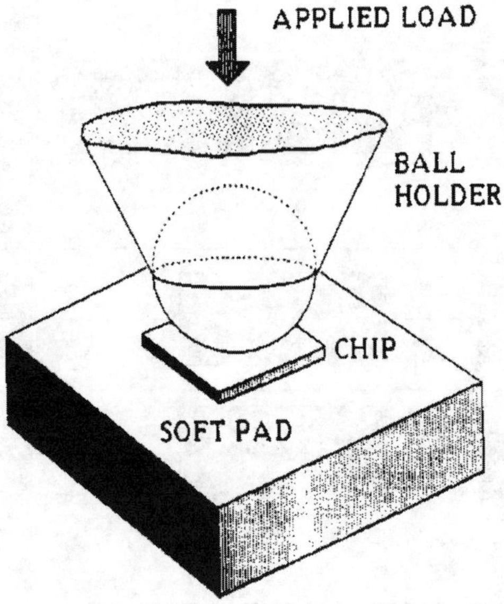

FIGURE 13.6 Technique used to measure fracture strength of semiconductor die.

Figure 13.6 shows one technique used to measure the fracture strength of a GaAs or silicon die. A load is applied to a fixtured sphere, which then impacts the die at a precise force level. From the test, the critical value of the force to break the die is recorded. Then this force is converted to the maximum die stress via the following well-known [5] equation:

$$\sigma_t = \frac{3W}{2\pi m t^2}\left[(m+1)\ln\left(\frac{a}{r_o}\right) + (m-1)\left(1-\left(\frac{r_o}{a}\right)^2\right)\right]$$

After various material properties tests and all simulations and models have been completed, initial prototypes are built and tested. This next phase of tests typically assess long-term reliability of the product through thermal cycling, mechanical shock, variable random vibration, long-term storage, and high temperature and high humidity under biasing conditions. These, as well as other such tests, are the mainstay of common qualification programs. The levels of testing and cycles or hours experienced by the packaged device are often defined by the particular final application or system. For instance, a space-qualified product will require considerably more qualification assessment than a component or module going into a wireless handset that is expected to live 3 to 5 years. The temperature range of assessment for the space-qualified product may span from cryogenic temperatures to +150°C. The RF component for the wireless phone, on the other hand, may simply be tested from 0°C to 90°C.

In high power applications, often part of the reliability assessment involves powering up the device or module after it is mounted to a simulated customer board. The device is powered up and down at a specific duty cycle, through a number of cycles often as the ambient progresses through a series of thermal excursions. This represents what the RF packaged component would experience in the customer system, though usually at an accelerated power and/or temperature condition. Lifetime behavior can then be predicted, using standard prediction algorithms such as Black's equation, depending on the test results.

The RF product and packaging team submit a series of prototype lots through final, standard qualification/verification testing. If all results are positive, then samples are usually given to the customer at this time. These will undergo system accelerated life testing. The behavior of the system through this series of tests will be used to predict expected life cycle.

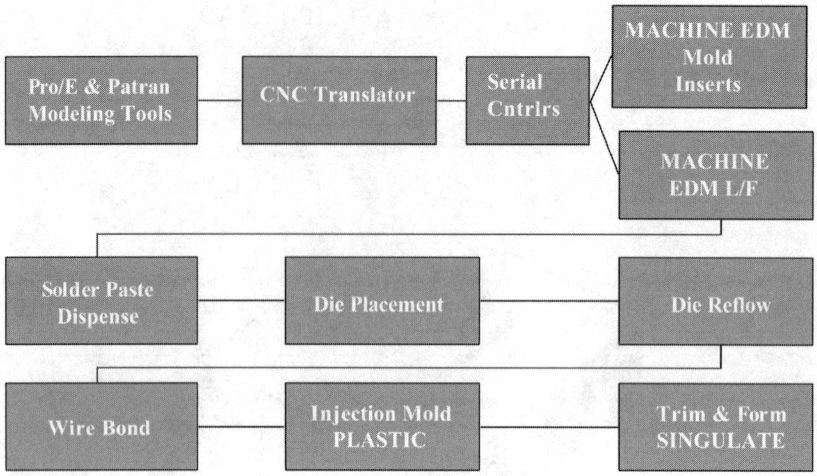

FIGURE 13.7 Rapid prototyping system.

13.6 Computer-Integrated Manufacturing

As mentioned above, there are tools available in the industry that can be used to rapidly develop prototypes directly from the package design files. These prototypes may be constructed of plastic or of various metals for examination and further assessment. Parametric Technologies offers such design and assembly software modules, although they are not by any means the only company with this type of software. The package design is done parameterically in Pro-E so that elements of the design can easily be changed and/or uploaded to form the next higher assembly. The package design elements then go through a series of algorithms to which processing conditions can be attached. These algorithms translate the information into CNC machine code which is used to operate equipment such as a wire EDM for the cutting of metals. Thus, a lead frame is fashioned automatically, in a construction that is a perfect match to the requirements of the die to be assembled. A process flow chart for this rapid prototyping scenario is shown in Fig. 13.7.

Computer-integrated manufacturing is also a highly effective tool utilized on the production floor, once the designed package has been accepted by the customer and is ready for production implementation. Here it is utilized for automated equipment operation, for statistical process control (SPC), for equipment shut down in out-of-control situations, etc. Coupled with neural networks, computer-integrated manufacturing can also be used for advanced automated process optimization techniques.

13.7 Conclusions

The development and design of packages for RF and microwave applications must involve a rigorous and systematic application of the proper tools and methodology to create a design that "works the first time" and every time for the predicted lifetime of the product. This encompasses an in-depth knowledge of the system requirements, the environmental conditions, and the mounting method and materials to be used for package assembly into the customer's system. Then modeling and simulation can take place. Often, in order to understand material properties and to use these more effectively in the models, material studies are done on specific parameters. These are then inserted into electrical, mechanical, and thermal models, which must be completed for effective package design. After a full set of models is completed, verification testing of the design can be done on the first prototypes. Rapid prototyping is made simple through techniques that automatically convert design parameters into machine code for operation of machining equipment. Computer-integrated manufacturing is a highly effective technique that can be utilized at various levels of the product introduction. In package design, it is often used for rapid prototyping and as a tool for better understanding the design. At the production level, it is often used for automated equipment operation and for statistical process control.

Verification testing of the prototypes may include IR scanning to assess thermal transfer. It may include instron testing to test the integrity of a solder interface or of a package construction. It may include power cycling under DC or RF conditions to insure that the packaged design will work in the customer application.

The final phase of assessment is the full qualification of the RF packaged device or module. This certifies to the engineer, and ultimately to the customer that the packaged product can live through a series of thermal cycles, through high temperature and high humidity conditions. It certifies that there will be no degradation under high temperature storage conditions. And it certifies that the product will still perform after appropriate mechanical shock or vibration have been applied. Typically, predictive lifetime assessment can be made using performance to accelerated test conditions during qualification and applying these results to standard reliability equations.

These package design elements, when integrated in a multidisciplined approach, provide the basis for successful package development at RF and microwave frequencies.

References

1. CINDAS = Center for Information and Data Analysis; Operated by Purdue University; Package Materials Database created under SRC (Semiconductor Research Corporation) funding.
2. G. Hawkins, H. Berg, M. Mahalingam, G. Lewis, and L. Lofgran "Measurement of silicon strength as affected by wafer back processing," International Reliability Physics Symposium, 1987.
3. T. Liang, J. Pla, and M. Mahalingam, Electrical Package Modeling for High Power RF Semiconductor Devices, Radio and Wireless Conference, IEEE, Aug. 9-12, 1998.
4. R.J. Roark, *Formulas for Stress and Strain*, 4th Edition, McGraw-Hill, New York, 219.

14

Thermal Analysis and Design of Electronic Systems

Avram Bar-Cohen
University of Minnesota

Karl J. Geisler
University of Minnesota

Allan D. Kraus
Allan D. Kraus Associates

14.1 Motivation

In the thermal control of RF devices, it is necessary to provide an acceptable *microclimate* for a diversity of devices and packages that vary widely in size, power dissipation, and sensitivity to temperature. Although the thermal management of all electronic components is motivated by a common set of concerns, this diversity often leads to the design and development of distinct thermal control systems for different types of electronic equipment. Moreover, due to substantial variations in the performance, cost, and environmental specifications across product categories, the thermal control of similar components may require widely differing thermal management strategies.

The prevention of catastrophic thermal failure (defined as an immediate, thermally induced, total loss of electronic function) must be viewed as the primary and foremost aim of electronics thermal control. Catastrophic failure may result from a significant deterioration in the performance of the component/system or from a loss of structural integrity at the relevant packaging levels. In early microelectronic systems, catastrophic failure was primarily *functional* and thought to result from changes in the bias voltage, *thermal runaway* produced by regenerative heating, and dopant migration, all occurring at elevated transistor junction temperatures. While these failure modes may still occur during the device development process, improved semiconductor simulation tools and thermally compensated devices have largely quieted these concerns and substantially broadened the operating temperature range of today's RF devices.

In microelectronic, microwave, and RF components, the levels of integration and device density on the chips, as well as frequencies of operation, continue to increase. The most critical heat-producing component for most RF systems is the power amplifier (PA) stage. Output power required from these stages ranges from less than 1 watt for some handheld commercial applications to greater than 1 kW

0-8493-1562-X/03/$0.00+$1.50

(multiple parallel stages) for certain military, avionics, and data link applications. Single transistor output power levels are as high as 100 W to 200 W for applications ranging from commercial base stations to avionics, satellite communications, and military. Amplified efficiencies for the highest output power requirements are typically in the 15 to 35% range. To facilitate effective thermal management for such high power levels, PA operation must be pulsed with low duty cycles (reducing thermal power dissipation requirements). Improved thermal performance can be translated into higher duty cycles and therefore into greater data transfer or more efficient use of bandwidth. For these kinds of applications, performance is already limited primarily by the maximum achievable heat flux. Improvements in that figure of merit automatically and immediately translate into improved system performance.

More generally, however, thermal design is aimed at preventing thermally induced physical failures through reduction of the temperature rise above ambient and minimization of temperature variations within the packaging structure(s). With RF integrated circuits or discrete RF high-performance devices, maximum frequency of operation, noise figure, power saturation levels, and nonlinear behavior are all affected by temperature. The use of many low-temperature materials and the structural complexity of chip packages and printed circuit boards has increased the risk of catastrophic failures associated with the vaporization of organic materials, the melting of solders, and thermal-stress fractures of leads, joints, and seals as well as the fatigue-induced delamination and fracture or creep-induced deformation of encapsulants and laminates. To prevent catastrophic thermal failure, the designer must know the maximum allowable temperatures, acceptable internal temperature differences, and the power consumption/dissipation of the various components. This information can be used to select the appropriate fluid, heat transfer mode, and inlet temperature for the coolant and to thus establish the thermal control strategy early in the design process.

After the selection of an appropriate thermal control strategy, attention can be turned to meeting the desired system-level reliability and the target failure rates of each component and subassembly. Individual solid-state electronic devices are inherently reliable and can typically be expected to operate, at room temperature, for some 100,000 years, i.e., with a base failure rate of 1 FIT (failures in 10^9 h). However, since the number of devices in a typical radio component is rapidly increasing and since an RF system may consist of many tens to several hundreds of such components, achieving a system Mean Time Between Failures of several thousand hours in military equipment and 40,000 to 60,000 hours in commercial systems is a most formidable task.

Many of the failure mechanisms, which are activated by prolonged operation of electronic components, are related to the local temperature and/or temperature gradients, as well as the thermal history of the package.[1] Device-related functional failures often exhibit a strong relationship between failure rate and operating temperature. This dependence can be represented in the form of an exponential Arrhenius relation, with unique, empirically determined coefficients and activation energy for each component type and failure mechanism. In the normal operating range of microelectronic components, a 10 to 20°C increase in chip temperature may double the component failure rate, and even a 1°C decrease may then lower the predicted failure rate associated with such mechanisms by 2 to 4%.[2]

Unfortunately, it is not generally possible to characterize thermally induced structural failures, which develop as a result of differential thermal expansion among the materials constituting a microwave package, in the form of an Arrhenius relation. Although these mechanical stresses may well increase as the temperature of the component is elevated, thermal stress failures are, by their nature, dependent on the details of the local temperature fields, as well as the assembly, attachment, and local operating history of the component. Furthermore, thermal stress generation in packaging materials and structures is exacerbated by power transients, as well as by the periodically varying environmental temperatures, experienced by most electronic systems, during both qualification tests and actual operation. However, stress variations in the elastic domain or in the range below the fatigue limit may have little effect on the component failure rate. Consequently, the minimization of elimination of thermally induced failures often requires careful attention to both the temperature and stress fields in the electronic components and necessitates the empirical validation of any proposed thermostructural design criteria.

14.1.1 Thermal Packaging Options

When the heat flux dissipated by the electronic component, device, or assembly is known and the allowable temperature rise above the local ambient condition is specified, the equations of the following sections can be used to determine which heat transfer process or combination of processes (if any) can be employed to meet the desired performance goals. Figure 14.1 shows the variation of attainable temperature differences with surface heat flux for a variety of heat transfer modes and coolant fluids.

Examination of Fig. 14.1 reveals that for a typical allowable temperature difference of 60°C between the component surface and the ambient, "natural" cooling in air — relying on both free convection and radiation — is effective only for heat fluxes below approximately 0.05 W/cm^2. Although forced convection cooling in air offers approximately an order-of-magnitude improvement in heat transfer coefficient, this thermal configuration is unlikely to provide heat removal capability in excess of 1 W/cm^2 even at an allowable temperature difference of 100°C.

To facilitate the transfer of moderate and high heat fluxes from component surfaces, the thermal designer must choose between the use of finned, air-cooled heat sinks and direct or indirect liquid cooling. Finned arrays and sophisticated techniques for improving convective heat transfer coefficients can extend the effectiveness of air cooling to progressively higher component heat fluxes, but often at ever-increasing weight, cost, and volume penalties. Alternately, reliance on heat transfer to liquids flowing at high velocity through so-called "cold plates" can offer a dramatic improvement in the transferable heat flux even at temperature differences as low as 10°C, when the conduction resistance in the cold plate wall is negligible.

A similar high heat flux capability is offered by boiling heat transfer to perfluorinated (FCs) and hydrofluoroether (HFEs) liquids. The high dielectric strength and low dielectric constant of these liquids make it possible to implement this approach for a wide range of components. Direct liquid contact allows the removal of component heat fluxes in excess of 10 W/cm^2 with saturated pool boiling at temperature differences typically less than 20°C. Natural convection (i.e., non-boiling) immersion cooling can also offer significant advantages and, as seen in Fig. 14.1, serves to bridge the gap between direct air cooling and cold plate technology.

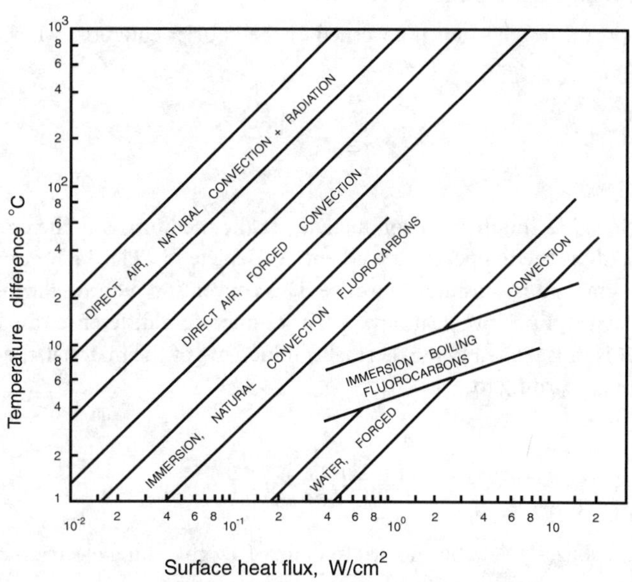

FIGURE 14.1 Temperature differences attainable as a function of heat flux for various heat transfer modes and various coolant fluids. (From Ref. 3, with permission.)

Unfortunately, when addressed within stringent cost targets, the cooling requirements of 21st-century microelectronic, microwave, and RF components cannot be met by today's thermal packaging technology. Rather, ways must be sought to improve on currently available technology, to leverage and combine the best features of existing thermal packaging hardware, and to introduce unconventional, perhaps even radical, thermal solutions into the electronic product family. In so doing, attention must be devoted to three primary issues:

- Highly effective air cooling — removing dissipated power from one or several high-performance components within minimal volumes and with low air-side pressure drops.
- Heat spreading — transporting heat from the relatively small area of the device to the substrate, card, or board, or to a relatively large heat sink or cold plate base.
- Interfacial heat transfer — transferring heat across the thermal resistances between the device and the next level of thermal packaging.

Attention now turns to a detailed discussion of basic heat transfer and the determination of the various types of thermal resistances often encountered in electronic equipment.

14.2 Thermal Modeling

To determine the temperature differences encountered in the flow of heat within electronic systems, it is necessary to recognize the relevant heat transfer mechanisms and their governing relations. In a typical system, heat removal from the active regions of the device(s) may require the use of several mechanisms, some operating in series and others in parallel, to transport the generated heat to the coolant or ultimate heat sink. Practitioners of the thermal arts and sciences generally deal with four basic thermal transport modes: conduction, convection, phase change, and radiation.

14.2.1 Conduction Heat Transfer

14.2.1.1 One-Dimensional Conduction

Steady thermal transport through solids is governed by the Fourier equation, which in one-dimensional form, is expressible as

$$q = -kA\frac{\mathrm{d}T}{\mathrm{d}x} \tag{14.1}$$

where q is the heat flow, k is the thermal conductivity of the medium, A is the cross-sectional area for the heat flow, and $\mathrm{d}T/\mathrm{d}x$ is the temperature gradient. As depicted in Fig. 14.2, heat flow produced by a negative temperature gradient is considered positive. This convention requires the insertion of the minus sign in Eq. (14.1) to assure a positive heat flow, q. The temperature difference resulting from the steady state diffusion of heat is thus related to the thermal conductivity of the material, the cross-sectional area, and the path length, L, according to

$$\left(T_1 - T_2\right)_{\mathrm{cd}} = q\frac{L}{kA} \tag{14.2}$$

The form of Eq. (14.2) suggests that, by analogy to Ohm's Law governing electrical current flow through a resistance, it is possible to define a thermal resistance for conduction, R_{cd}, as

$$R_{\mathrm{cd}} \equiv \frac{\left(T_1 - T_2\right)}{q} = \frac{L}{kA} \tag{14.3}$$

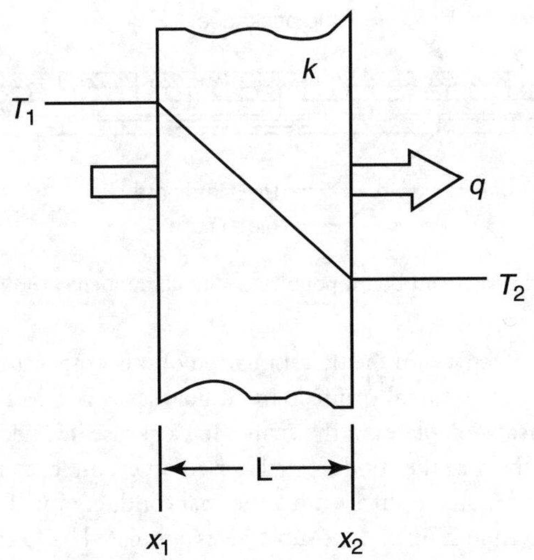

FIGURE 14.2 One-dimensional conduction through a slab. (From Ref. 4, with permission.)

14.2.1.2 One-Dimensional Conduction with Internal Heat Generation

Situations in which a solid experiences internal heat generation, such as that produced by the flow of an electric current, give rise to more complex governing equations and require greater care in obtaining the appropriate temperature differences. The axial temperature variation in a slim, internally-heated conductor whose edges (ends) are held at a temperature T_o, is found to equal

$$T = T_o + q_g \frac{L^2}{2k}\left[\left(\frac{x}{L}\right) - \left(\frac{x}{L}\right)^2\right] \tag{14.4}$$

When the volumetric heat generation rate, q_g, in W/m³, is uniform throughout, the peak temperature is developed at the center of the solid and is given by

$$T_{max} = T_o + q_g \frac{L^2}{8k} \tag{14.5}$$

Alternatively, since q_g is the volumetric heat generation, $q_g = q/LW\delta$, the center-edge temperature difference can be expressed as

$$T_{max} - T_o = q\frac{L^2}{8kLW\delta} = q\frac{L}{8kA} \tag{14.6}$$

where the cross-sectional area, A, is the product of the width, W, and the thickness, δ. An examination of Eq. (14.6) reveals that the thermal resistance of a conductor with a distributed heat input is only one quarter that of a structure in which all of the heat is generated at the center.

In the design of airborne electronic system and equipment to be operated in a corrosive or damaging environment, it is often necessary to conduct the heat dissipated by the components down into the substrate or printed circuit board and, as shown in Fig. 14.3, across the substrate/PCB to a cold plate or sealed heat exchanger. For a symmetrically cooled substrate/PCB with approximately uniform heat dissipation on the surface, a first estimate of the peak temperature at the center of the board, can be obtained using Eq. (14.6).

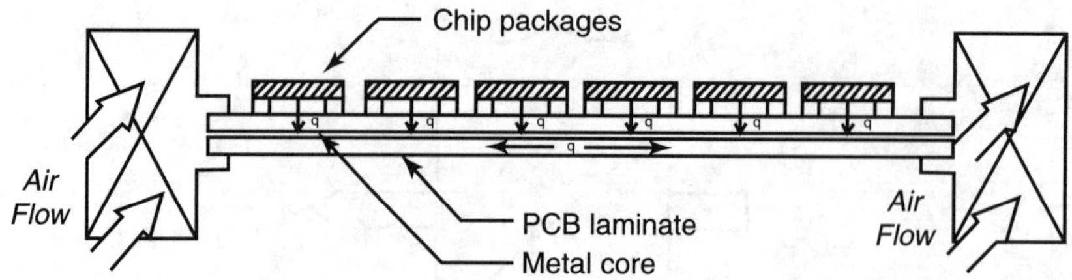

FIGURE 14.3 Edge-cooled printed circuit board populated with components. (From Ref. 4, with permission.)

This relation can be used effectively in the determination of the temperatures experienced by conductively cooled substrates and conventional printed circuit boards, as well as PCBs with copper lattice on the surface, metal cores, or heat sink plates in the center. In each case it is necessary to evaluate or obtain the effective thermal conductivity of the conducting layer. As an example, consider an alumina substrate, 0.20 m long, 0.15 m wide, and 0.005 m thick with a thermal conductivity of 20 W/mK, whose edges are cooled to 35°C by a cold plate. Assuming that the substrate is populated by 15 compounds, each dissipating 2 W, the substrate center temperature will be found to be equal to 85°C when calculated using Eq. (14.6).

14.2.1.3 Spreading Resistance

In component packages that provide for lateral spreading of the heat generated in the device(s), the increasing cross-sectional area for heat flow at successive "layers" adjacent to the device reduces the internal thermal resistance. Unfortunately, however, there is an additional resistance associated with this lateral flow of heat. This, of course, must be taken into account in the determination of the overall component package temperature difference.

For the circular and square geometries common in many applications, Negus et al.[5] provided an engineering approximation for the spreading resistance of a small heat source on a thick substrate or heat spreader (required to be 3 to 5 times thicker than the square root of the heat source area), which can be expressed as

$$R_{sp} = \frac{0.475 - 0.62\epsilon + 0.13\epsilon^3}{k\sqrt{A_c}} \tag{14.7}$$

where ϵ is the ratio of the heat source area to the substrate area, k is the thermal conductivity of the substrate, and A_c is the area of the heat source.

For relatively thin layers on thicker substrates, such as encountered in the use of thin lead frames, or heat spreaders interposed between the device and substrate, Eq. (14.7) cannot provide an acceptable prediction of R_{sp}. Instead, use can be made of the numerical results plotted in Fig. 14.4 to obtain the requisite value of the spreading resistance.

14.2.1.4 Interface/Contact Resistance

Heat transfer across the interface between two solids is generally accompanied by a measurable temperature difference, which can be ascribed to a contact or interface thermal resistance. For perfectly adhering solids, geometrical differences in the crystal structure (lattice mismatch) can impede the flow of photons and electrons across the interface, but this resistance is generally negligible in engineering design. When dealing with real interfaces, the asperities present on each of the surfaces, as shown in an artist's conception in Fig. 14.5, limit actual contact between the two solids to a very small fraction of the apparent interface area. The flow of heat across the gap between two solids in nominal contact is, thus, seen to involve solid conduction in the areas of actual contact and fluid conduction across the "open" spaces. Radiation across the gap can be important in a vacuum environment or when the surface temperatures are high.

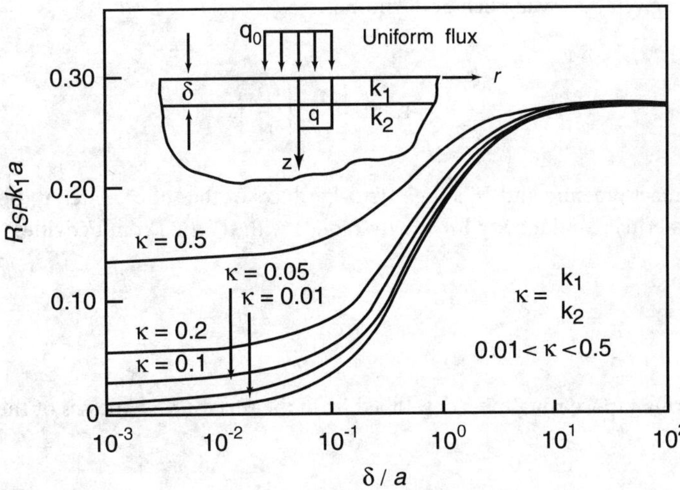

FIGURE 14.4 The thermal resistance for a circular heat source on a two-layer substrate. (From Ref. 6, with permission.)

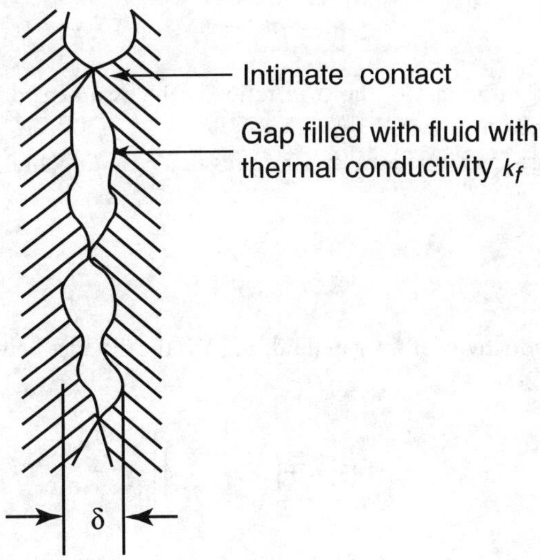

FIGURE 14.5 Physical contact between two nonideal surfaces. (From Ref. 4, with permission.)

The total contact conductance, h_{co}, is taken as the sum of the solid-to-solid conductance, h_c, and the gap conductance, h_g.

$$h_{co} = h_c + h_g \tag{14.8}$$

The contact resistance based on the apparent contact area, A_a may be defined as

$$R_{co} \equiv \frac{1}{h_{co} A_a} \tag{14.9}$$

In Eq. (14.8), h_c is given by Yovanovich and Antonetti[6] as

$$h_c = 1.25k_s\left(\frac{m}{\sigma}\right)\left(\frac{P}{H}\right)^{0.95} \tag{14.10}$$

where P is the contact pressure and H is the micro-hardness of the softer material (both in Pa), k_s is the harmonic mean thermal conductivity for the two solids with thermal conductivities, k_1 and k_2,

$$k_s = \frac{2k_1k_2}{k_1 + k_2}$$

σ is the effective rms surface roughness developed from the surface roughnesses of the two materials, σ_1 and σ_2,

$$\sigma = \sqrt{\sigma_1^2 + \sigma_2^2}$$

and m is the effective absolute surface slope composed of the individual slopes of the two materials, m_1 and m_2,

$$m = \sqrt{m_1^2 + m_2^2}$$

In the absence of detailed information, the σ/m ratio can be assumed to fall into the range of 5 to 9 microns for relatively smooth surfaces.[7]

For normal interstitial gases around atmospheric pressure, h_g in Eq. (14.8) is given by

$$h_g = \frac{k_g}{Y} \tag{14.11}$$

where k_g is the thermal conductivity of the gap fluid, and Y is the distance between the mean planes given by

$$Y = 1.185\left[-\ln\left(3.132\frac{P}{H}\right)\right]^{0.547}\sigma$$

Equations (14.10) and (14.11) can be added, and then in accordance with (14.9), the total contact resistance becomes

$$R_{co} \equiv \left\{\left[1.25k_s\left(\frac{m}{\sigma}\right)\left(\frac{P}{H}\right)^{0.95} + \frac{k_g}{Y}\right]A_a\right\}^{-1} \tag{14.12}$$

14.2.1.5 Transient Heating or Cooling

An externally heated solid of relatively high thermal conductivity, that is experiencing no external cooling, will undergo a constant rise in temperature according to

$$\frac{dT}{dt} = \frac{q}{mc} \tag{14.13}$$

where q is the rate of internal heat generation, m is the mass of the solid, and c is the specific heat of the solid. Equation (14.13) assumes that all of the mass can be represented by a single temperature and this relation is frequently termed the "lumped capacity" solution for transient heating.

Expanding on the analogy between thermal and electrical resistances, the product of mass and specific heat can be viewed as analogous to electrical capacitance and thus to constitute the "thermal capacitance."

When the same solid is externally cooled, the temperature rises asymptotically toward the steady-state temperature, which is itself determined by the external resistance to the heat flow, R_{ex}. Consequently, the time variation of the temperature of the solid is expressible as

$$T(t) = T(t=0) + qR_{ex}\left[1 - e^{-t/mcR_{ex}}\right] \tag{14.14}$$

The lump capacitance model is accurate when the ratio of the internal conduction resistance of a solid to the external thermal resistance is small. This ratio is represented by the Biot number, and the criterion for applicability of the lumped capacitance model is given as

$$\mathrm{Bi} = \frac{hL_c}{k} < 0.1 \tag{14.15}$$

where the characteristic length, L_c, is typically defined as the ratio of the solid's volume to its surface area. More generally, L_c should be taken as the distance over which the solid experiences its maximum temperature difference.[8]

14.2.2 Convective Heat Transfer

14.2.2.1 The Heat Transfer Coefficient

Convective thermal transport from a surface to a fluid in motion can be related to the heat transfer coefficient, h, the surface-to-fluid temperature difference and the "wetted" surface area, A, in the form

$$q = hA(T_s - T_{fl}) \tag{14.16}$$

The differences between convection to a rapidly moving fluid, a slowly flowing or stagnant fluid, as well as variations in the convective heat transfer rate among various fluids, are reflected in the values of h. For a particular geometry and flow regime, h may be found from available empirical correlations and/or theoretical relations. Use of Eq. (14.16) makes it possible to define the convective thermal resistance, as

$$R_{cv} \equiv \frac{1}{hA} \tag{14.17}$$

14.2.2.2 Dimensionless Parameters

Common dimensionless quantities that are used in the correlation of heat transfer data are the *Nusselt number,* Nu, which relates the convective heat transfer coefficient to the conduction in the fluid where the subscript, fl, pertains to a fluid property,

$$\mathrm{Nu} \equiv \frac{h}{k_{fl}/L} = \frac{hL}{k_{fl}} \tag{14.18}$$

the *Prandtl number,* Pr, which is a fluid property parameter relating the diffusion of momentum to the conduction of heat,

$$\mathrm{Pr} \equiv \frac{c_p \mu}{k_{fl}} \tag{14.19}$$

the *Grashof number,* Gr, which accounts for the buoyancy effect produced by the volumetric expansion of the fluid,

$$\mathrm{Gr} \equiv \frac{\rho^2 \beta g L^3 \Delta T}{\mu^2} \tag{14.20}$$

and the *Reynolds number,* Re, which relates the momentum in the flow to the viscous dissipation,

$$\mathrm{Re} \equiv \frac{\rho V L}{\mu} \tag{14.21}$$

14.2.2.3 Natural Convection

Despite increasing performance demands and advances in thermal management technology, direct air-cooling of electronic equipment continues to command substantial attention. Natural convection is the quietest, least expensive, and most reliable implementation of direct fluid cooling. In more demanding systems, natural convection cooling with air is often investigated as a baseline design to justify the application of more sophisticated techniques.

In natural convection, fluid motion is induced by density differences resulting from temperature gradients in the fluid. The heat transfer coefficient for this regime can be related to the buoyancy and the thermal properties of the fluid through the *Rayleigh number,* Ra, which is the product of the Grashof and Prandtl numbers,

$$\mathrm{Ra} = \frac{\rho^2 \beta g c_p}{\mu k_{fl}} L^3 \Delta T \tag{14.22}$$

where the fluid properties, ρ, β, c_p, μ, and k are evaluated at the fluid bulk temperature, and ΔT is the temperature difference between the surface and the fluid.

Empirical correlations for the natural convection heat transfer coefficient generally take the form

$$h = C\left(\frac{k_{fl}}{L}\right)(\mathrm{Ra})^n \tag{14.23}$$

where n is found to be approximately 0.25 for $10^3 < \mathrm{Ra} < 10^9$, representing laminar flow, 0.33 for $10^9 < \mathrm{Ra} < 10^{12}$, the region associated with the transition to turbulent flow, and 0.4 for $\mathrm{Ra} > 10^{12}$ when strong turbulent flow prevails. The precise value of the correlating coefficient, C, depends on fluid, the geometry of the surface, and the Rayleigh number range. Nevertheless, for natural convection in air from common plate, cylinder, and sphere configurations, it has been found to vary in the relatively narrow range of 0.45 to 0.65 for laminar flow and 0.11 to 0.15 for turbulent flow past the heated surface.[3]

Vertical Channels — Vertical channels formed by parallel printed circuit boards (PCBs) or longitudinal fins are a frequently encountered configuration in natural convection cooling of electronic equipment. The historical work of Elenbaas,[9] a milestone of experimental results and empirical correlations, was the first to document a detailed study of natural convection in smooth, isothermal parallel plate channels. In subsequent years, this work was confirmed and expanded both experimentally and numerically by a number of researchers, including Bodoia[10], Sobel et al.,[11] Aung,[12] Aung et al.,[13] Miyatake and Fujii,[14] and Miyatake et al.[15]

These studies revealed that the value of the Nusselt number lies between two extremes associated with the separation between the plates or the channel width. For wide spacing, the plates appear to have little influence upon one another and the Nusselt number in this case achieves its *isolated plate limit.* On the other hand, for closely spaced plates or for relatively long channels, the fluid attains its *fully developed* velocity profile and the Nusselt number reaches its *fully developed limit.* Intermediate values of the Nusselt number can be obtained from a family of composite expressions developed by Bar-Cohen and Rohsenow[16] and verified by comparison to numerous experimental and numerical studies.

For an isothermal channel, at the fully developed limit, the Nusselt number takes the form

$$Nu = \frac{El}{C_1} \tag{14.24}$$

where El is the Elenbaas number, defined as

$$El \equiv \frac{c_p \rho^2 g\beta\left(T_w - T_{amb}\right)b^4}{\mu k L} \tag{14.25}$$

where b is the channel spacing, L is the channel length, and $(T_w - T_{amb})$ is the temperature difference between the channel wall and the ambient, or channel inlet. For an asymmetric channel, or one in which one wall is heated and the other is insulated, the appropriate values of C_1 is 12, while for symmetrically heated channels, $C_1 = 24$.

For an isoflux channel, at the fully developed limit, the Nusselt number has been shown to take the form

$$Nu = \sqrt{\frac{El'}{C_1}} \tag{14.26}$$

where the modified *Elenbaas number, El′*, is defined as

$$El' \equiv \frac{c_p \rho^2 g\beta q'' b^5}{\mu k^2 L} \tag{14.27}$$

where q'' is the heat flux leaving the channel wall(s). When this Nusselt number is based on the maximum wall temperature ($x = L$), the appropriate values of C_1 are 24 and 48 for the asymmetric and symmetric cases, respectively. When based on the mid-height ($x = l/2$) wall temperature, the asymmetric and symmetric C_1 values are 6 and 12, respectively.

In the limit where the channel spacing is very large, the opposing channel walls do not influence each other either hydrodynamically or thermally. This situation may be accurately modeled as heat transfer from an isolated vertical surface in an infinite medium. Natural convection from an isothermal plate can be expressed as

$$Nu = C_2 El^{1/4} \tag{14.28}$$

where McAdams[17] suggests a C_2 value of 0.59. However, more recent research suggests that the available data could be better correlated with $C_2 = 0.515$.[8] Natural convection from an isoflux plate is typically expressed as

$$Nu = C_2 El'^{1/5} \tag{14.29}$$

with a leading coefficient of 0.631 when the Nusselt number is based on the maximum ($x = L$) wall temperature and 0.73 when the Nusselt number is based on the mid-height ($x = l/2$) wall temperature.

Composite Equations — When a function is expected to vary smoothly between two limiting expressions, which are themselves well defined, and when intermediate values are difficult to obtain, an approximate composite relation can be obtained by appropriately summing the two limiting expressions. Using the Churchill and Usagi[18] method, Bar-Cohen and Rohsenow[19] developed composite Nusselt number relations for natural convection in parallel plate channels of the form

$$Nu = \left[\left(Nu_{fd}\right)^{-n} + \left(Nu_{ip}\right)^{-n} \right]^{-1/n} \tag{14.30}$$

where Nu_{fd} and Nu_{ip} are Nusselt numbers for the fully developed and isolated plate limits, respectively. The correlating exponent n was given a value of 2 to offer good agreement with Elenbaas'[9] experimental results.

For an isothermal channel, combining Eqs. (14.24) and (14.28) yields a composite relation of the form

$$Nu = \left[\frac{C_3}{El^2} + \frac{C_4}{\sqrt{El}} \right]^{-1/2} \tag{14.31}$$

while for an isoflux channel, Eqs. (14.26) and (14.29) yield a result of the form

$$Nu = \left[\frac{C_3}{El'} + \frac{C_4}{El'^{2/5}} \right]^{-1/2} \tag{14.32}$$

Values of the coefficients C_3 and C_4 appropriate to various cases of interest appear in Table 14.1. It is to be noted that the tabulated values reflect a change in the value of the coefficient from 0.59 originally used by Bar-Cohen and Rohsenow[19] in the isolated, isothermal plate limit to the more appropriate 0.515 value.

In electronic cooling applications where volumetric concerns are not an issue, it is desirable to space PCBs far enough apart that the isolated plate Nusselt number prevails along the surface. In lieu of choosing an infinite plate spacing, the composite Nusselt number may be set equal to 99%, or some other high fraction of its associated isolated plate value. The composite Nusselt number relation may then be solved for the appropriate channel spacing.

For an isothermal channel, the channel spacing that maximizes the rate of heat transfer from individual PCBs takes the form

$$b_{max} = \frac{C_5}{P^{1/4}} \tag{14.33}$$

where

$$P = \frac{c_p \rho^2 g \beta \left(T_w - T_{amb}\right)}{\mu k L} = \frac{El}{b^4}$$

TABLE 14.1 Appropriate Values for the C_i coefficients appearing in Eqs. (14.24)–(14.39).

Case	C_1	C_2	C_3	C_4	C_5	C_6	C_7
			Isothermal				
Symmetric heating	24	0.515	576	3.77	4.43	0.00655	2.60
Asymmetric heating	12	0.515	144	3.77	3.51	0.0262	2.06
			Isoflux				
Symmetric heating							
Maximum temp.	48	0.63	48	2.52	9.79	0.105	2.12
Midheight temp.	12	0.73	12	1.88	6.80	0.313	1.47
Asymmetric heating							
Maximum temp.	24	0.63	24	2.52	7.77	0.210	1.68
Midheight temp.	6	0.73	6	1.88	5.40	0.626	1.17

while for an isoflux channel, the channel spacing that minimizes the PCB temperature for a given heat flux takes the form

$$b_{\text{max}} = \frac{C_5}{R^{1/5}} \tag{14.34}$$

where

$$R = \frac{c_p \rho^2 g\beta q''}{\mu k^2 L} = \frac{El'}{b^5}$$

Values of the coefficient C_5 appropriate to various cases of interest appear in Table 14.1.

Optimum Spacing — In addition to being used to predict heat transfer coefficients, the composite relations presented may be used to optimize the spacing between plates. For isothermal arrays, the optimum spacing maximizes the total heat transfer from a given base area or the volume assigned to an array of plates or printed circuit boards. In the case of isoflux parallel plate arrays, the total array heat transfer for a given base area may be maximized by increasing the number of plates indefinitely, though the plate will experience a commensurate increase in temperature. Thus, it is more appropriate to define the optimum channel spacing for an array of isoflux plates as the spacing that will yield the maximum volumetric heat dissipation rate per unit temperature difference. Despite this distinction, the optimum spacing is found in the same manner.

The total heat transfer rate from an array of vertical, single-sided plates can be written as

$$\frac{Q_T}{LsWk\Delta t} = \left(\frac{Nu}{b(b+d)}\right) \tag{14.35}$$

where the number of plates, $m = W/(b + d)$, d is the plate thickness, W is the width of the entire array, and s is the depth of the channel. The optimum spacing may be found by substituting the appropriate composite Nusselt number equation into Eq. (14.35), taking the derivative of the resulting expression with respect to b, and setting the result equal to zero. Use of an isothermal composite Nusselt number in Eq. (14.35) yields a relation of the form

$$\left(2b+3d-C_6P^{3/2}b^7\right)_{\text{opt}}=0 \tag{14.36}$$

or

$$b_{\text{opt}}=\frac{C_7}{P^{1/4}}\ (d=0) \tag{14.37}$$

where d, the plate thickness, is negligible. Use of an isoflux composite Nusselt number yields

$$\left(b+3d-C_6R^{3/5}b^4\right)_{\text{opt}}=0 \tag{14.38}$$

or

$$b_{\text{opt}}=\frac{C_7}{R^{1/5}}\ \left(d=0\right) \tag{14.39}$$

Values of the coefficients C_6 and C_7 appropriate to various cases of interest appear in Table 14.1.

Limitations — These smooth-plate relations have proven useful in a wide variety of applications and have been shown to yield very good agreement with measured empirical results for heat transfer from arrays of PCBs. However, when applied to closely spaced printed circuit boards, where the spacing is of the order of the component height, these equations tend to under-predict heat transfer in the channel due to the presence of between-package "wall flow" and the non-smooth nature of the channel surfaces.[20]

14.2.2.4 Forced Convection

For forced flow in long or very narrow parallel-plate channels, the heat transfer coefficient attains an asymptotic value (a fully developed limit), which for symmetrically heated channel surfaces is approximately equal to

$$h=\frac{4k_{\text{fl}}}{d_{\text{e}}} \tag{14.40}$$

where d_e is the *hydraulic diameter* defined in terms of the flow area, A, and the wetted perimeter of the channel, P_w

$$d_{\text{e}}\equiv\frac{4A}{P_{\text{w}}}$$

In the inlet zones of such parallel-plate channels and along isolated plates, the heat transfer coefficient varies with the distance from the leading edge. The low-velocity, or laminar flow, average convective heat transfer coefficient for $\text{Re} < 2\times10^5$ is given by[3]

$$h=0.664\left(\frac{k}{L}\right)\text{Re}^{1/2}\,\text{Pr}^{1/3} \tag{14.41}$$

where k is the fluid thermal conductivity and L is the characteristic dimension of the surface. This heat transfer coefficient decreases asymptotically toward the fully developed value given by Eq. (14.40).

A similar relation applies to flow in tubes, pipes, ducts, channels and/or annuli with the equivalent diameter, d_e, serving as the characteristic dimension in both the Nusselt and Reynolds numbers. For laminar flow, $\text{Re}\le 2100$

$$\frac{\bar{h}d_e}{k} = 1.86\left[\mathrm{Re}\,\mathrm{Pr}\left(\frac{d_e}{L}\right)\right]^{1/3}\left(\frac{\mu}{\mu_w}\right)^{0.14} \tag{14.42}$$

which is attributed to Sieder and Tate[21] and where μ_w is the viscosity of the convective medium at the wall temperature. Observe that Eqs. (14.41) and (14.42) show that the heat transfer coefficient from the surface to the fluid is highest for short channels and decreases as L increases.

In higher velocity turbulent flow, the dependence of the convective heat transfer coefficient on the Reynolds number increases and, in the range $\mathrm{Re} \geq 3 \times 10^5$, is typically given by[3]

$$h = 0.036\left(\frac{k}{L}\right)(\mathrm{Re})^{0.80}(\mathrm{Pr})^{1/3} \tag{14.43}$$

In pipes, tubes, channels, ducts and/or annuli, transition to turbulent flow occurs at an equivalent diameter-based Reynolds number of approximately 10,000. Thus, the flow regime bracketed by

$$2100 \leq \mathrm{Re} \leq 10{,}000$$

is usually referred to as the transition region. Hausen[22] has provided the correlation

$$\frac{hd_e}{k} = 0.116\left[\mathrm{Re} - 125\right](\mathrm{Pr})^{1/3}\left(1 + \frac{d_e}{L}\right)^{2/3}\left(\frac{\mu}{\mu_w}\right) \tag{14.44}$$

and Sieder and Tate[21] give for turbulent flow

$$\frac{hd_e}{k} = 0.23(\mathrm{Re})^{0.80}(\mathrm{Pr})^{1/3}\left(\frac{\mu}{\mu_w}\right) \tag{14.45}$$

Additional correlations for the coefficient of heat transfer in forced convection for various configurations may be found in the heat transfer textbooks.[8,23–25]

14.2.3 Phase Change Heat Transfer

When heat exchange is accompanied by evaporation of a liquid or condensation of a vapor, the resulting flow of vapor toward or away from the heat transfer surface and the high rates of thermal transport associated with the latent heat of the fluid can provide significantly higher heat transfer rates than single-phase heat transfer alone.

14.2.3.1 Boiling

Boiling heat transfer displays a complex dependence on the temperature difference between the heated surface and the saturation temperature (boiling point) of the liquid. In nucleate boiling, the primary region of interest, the ebullient heat transfer rate is typically expressed in the form of the Rohsenow[26] equation

$$q = \mu_f h_{fg}\sqrt{\frac{g(\rho_f - \rho_g)}{\sigma}}\left[\frac{c_{pf}}{C_{sf}Pr_f^{1.7}h_{fg}}\right]^{1/r}(T_S - T_{sat})^{1/r} \tag{14.46}$$

where $1/r$ is typically correlated with a value of 3, and C_{sf} is a function of characteristics of the surface/fluid combination. Rohsenow recommended that the fluid properties in Eq. (14.46) be evaluated at the liquid saturation temperature.

For pool boiling of the dielectric liquid FC-72 (T_{sat} = 56°C at 101.3 kPa) on a plastic-pin-grid-array (PPGA) chip package, Watwe et al.[27] obtained values of 7.47 for $1/r$ and 0.0075 for C_{sf}. At a surface heat flux of 10 W/cm², the wall superheat at 101.3 kPa is nearly 30°C, corresponding to a average surface temperature of approximately 86°C.

The departure from nucleate boiling, or "Critical Heat Flux" (CHF), places an upper limit on the use of the highly efficient boiling heat transfer mechanism. CHF can be significantly influenced by system parameters such as pressure, subcooling, heater thickness and properties, and dissolved gas content. Watwe et al.[27] presented the following equation to predict the pool boiling critical heat flux of dielectric coolants from microelectronic components and under a variety of parametric conditions.

$$\mathrm{CHF}=\left\{\frac{\pi}{24}h_{fg}\sqrt{\rho_g}\left[\sigma_f g\left(\rho_f-\rho_g\right)\right]^{1/4}\right\}\left(\frac{\delta\sqrt{\rho_h c_{ph} k_h}}{\delta\sqrt{\rho_h c_{ph} k_h}+0.1}\right)$$
$$\times\left\{1+\left[0.3014-0.01507L'(P)\right]\right\}\left\{1+0.03\left[\left(\frac{\rho_f}{\rho_g}\right)^{0.75}\frac{c_{pf}}{h_{fg}}\right]\Delta T_{sub}\right\} \tag{14.47}$$

The first term on the right-hand side of Eq. (14.47) is the classical Kutateladze-Zuber prediction, which is the upper limit on the saturation value of CHF on very large horizontal heaters. The second term represents the effects of heater thickness and thermal properties on the critical heat flux. The third term in Eq. (14.47) accounts for the influence of the heater size, where

$$L'=L\sqrt{\frac{g\left(\rho_f-\rho_g\right)}{\sigma_f}} \tag{14.48}$$

This third term is only to be included when its value is larger than unity (i.e., $0.3014 - 0.01507L' > 0$) as small heaters show an increase in CHF over larger heaters. The last term is an equation representing the best-fit line through the experimental data of Watwe et al.[27] and represents the influence of subcooling on CHF. The pressure effect on CHF is embodied in the Kutateladze-Zuber and the subcooling model predictions, which make up Eq. (14.47), via the thermophysical properties. Thus, Eq. (14.47) can be used to estimate the combined influences of various system and heater parameters on CHF. The critical heat flux, under saturation conditions at atmospheric pressure, for a typical dielectric coolant like FC-72 and for a 1-cm component is approximately 15 W/cm². Alternately, at 2 atm and 30°C of subcooling CHF for FC-72 could be expected to reach 22 W/cm².

14.2.3.2 Condensation

Closed systems involving an evaporative process must also include some capability for vapor condensation. Gerstmann and Griffith[28] correlated film condensation on a downward-facing flat plate as

$$\mathrm{Nu}=0.81Ra^{0.193} \qquad 10^{10}>Ra>10^{8} \tag{14.49}$$

$$\mathrm{Nu}=0.69Ra^{0.20} \qquad 10^{8}>Ra>10^{6} \tag{14.50}$$

where,

$$Nu \equiv \frac{h}{k}\left(\frac{\sigma}{g(\rho_f - \rho_g)}\right)^{1/2} \tag{14.51}$$

$$Ra \equiv \frac{g\rho_f(\rho_f - \rho_g)h_{fg}}{k\mu\Delta T}\left(\frac{\sigma}{g(\rho_f - \rho_g)}\right)^{3/2} \tag{14.52}$$

The Nusselt number for laminar film condensation on vertical surfaces was correlated by Nusselt[29] and later modified by Sadasivan and Lienhard[30] as:

$$\mathrm{Nu} = \frac{hL}{k_f} = 0.943\left[\frac{g\Delta\rho_{fg}L^3h'_{fg}}{k_f v_f(T_{sat} - T_c)}\right]^{1/4} \tag{14.53}$$

where

$$h'_{fg} = h_{fg}(1 + C_c Ja)$$

$$C_c = 0.683 - \frac{0.228}{Pr_l}$$

$$Ja = \frac{c_{pf}(T_{sat} - T_c)}{h_{fg}}$$

14.2.3.3 Phase Change Materials

In recent years there has been growing use of solid-liquid phase change materials (PCM) to help mitigate the deleterious effects of transient "spikes" in the power dissipation and/or environmental load imposed on RF modules. This is of particular importance for outdoor modules, where PCMs can serve to smooth diurnal variations in the air temperature and solar radiations. To determine the mass of PCM needed to absorb a specified thermal load at a constant (melting) temperature, it is necessary to obtain the latent heat of fusion of that material and insert it in the following relation

$$m = \frac{Q}{h_{fs}} \tag{14.54}$$

14.2.4 Flow Resistance

The transfer of heat to a flowing gas or liquid that is not undergoing a phase change results in an increase in the coolant temperature from an inlet temperature of T_{in} to an outlet temperature of T_{out}, according to

$$T_{out} - T_{in} = \frac{q}{\dot{m}c_p} \tag{14.55}$$

Based on this relation, it is possible to define an effective flow resistance, R_{fl}, as

$$R_{fl} \equiv \frac{1}{\dot{m}c_p} \tag{14.56}$$

where $\dot{m}$, the mass flow rate, is given in kg/s.

In multicomponent systems, determination of individual component temperatures requires knowledge of the fluid temperature adjacent to the component. The rise in fluid temperature relative to the inlet value can be expressed in a flow thermal resistance, as done in Eq. (14.56). When the coolant flow path traverses many individual components, care must be taken to use R_{fl} with the total heat absorbed by the coolant along its path, rather than the heat dissipated by an individual component. For system-level calculations aimed at determining the average component temperature, it is common to base the flow resistance on the average rise in fluid temperature, that is, one-half the value indicated by Eq. (14.56).

14.2.5 Radiative Heat Transfer

Unlike conduction and convection, radiative heat transfer between two surfaces or between a surface and its surroundings is not linearly dependent on the temperature difference and is expressed instead as

$$q = \sigma AF\left(T_1^4 - T_2^4\right) \tag{14.57}$$

where F includes the effects of surface properties and geometry and σ is the Stefan-Boltzman constant, $\sigma = 5.67 \times 10^{-8}$ W/m^{2}K^4. For modest temperature differences, this equation can be linearized to the form

$$q = h_r A\left(T_1 - T_2\right) \tag{14.58}$$

where h_r is the effective "radiation" heat transfer coefficient

$$h_r = \sigma F\left(T_1^2 + T_2^2\right)\left(T_1 + T_2\right) \tag{14.59}$$

and, for small $\Delta T = T_1 - T_2$, h_r is approximately equal to

$$h_r = 4\sigma F\left(T_1 T_2\right)^{3/2} \tag{14.60}$$

where T_1 and T_2 must be expressed in absolute degrees Kelvin. It is of interest to note that for temperature differences of the order of 10 K with absolute temperatures around room temperature, the radiative heat transfer coefficient, h_r, for an ideal (or "black") surface in an absorbing environment is approximately equal to the heat transfer coefficient in natural convection of air.

Noting the form of Eq. (14.58), the radiation thermal resistance, analogous to the convective resistance, is seen to equal

$$R_r \equiv \frac{1}{h_r A} \tag{14.61}$$

14.2.6 Environmental Heat Transfer

In applying the foregoing thermal transport relations to microwave equipment located outdoors, attention must be devoted to properly characterizing the atmospheric conditions and including both incoming solar radiation and outgoing night-sky radiation in the heat balance relations. While best results will be obtained by using precise environmental specifications for the "microclimate" at the relevant location, more general specifications may be of use in early stages of product design. The external environment can vary in temperature from –50°C to +50°C, representing the polar regions at one extreme and the subtropical deserts at the other, and experience a range in air pressure from 76 kPa (11 psi), at high plateaus, to 107 kPa (15.5 psi), in deep rift valleys. Incident solar fluxes at noon can reach 1 kW/m^2 on a horizontal surface, but more typically may average 0.5 kW/m^2, of both direct and diffuse radiation, during the peak solar hours. The outgoing long-wave radiation from an outdoor module exposed to the clear nighttime sky falls in the range of 0.01 to 0.1 kW/m^2.[31] It may be anticipated that convective heat transfer coefficients on large exposed surfaces at sea level will attain values of 6 W/m^{2}K for still air to 75 W/m^{2}K, at wind velocities approaching 100 km/h. To determine the surface temperature of an outdoor module, use can be made of the heat balance relation equating the incoming heat — from the microwave components and solar load — with the outgoing heat — by radiation and convection, as

$$q_{rf} + q_{solar} = q_{rad} + q_{conv} \tag{14.62}$$

or

$$\begin{aligned} q_{rf} &= q_{rad} + q_{conv} - q_{solar} \\ &= \sigma A_{surf} F\left(T_{surf}^4 - T_{sky}^4\right) + h_{conv} A_{surf}\left(T_{surf} - T_{amb}\right) - \alpha A_{surf} S \end{aligned} \tag{14.63}$$

or

$$T_{surf} = \frac{\dfrac{q_{rf}}{A_{surf}} + \alpha S}{\sigma F\left(T_{surf}^2 + T_{sky}^2\right)\left(T_{surf} + T_{sky}\right) + h_{conv}} + T_{sky} \tag{14.64}$$

where S is the solar incidence (W/m^2) and T_{sky} is the effective sky temperature (K) (typically equal to ambient temperature during the day and up to 20 K below the air temperature on a dry, clear night).

14.3 Thermal Resistance Networks

The expression of the governing heat transfer relations in the form of thermal resistances greatly simplifies the first-order thermal analysis of electronic systems. Following the established rules for resistance networks, thermal resistances that occur sequentially along a thermal path can be simply summed to establish the overall thermal resistance for that path. In similar fashion, the reciprocal of the effective overall resistance of several parallel heat transfer paths can be found by summing the reciprocals of the individual resistances. In refining the thermal design of an electronic system, prime attention should be devoted to reducing the largest resistances along a specified thermal path and/or providing parallel paths for heat removal from a critical area.

While the thermal resistances associated with various paths and thermal transport mechanisms constitute the "building blocks" in performing a detailed thermal analysis, they have also found widespread application as "figures-of-merit" in evaluating and comparing the thermal efficacy of various packaging techniques and thermal management strategies.

14.3.1 Chip Module Thermal Resistance

14.3.1.1 Definition

The thermal performance of alternative chip packaging techniques is commonly compared on the basis of the overall (junction-to-coolant or junction-to-ambient) thermal resistance, R_{ja}. This packaging figure-of-merit is generally defined in a purely empirical fashion,

$$R_{ja} \equiv \frac{T_j - T_{fl}}{q_c} \tag{14.65}$$

where T_j and T_{fl} are the junction and coolant (fluid) temperatures, respectively, and q_c is the chip heat dissipation.

Unfortunately, however, most measurement techniques are incapable of detecting the actual junction temperature, that is, the temperature of the small volume at the interface of p-type and n-type semiconductors. Hence, this term generally refers to the average temperature or a representative temperature on the chip.

Examination of various packaging techniques reveals that the junction-to-coolant thermal resistance is, in fact, composed of an internal, largely conductive resistance and an external, primarily convective resistance. As shown in Fig. 14.6, the internal resistance, R_{jc}, is encountered in the flow of dissipated heat from the active chip surface through the materials used to support and bond the chip and on to the case of the integrated circuit package. The flow of heat from the case directly to the coolant, or indirectly through a fin structure and then to the coolant, must overcome the external resistance, R_{ex}.

14.3.1.2 Internal Thermal Resistance

As previously discussed, conductive thermal transport is governed by the Fourier Equation, which can be used to define a conduction thermal resistance, as in Eq. (14.3). In flowing from the chip to the package surface or case, the heat encounters a series of resistances associated with individual layers of materials, starting with the chip (silicon, galium arsenide, indium phosphide, etc.) and continuing thru solder, copper, alumina, and epoxy, as well as the contact resistances that occur at the interfaces between pairs of materials. Although the actual heat flow paths within a chip package are rather complex and may shift to accommodate varying external cooling situations, it is possible to obtain a first-order estimate

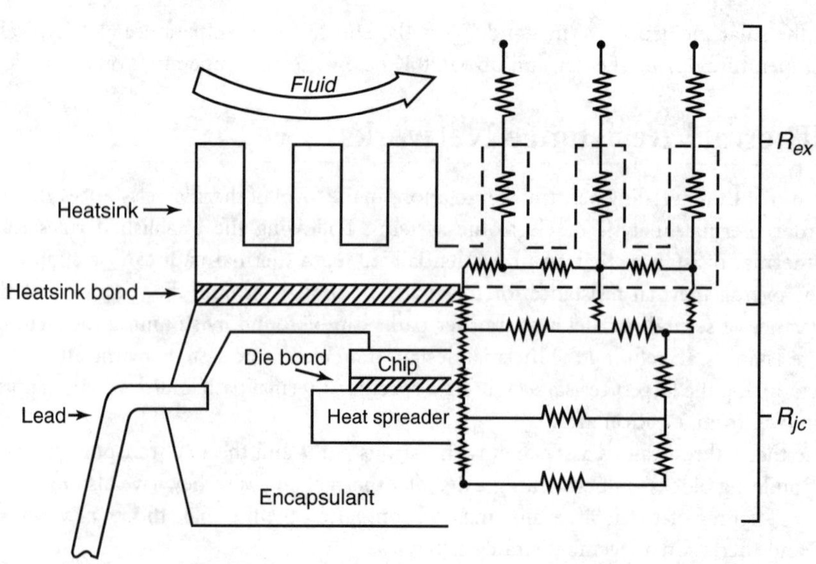

FIGURE 14.6 Primary thermal resistances in a single chip package. (From Ref. 4, with permission.)

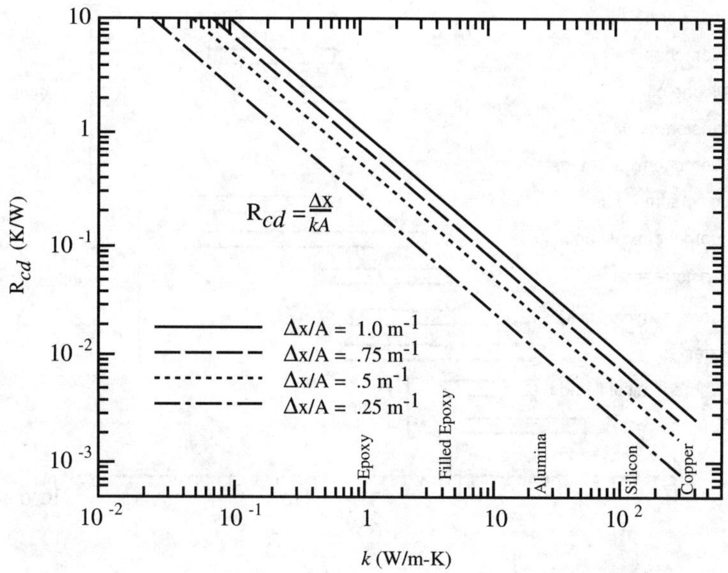

FIGURE 14.7 Conductive thermal resistances for packaging materials. (From Ref. 4, with permission.)

of the internal resistance by assuming that power is dissipated uniformly across the chip surface and that heat flow is largely one-dimensional. To the accuracy of these assumptions, Eq. (14.67)

$$R_{jc} = \frac{T_j - T_c}{q_c} = \sum \frac{\Delta x}{kA} \tag{14.66}$$

can be used to determine the internal chip module resistance, where the summed terms represent the conduction thermal resistances posed by the individual layers, each with thickness Δx. As the thickness of each layer decreases and/or the thermal conductivity and cross-sectional area increase, the resistance of the individual layers decreases. Values of R_{cd} for packaging materials with typical dimensions can be found via Eq. (14.67) or Fig. 14.7, to range from 2 K/W for a 1000 mm^2 by 1-mm thick layer of epoxy encapsulant to 0.0006 K/W for a 100 mm^2 by 25-micron (1-mil) thick layer of copper. Similarly, the values of conduction resistance for typical "soft" bonding materials are found to lie in the range of approximately 0.1 K/W for solders and 1 to 3 K/W for epoxies and thermal pastes for typical $\Delta x/A$ ratios of 0.25 to 1.0.

Comparison of theoretical and experimental values of R_{jc} reveals that the resistances associated with compliant, low thermal conductivity bonding materials and the spreading resistances, as well as the contact resistances at the lightly loaded interfaces within the package, often dominate the internal thermal resistance of the chip package. It is, thus, not only necessary to correctly determine the bond resistance but to also add the values of R_{sp}, obtained from Eq. (14.7) and/or Fig. 14.4, and R_{co} from Eq. (14.9) or (14.12) to the junction-to-case resistance calculated from Eq. (14.67). Unfortunately, the absence of detailed information on the voidage in the die-bonding and heat-sink attach layers and the present inability to determine, with precision, the contact pressure at the relevant interfaces, conspire to limit the accuracy of this calculation.

14.3.1.3 External Resistance

An application of Eq. (14.41) or (14.43) to the transfer of heat from the case of a chip module to the coolant shows that the external resistance, $R_{ex} = 1/hA$, is inversely proportional to the wetted surface area and to the coolant velocity to the 0.5 to 0.8 power and directly proportional to the length scale in the flow direction to the 0.5 to 0.2 power. It may, thus, be observed that the external resistance can be strongly

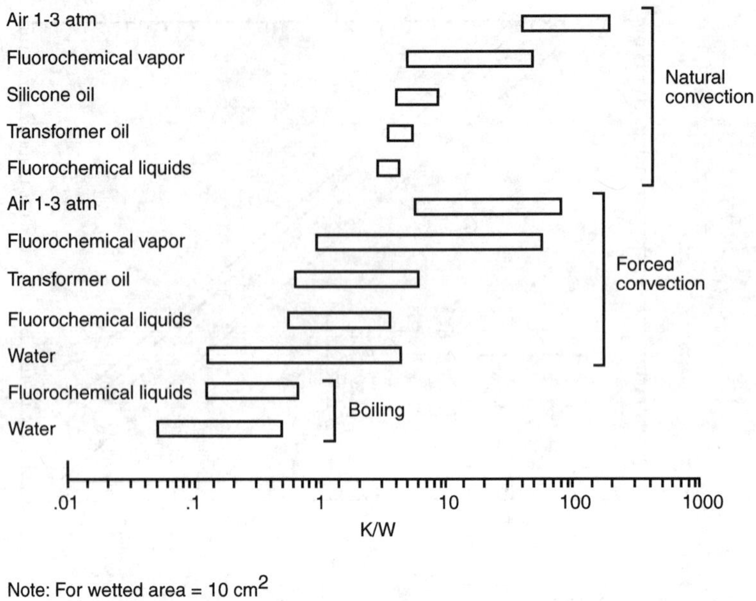

FIGURE 14.8 Typical external (convective) thermal resistances for various coolants and cooling modes. (From Ref. 4, with permission.)

influenced by the fluid velocity and package dimensions and that these factors must be addressed in any meaningful evaluation of the external thermal resistances offered by various packaging technologies.

Values of the external resistance for a variety of coolants and heat transfer mechanisms are shown in Fig. 14.8 for a typical component wetted area of 10 cm^2 and a velocity range of 2 to 8 m/s. They are seen to vary from a nominal 100 K/W for natural convection in air, to 33 K/W for forced convection in air, to 1 K/W in fluorocarbon liquid forced convection and to less than 0.5 K/W for boiling in fluorocarbon liquids. Clearly, larger chip packages will experience proportionately lower external resistances than the displayed values. Moreover, conduction of heat through the leads and package base into the printed circuit board or substrate will serve to further reduce the effective thermal resistance.

In the event that the direct cooling of the package surface is inadequate to maintain the desired chip temperature, it is common to attach finned heat sinks, or compact heat exchangers, to the chip package. These heat sinks can considerably increase the wetted surface area, but may act to reduce the convective heat transfer coefficient by obstructing the flow channel. Similarly, the attachment of a heat sink to the package can be expected to introduce additional conductive resistances, in the adhesive used to bond the heat sink and in the body of the heat sink. Typical air-cooled heat sinks can reduce the external resistance to approximately 10 to 15 K/W in natural convection and to as low as 3 to 5 K/W for moderate forced convection velocities.

When a heat sink or compact heat exchanger is attached to the package, the external resistance accounting for the bond-layer conduction and the total resistance of the heat sink, R_{hs}, can be expressed as

$$R_{ex} = \frac{T_c - T_{fl}}{q_c} = \sum \left(\frac{\Delta x}{kA} \right)_b + R_{hs} \tag{14.67}$$

where R_{hs}

$$R_{hs} = \left[\frac{1}{nhA_f \eta} + \frac{1}{h_b A_b} \right]^{-1} \tag{14.68}$$

is the parallel combination of the resistance of the *n* fins

$$R_f = \frac{1}{nhA_f\eta} \tag{14.69}$$

and the *bare* or base surface not occupied by the fins

$$R_b = \frac{1}{h_b A_b} \tag{14.70}$$

Here, the base surface is $A_b = A - A_f$ and use of the heat transfer coefficient, h_b, is meant to recognize that the heat transfer coefficient that is applied to the base surfaces is not necessarily equal to that applied to the fins.

An alternative expression for R_{hs} involves an *overall surface efficiency,* η_o, defined by

$$\eta_o = 1 - \frac{nA_f}{A}(1-\eta) \tag{14.71}$$

where *A* is the total surface composed of the base surface and the finned surfaces of *n* fins

$$A = A_b + nA_f \tag{14.72}$$

In this case, it is presumed that $h_b = h$ so that

$$R_{hs} = \frac{1}{h\eta_o A} \tag{14.73}$$

In an optimally designed fin structure, η can be expected to fall in the range of 0.50 to 0.70.[4] Relatively thick fins in a low velocity flow of gas are likely to yield fin efficiencies approaching unity. This same unity value would be appropriate, as well, for an unfinned surface and, thus, serve to generalize the use of Eq. (14.68) to all package configurations.

14.3.1.4 Total Resistance — Single Chip Packages

To the accuracy of the assumptions employed in the preceding development, the overall single chip package resistance, relating the chip temperature to the inlet temperature of the coolant, can be found by summing the internal, external, and flow resistances to yield

$$\begin{aligned} R_{jaj} &= R_{jc} + R_{ex} + R_{fl} \\ &= \sum \frac{\Delta x}{kA} + R_{int} + R_{sp} \\ &= \frac{1}{\eta hA} + \left(\frac{Q}{q}\right)\left(\frac{1}{2\rho Qc_p}\right) \end{aligned} \tag{14.74}$$

In evaluating the thermal resistance by this relationship, care must be taken to determine the effective cross-sectional area for heat flow at each layer in the module and to consider possible voidage in any solder and adhesive layers.

As previously noted in the development of the relationships for the external and internal resistances, Eq. (14.75) shows R_{ja} to be a strong function of the convective heat transfer coefficient, the flowing heat capacity of the coolant, and geometric parameters (thickness and cross-sectional area of each layer). Thus, the introduction of a superior coolant, use of thermal enhancement techniques that increase the local heat transfer coefficient, or selection of a heat transfer mode with inherently high heat transfer coefficients (boiling, for example) will all be reflected in appropriately lower external and total thermal resistances. Similarly, improvements in the thermal conductivity and reduction in the thickness of the relatively low conductivity bonding materials (such as soft solder, epoxy, or silicone) would act to reduce the internal and total thermal resistances.

14.3.1.5 Weighted-Average Modification of R_{jc}

The commonly used junction-to-case thermal resistance, relying on just a single case temperature, can be used with confidence only when the package case is nearly isothermal. In a more typical packaging configuration, when substantial temperature variations are encountered among and along the external surfaces of the package,[32–34] the use of the reported R_{jc} can lead to erroneous chip temperature predictions. This is especially of concern in the analysis and design of plastic chip packages, due to the inherently high thermal resistance of the plastic encapsulant and the package anisotropies introduced by the large differences in the conductivity between the lead frame and/or heat spreader and the plastic encapsulant. Since R_{jc} is strictly valid only for an isothermal package surface, a method must be found to address the individual contributions of the various surface segments according to their influence on the junction temperature.

Following Krueger and Bar-Cohen,[35] it is convenient to introduce the expanded R_{jc} methodology with a thermal model of a chip package that can be approximated by a network of three thermal resistances connected in parallel, from the chip to the top, sides, and bottom of the package, respectively. This type of compact model is commonly referred to as a "star network" and, in this model, the heat flow from the chip is

$$q = q_1 + q_2 + q_3 \tag{14.75}$$

or

$$q = \frac{T_j - T_1}{R_1} + \frac{T_j - T_2}{R_2} + \frac{T_j - T_3}{R_3} \tag{14.76}$$

This compact model of an electronic device is shown schematically in Fig. 14.9.

Equation (14.77) can be rearranged to yield the dependence of the chip (or junction) temperature on the temperature of the three surface segments as

$$T_j = \left(\frac{R_2R_3}{R_s}\right)T_1 + \left(\frac{R_3R_1}{R_s}\right)T_2 + \left(\frac{R_1R_2}{R_s}\right)T_3 + \left(\frac{R_1R_2R_3}{R_s}\right)q \tag{14.77}$$

where $R_s = R_1R_2 + R_1R_3 + R_2R_3$

Equation (14.78) may be generalized to admit n-distinct elements along the package surface, or

$$T_j = \sum_{k=1}^{n} I_k T_k + I_{n+1} q \tag{14.78}$$

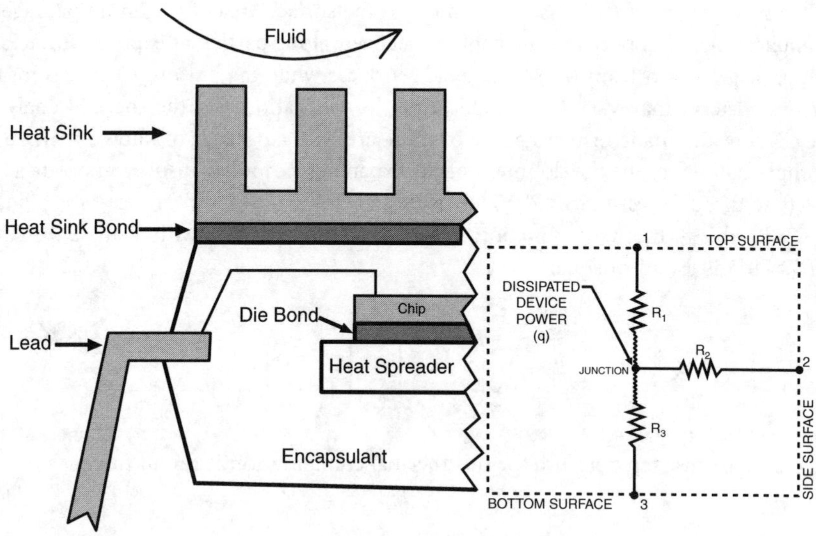

FIGURE 14.9 Geometry of a 28-lead PLCC device. (a) The compact model schematic, and (b) the actual device cross-section.[35]

A comparison of Eqs. (14.78) and (14.79) shows that the coefficients of the specified surface temperatures, the I_k's are totally determined by the internal resistances of the chip package

$$I_1 = \frac{R_2 R_3}{R_s} \qquad I_2 = \frac{R_3 R_1}{R_s}$$
$$I_3 = \frac{R_1 R_2}{R_s} \qquad I_4 = \frac{R_1 R_2 R_3}{R_s} \tag{14.79}$$

The temperature coefficients needed to generate a junction temperature relation of the form shown in Eq. (14.79) can thus be determined from previously calculated internal resistance or, in the absence of such values, by extraction from empirical data or numerical results for the junction temperature. Furthermore, it is to be noted that the sum of the coefficients of the various surface temperature is equal to unity and that the power dissipation coefficient, $I_{n+1}q$, is, in fact, the familiar R_{jc} isothermal, junction-to-case thermal resistance. Consequently, Eq. (14.79) may be rewritten as

$$T_j = \sum_{k=1}^{n} I_k T_k + R_{jc} q \tag{14.80}$$

or, returning to R_{jc}

$$R_{jc} = \frac{T_j - \sum_{k=1}^{n} I_k T_k}{q} = \frac{T_j - \bar{T}_c}{q} \tag{14.81}$$

where $\bar{T}_c$ is the average case temperature

$$\bar{T}_c = \frac{\sum_{k=1}^{n} A_k T_k}{A_T} \tag{14.82}$$

where A_k is the surface area of the kth surface and A_T is the surface area of the entire package.

In many applications, chip packages are cooled selectively along particular exposed surfaces. One such example is a package cooled from the top and side surfaces while the bottom surface is insulated. The thermally active surfaces may vary from application, to application and the thermal analyst needs to quantify the effect of thermally insulating one or more areas on a package of known thermal resistance. For the assumptions used in the development of the expanded R_{jc} model, insulation of surface-m results in zero heat flow through resistance, R_m. This causes the temperature of surface-m to equal the chip temperature. With this in mind, the junction temperature for a package with a single insulated surface given by Eq. (14.81) is found to equal

$$T_j = \sum_{k \neq m} \left(\frac{I_k}{1 - I_m} \right) T_k + \left(\frac{I_{n+1}}{1 - I_m} \right) q \tag{14.83}$$

The weighted average case temperature for this thermal configuration is found to equal

$$\overline{T}_c = \sum_{k \neq m} \left(\frac{I_k}{1 - I_m} \right) T_k \tag{14.84}$$

and the modified junction to case resistance, R^*_{jc} is

$$R^*_{jc} = \frac{R_{jc}}{1 - I_m} \tag{14.85}$$

14.3.2 Multichip Modules

The thermostructural complexity of multichip modules in current use hampers effective thermal characterization and introduces significant uncertainty in any attempt to compare the thermal performance of these packaging configurations. Variations in heat generation patterns across the active chips (reflecting differences in functional requirements and the duty cycle among the macrocells constituting a particular chip), as well as nonuniformities in heat dissipation among the chips assembled in a single module, further complicate this task. While nonthermal issues (e.g., electromagnetic crosstalk) may often be the dominant limiting factors in RF multichip modules, the high power associated with microwave devices makes it essential that the thermal performance of this packaging configuration be analyzed and reported in a consistent manner. To establish a common basis for comparison of multichip modules, it is possible to neglect the on-chip and chip-to-chip variations and consider that the heat generated by each chip flows through a unit cell of the module structure to the external coolant.[36,37] For a given structure, increasing the area of the unit cell allows heat to spread from the chip to a larger cross-section, reducing the heat flux at some of the thermally critical interfaces and at the convectively cooled surfaces. Consequently, the thermal performance of a multichip module can be best represented by the area-specific thermal resistance, i.e., the temperature difference between the chip and the coolant divided by the substrate heat flux, expressed in units of K/(W/cm^2). This figure of merit is equivalent to the inverse of the overall heat transfer coefficient, U, commonly used in the compact heat exchanger literature. Despite significant variation in design and fabrication, the leading edge water-cooled and air-cooled modules of the late 1980s provided a specific thermal resistance of approximately 20°C for every watt per square centimeter at the substrate. A decade later, the thermally best multichip modules of the 1990s offered specific thermal resistance values between 5 and 10 K/(W/cm^2).

14.3.3 Radar System Applications

The frequent demand for high radiated power in civilian and military radar systems, for the ranging and detection of remote objects, has led to the development of many high-energy microwave systems. Due to the inefficiencies inherent in electrical-to-microwave energy conversion and the management of radio-frequency (RF) energy, the operation of such radar equipment often results in significant heat dissipation. To avoid catastrophic failures, to achieve the reliability targets of the system, and to satisfy the frequency stability requirements of the RF tubes, system requirements commonly specify temperature control to within several degrees Celsius around 150°C. These thermal requirements necessitate the use of aggressive thermal management techniques, including the use of high flow rate liquid forced convection, pool and flow boiling, and high pressure-drop, air-cooled compact heat exchangers.[3]

In traditional, mechanically steered radar systems, much of the total power dissipation occurs in the power tubes (e.g., klystron, gyrotrons, amplitrons), with secondary dissipation in the rotary joints and wave guides. Heat release along the collector surfaces of the tubes is proportional to the RF power of the system and can range from 1 kW to as much as 20 kW, with a peak local flux of several watts/cm^2 to, perhaps, several thousand watts/cm^2, at operating temperatures of 150°C.[3] Similar, though less severe thermal problems are encountered in the RF rotary joints and waveguides, where heat fluxes of 5 W/cm^2 to 10 W/cm^2, with allowable temperatures of 150°C, may be encountered.

Growing applications of active antenna array elements, utilizing amplifier diodes and solid-state phase shifters to provide electronic steering of the radiated RF beam, have introduced new dimensions into the thermal control of microwave equipment. In such "phased array" radar systems, high power tubes and waveguides can be eliminated with low power RF delivered to each antenna element. The low conversion efficiency of the amplifier diodes results in diode heat dissipation between 1 W and 10 W and local heat fluxes comparable to power transistors. In "phased array" antenna, precise shaping of the radiated wave often requires relatively cool (<90°C) ferrite phase shifters and no significant temperature variations (<10°C) across the elements used in the array.

References

1. M. Pecht, P. Lall, and E. Hakim, The influence of temperature on integrated circuit failure mechanisms, *Advances in Thermal Modeling of Electronic Components and Systems*, vol. 3, A. Bar-Chen and A.D. Kraus, Eds., ASME Press, New York, 1992.
2. R.A. Morrison, Improved avionics reliability through phase change conductive cooling, *Proceedings, IEEE National Telesystems Conference*, 1982, B5.6.1–B5.6.6.
3. A.D. Kraus and A. Bar-Cohen, *Thermal Analysis and Control of Electronic Equipment*, McGraw-Hill, New York, 1983.
4. A.D. Kraus and A. Bar-Cohen, *Design and Analysis of Heat Sinks*, John Wiley & Sons, New York, 1995.
5. K.J. Negus, M.M. Yovanovich, and J.V. Beck, On the non-dimensionalization of constriction resistance for semi-infinite heat flux tubes, *J. Heat Trans.*, 1111, 804–807, 1989.
6. M.M. Yovanovich and V.W. Antonetti, Application of thermal contact resistance theory to electronic packages, in *Advances in Thermal Modeling of Electronic Components and Systems*, A. Bar-Cohen and A.D. Kraus (eds.), Hemisphere Publishing Co, New York, 1988, 79, 28.
7. M.M. Yovanovich, personal communication, 1990.
8. F.P. Incropera and D.P. Dewitt, *Introduction to Heat Transfer*, John Wiley and Sons, New York, 1996.
9. W. Elenbaas, Heat dissipation of parallel plates by free convection, *Physica*, 9, 1, 665–671, 1942.
10. J.R. Bodoia and J.F. Osterle, The development of free convection between heated vertical plates, *J. Heat Transfer*, 84, 40–44, 1964.
11. N. Sobel, F. Landis, and W.K. Mueller, Natural convection heat transfer in short vertical channels including the effect of stagger, *Proc. 3d Int. Heat Transfer Conf.*, Chicago, 2, 1966, 121–125.

12. W. Aung, Fully developed laminar free convection between vertical plates heated asymmetrically, *Int. J. Heat Mass Transfer,* 15, 40–44, 1972.
13. W. Aung, L.S. Fletcher, and V. Sernas, Developing laminar free convection between vertical flat plates with asymmetric heating, *Int. J. Heat Mass Transfer,* 15, 2293–2308, 1972.
14. O. Miyatake and T. Fujii, Free convection heat transfer between vertical parallel plates — one plate isothermally heated and the other thermally insulated, *Heat Transfer Jpn. Res.,* 3, 30–38, 1972.
15. O. Miyatake, T. Fujii, M. Fujii, and H. Tanaka, Natural convection heat transfer between vertical parallel plates — one plate with a uniform heat flux and the other thermally insulated, *Heat Transfer Jpn. Res.,* 4, 25–33, 1973.
16. A. Bar-Cohen and W.M. Rohsenow, Thermal optimum spacing of vertical, natural convection cooled, vertical plates, *J. Heat Trans.,* 106, 116–122, 1984.
17. W.H. McAdams, *Heat Transmission,* McGraw-Hill, New York, 1954.
18. S.W. Churchill and R. Usagi, A. general expression for the correlation of rates of transfer and other phenomena, *AIChE J.,* 18, 6, 1121–1138, 1972.
19. A. Bar-Cohen and W.M. Rohsenow, Thermally optimum spacing of vertical, natural convection cooled, parallel plates, *Heat Transfer in Electronic Equipment, ASME HTD 20, ASME WAM,* Washington, D.C., 11–18, 1981.
20. A. Bar-Cohen, Bounding relations for natural convection heat transfer from vertical printed circuit boards, *Proceedings of the IEEE,* September 1985.
21. E.N. Sieder and G.E. Tate, Heat transfer and pressure drops of liquids in tubes, *Ind. Eng. Chem.,* 28, 1429–1435, 1936.
22. H. Hausen, Darstelling des wärmeuberganges in rohren durch verallgemeinerte potenzbeziehungen, *VDI Z,* 4, 91–98, 1943.
23. A. Bejan, *Heat Transfer,* John Wiley & Sons, New York, 1993.
24. J.P. Holman, *Heat Transfer,* McGraw-Hill, New York, 1990.
25. J.H. Lienhard, *A Heat Transfer Textbook,* Prentice-Hall, Englewood Cliffs, NJ, 1987.
26. W.M. Rohsenow, A method of correlating heat transfer data for surface boiling of liquids, *Transactions of ASME* (reprinted in *3rd ASME/JSME Thermal Engineering Joint Conference,* 1, 503–512), 74, 969–976, 1951.
27. A.A. Watwe, A. Bar-Cohen, and A. McNeil, Combined pressure and subcooling effects on pool boiling from a PPGA chip package, *ASME J. Electronics Packaging,* March 1997.
28. F. Gertsmann and P. Griffith, Laminar film condensation on the underside of horizontal and inclined surfaces, *Int. J. Heat Mass Trans.,* 10, 567–580, 1966.
29. W.Z. Nusselt, Die oberflächencondensation der wasserdamfes, *Z. Ver. Deut. Ing.,* 60, 541–569, 1916.
30. P. Sadasivan and J.H. Lienhard, Sensible heat correction in laminar film boiling and condensation, *J. Heat Trans.,* 109, 545–546, 1987.
31. C. Rambach and A. Bar-Cohen, Nocturnal heat rejection by skyward radiation, in *Future Energy Production Systems,* J.C. Denton and N. Afgan, eds., Hemisphere Publishing, New York, 1976, 713–726.
32. J.A. Andrews, Package thermal resistance model dependency on equipment design, *IEEE CHMT Trans.,* 11, 4, 528–537, 1988.
33. S.S. Furkay, Thermal characterization of plastic and surface mount components, *IEEE CHMT Trans.,* 11, 4, 521–527, 1988.
34. E.A. Wilson, Factors influencing the interdependence of R_{jc} and R_{ca}, *Proceedings of the Second International Electronic Packaging Society Meeting,* 247–255, 1981.
35. W.B. Krueger and A. Bar-Cohen, Thermal characterization of a PLCC — expanded R_{jc} methodology, *IEEE CHMT Trans.,* 15, 5, 691–698, 1992.
36. A. Bar-Cohen, Thermal management of air- and liquid-cooled multichip modules, *IEEE CHMT Trans.,* CHMT-10, 2, 159–175, 1987.
37. A. Bar-Cohen, Addendum and correction to thermal management of air- and liquid-cooled multichip modules, *IEEE CHMT Trans.,* CHMT-11, 3, 333–334, 1988.

15

Low Voltage/Low Power Microwave Electronics

Mike Golio
Motorola

15.1 Introduction

The development and manufacture of low voltage/low power RF and microwave electronics has been the subject of intense focus over the past few decades. The development of active circuits that consume lower power at lower voltage levels is consistent with the general trends of semiconductor scaling. From the simple relationship

$$E = V/d \tag{15.1}$$

where E is the electric field, V is the voltage, and d is the separation distance, it can be seen that smaller devices (reduced d) require less voltage to achieve the same electric field levels. Since to first order it is electric field in a material that dominates electron transport properties, electric field levels must be preserved when devices are scaled. Therefore, for each reduction in critical dimension (gate length or base width) reductions in operating voltage have been realized.

Because semiconductor physics implies an inverse relationship between device size and required voltage level, and because transistor scaling has been a major driver for integrated circuit (IC) technology since the earliest transistors were introduced, the trend toward reduced power supply voltage has been ongoing since the beginning of the semiconductor industry. The explosive growth of portable wireless products, however, has brought a sense of urgency and technical focus to this issue. Portable product requirements have forced technologists to consider battery drain and methods to extend battery lifetime.

Since the total global volume requirements for digital ICs far exceeds the requirements for RF and microwave ICs, digital applications have been the primary drivers for most IC device technologies. The continuing advances of CMOS technology in particular have supported the low voltage/low power trend. Bipolar, BiCMOS, and more recently SiGe HBT developments have similarly supported this trend. For digital circuitry, the low voltage/low power trend is essentially without negative consequences. Each reduction in device size and power supply requirement has led to significant performance improvements. The reduced device size leads to smaller, lighter, cheaper, and faster circuits while the reduced voltage

0-8493-1562-X/03/$0.00+$1.50

leads to reduced power consumption and improved reliability. The combination of these advantages provides the potential for higher levels of integration, which leads to the use of more complex and sophisticated architectures and systems.

For RF and microwave circuitry, however, the push toward lower voltage and lower power consumption is counter to many other radio requirements and presents challenges not faced by digital circuit designers. Although smaller, lighter, cheaper, and faster are generally desirable qualities for RF as well as digital applications, reduced voltage can produce negative consequences for RF circuit performance. In particular, reduced voltage leads to reduced dynamic range, which can be especially critical to the development of linear power amplifiers.

An imperative to develop and produce low voltage/low power RF products requires that every aspect of radio component development be considered. The ultimate success of a low voltage/low power RF product strategy is affected by all of the following issues:

- Materials technology (GaAs, Si, epitaxiy, implant, heterojunctions).
- Device technology (BJT, HBT, MESFET, MOSFET, HEMT).
- Circuit technology (matching and topology challenges for low voltage).
- Radio system design issues (impact of system and radio architecture on performance).

15.2 Motivations for Reduced Voltage

Commercial handheld products are clearly a significant driver for the low voltage/low power imperative. Reduced power consumption translates into longer battery lifetime and/or smaller, lighter products through reduction of battery size. Consumers value both of these attributes highly, which has led to the widespread acceptance of portable pagers, cell phones, and wireless personal area network products. An additional collateral benefit to reduced voltage operation is improved product reliability.

For vehicle-based products, the low power imperative is less compelling since battery size and lifetime issues are not relevant to these applications. Smaller and lighter electronics, however, are still valued in these applications. Power supplies represent a fixed payload that must be transported throughout the lifetime of the vehicle. In the case of jet aircraft, a single additional pound of payload represents tens of thousands of dollars of jet fuel over the life of the aircraft. A reduction in the size and weight of power supplies also represents payload that can be displaced with additional safety, communication, or navigation equipment. As in the case of handheld equipment, reduced power consumption will contribute to improved product reliability.

Even for many fixed site applications, reduced power consumption can provide benefits. Particularly in urban environments, base stations must be located in high-cost real estate areas (at the top of skyscrapers, for example). Rental of this space is often billed on a per square foot basis, so reduction in size carries some cost advantage. Increased reliability also leads to reduced maintenance costs.

Advantages of low voltage/low power RF circuitry are also significant for military applications. Supply line requirements during a military operation are often a determining factor in the outcome of battle. Transport vehicles are limited in number and capacity, so extended battery lifetime translates directly into decreased battlefield transport capacity requirements for battery resupply. This reduction in battery resupply transport requirements frees up valuable resources to transport additional lifesaving materials. The advantage is also directly felt by the foot soldier who is able to achieve greater mobility and/or additional electronic functionality (communication or navigation equipment).

As mentioned in the introduction, low voltage/low power trends have been almost without negative impact for most digital applications. Since most portable radio products are comprised of both digital ICs and RF parts, manufacturers would prefer to run all of the electronics from a single power supply. RF and microwave circuitry has had to follow the trend to low voltage, even when that trend brings on increasing challenges. This requirement to conform is exacerbated by the fact that the volume of digital circuitry required from semiconductor factories is far higher than the volume of RF circuitry. Applications that would benefit from high voltage operation comprise a nearly insignificant portion of the total

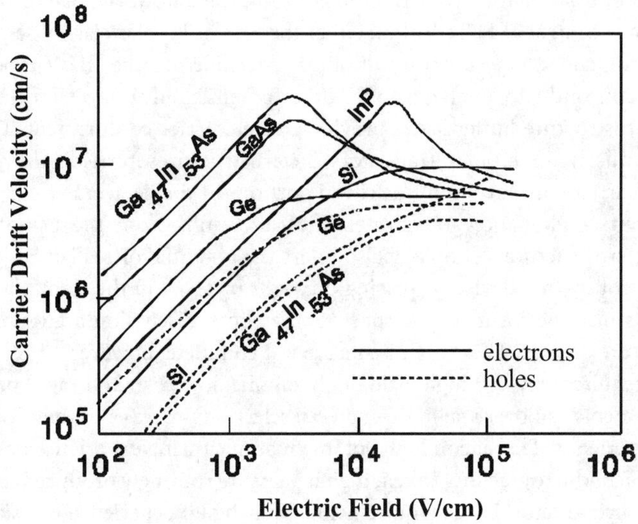

FIGURE 15.1 A plot of the carrier velocity in bulk semiconductor material as a function of the applied electric field for several commonly used semiconductors.

semiconductor market so get little attention from most semiconductor manufacturers. The RF designer often has little choice but to use a low voltage IC process.

15.3 Semiconductor Materials Technology

Low voltage/low power operation requires improved efficiency, low parasitic resistance, low parasitic capacitance, and precise control of on-voltage. The selection of the semiconductor material impacts all of these characteristics.

Improvements in efficiency and reductions in parasitic resistance can be achieved by using materials that exhibit increased carrier mobility and velocity. Figure 15.1 plots the carrier velocity in bulk semiconductor material as a function of the applied electric field for several commonly used semiconductors. The carrier (electron or hole) mobility is given by

$$\mu_0 = \frac{v}{E} \qquad \text{for small E} \tag{15.2}$$

where v is the carrier velocity in the material and E is the electric field. The figure illustrates that III-V semiconductor materials such as GaAs, GaInAs, and InP exhibit superior electron velocity and mobility characteristics relative to silicon. For this reason, devices fabricated in III-V materials are expected to exhibit some advantages in achieving high efficiency and low parasitic resistance.

Low parasitic capacitance is also a desirable characteristic to insure low loss and minimal battery drain. Because GaAs and InP offer semi-insulating substrates, they exhibit performance advantages over the much more conductive, and hence poorer performing silicon substrates. Advanced materials strategies such as silicon-on-insulator offer potential advantages even over III-V materials, but at increased cost and complexity.

Precise control of on-voltage becomes increasingly important as supply voltages diminish. Low voltage operation means that the total voltage swing that can be realized by the RF signal is reduced. Bipolar transistors offer clear, significant advantages over FETs with respect to this characteristic since bipolar on-voltage characteristics are determined primarily by material band gaps while FET on-voltage characteristics are determined by material process parameters such as thickness and charge densities. When other considerations indicate that a FET structure is appropriate, however, heterostructure buffer layers improve the on-voltage control of FETs. Within either the FET or bipolar class of semiconductor devices, Silicon process control tends to be superior to Gallium Arsenide or other III-V process control.

Heterostructure semiconductor materials offer other potential advantages for low voltage design. In FET structures, heterostructure buffer layers provide greater carrier confinement. This can reduce substrate parasitics. Similarly, in bipolar transistors, heterostructures provide the potential to reduce unwanted minority carrier concentrations in critical portions of the device. In both the FET and bipolar case improved output conductance and greater linearity results from the use of properly designed heterostructures. Heterostructure technology also offers the potential of realizing high current and high breakdown simultaneously in a device by placing the right material in the right place.

From a purely technical performance perspective, the above discussion argues primarily for the use of III-V heterostructure devices. These arguments are not complete, however. Most commercial wireless products also have requirements for high yield, high volume, low cost, and rapid product development cycles. These requirements can overwhelm the material selection process and they favor mature processes and high volume experience. The silicon high volume manufacturing experience base is far greater than that of any III-V semiconductor facility. Silicon digital parts are routinely produced in volumes that dwarf total global III-V semiconductor device requirements. This digital experience is easily leveraged into the production of silicon RF and microwave parts, giving silicon materials a clear advantage in these areas.

No short cut for the development of experience and maturity with advanced material structures is available. As demand for diminishing power consumption RF products continues to grow, however, experience will be gained and the fundamental physical advantages of III-V heterostructure devices will lead to increasing presence in commercial wireless products.

15.4 Semiconductor Device Technology

Device technology decisions influence many of the same performance metrics that are affected by material decisions. Efficiency, parasitic resistance, and on-voltage control are all affected by the choice of device. Additional device considerations may depend on the particular circuit application being considered. For an oscillator application, for example, the 1/f noise characteristics of the transistor are likely to be critical to achieving the required circuit level performance. These same transistor characteristics may be of little interest for many other circuit applications.

Table 15.1 evaluates most modern transistor technologies against a list of performance characteristics that are of interest for portable power amplifier applications. The performance metrics considered in the table include: parasitic loss, ability to easily use a single polarity power supply, power added efficiency, linearity, cost, and maturity. As is readily seen from the table, no particular device type excels in all areas and all devices exhibit at least some significant disadvantages. The relative values of each of the metrics are not equal and are not identical for all applications. Meeting linearity requirements, for example, is central to the power amplifier (PA) design process for many digitally modulated portable radios, while the linearity requirements for many analog radios is easily achieved and therefore less valued.

Parasitic loss is clearly important to reduce undesirable current drain. The III-V device technologies offer advantages over silicon technologies in this area because of their material advantages.

Single polarity power supplies are desirable since other radio components do not require a negative supply. The generation of a negative voltage required to bias a transistor adds to the parts count, cost, and current drain of the radio. HBTs, BJTs, and LDMOS parts have a clear advantage over conventional MESFETs and HEMTs in this area. Enhancement mode HEMTs offer a potential solution to this problem, but at a cost to manufacturing ability and dynamic range.

Linearity requirements are typically driven by government regulations to keep wireless products from interfering with each other. Linearity specifications are not consistent for all radio architectures and are sometimes the subject of erroneous discussions in the technical literature. The linearity of the Gummel plot [$\log(I_c)$ vs. $\log(I_b)$] of a BJT or HBT, for example, is not equivalent or translated into RF linearity.

TABLE 15.1 Semiconductor Devices are Evaluated against Several Important Device Characteristics for Portable Power Amplifer Applications

	MESFET/HEMT	III-V HBT	Si BJT	Si:Ge HBT	Power MOSFET (LDMOS)
		Low Voltage Related Characteristics			
Parasitic loss	Very good	Very good	Moderate	Moderate	Moderate
Single-polarity supply	No[a]	Yes	Yes	Yes	Yes
Power added efficiency	Excellent	Very good	Moderater	Moderate	Good for f<~2 GHz
Linearity	Excellent[b]	Very good	Moderate	Very good	Moderate
Power density	Moderate	Excellent	Very good	Excellent	Moderate
		General Characteristics			
Cost	Moderate to high	Moderate to high	Low to moderate	Moderate to high	Low to moderate
Maturity	Good	Good	Excellent	Moderate	Very good

[a] e-mode MESFETs and HEMTs do make use of single-polarity supply voltage, but at a cost of more difficult manufacturing and with a more limited dynamic range.

[b] MESFET devices manufactured using epi-taxial material (as opposed to ion-implanted) as well as HEMTs exhibit excellent linearity.

It is also difficult to measure linearity and determine an absolute performance limit for a particular device. RF circuit tuning performed to optimize power, gain, or efficiency is not necessarily optimum for achieving high linearity. MESFETs fabricated using epitaxial material (as opposed to ion-implanted material) and HEMTs exhibit slight advantages in linearity over other devices. HBTs and BJTs have also demonstrated competitive linearity characteristics.

Once power, gain, and linearity specifications are met, efficiency often becomes the key distinguishing figure of merit for modern wireless power amplifiers. Improvements in PA efficiency are translated directly into reduced battery drain and longer battery lifetime. HEMTs hold certain advantages in this area while silicon BJTs have not produced efficiency values that are as competitive. Both LDMOS and HBT parts have also produced competitive efficiencies for certain applications.

High power density is desirable since it leads to reduced part size. Smaller devices not only reduce the size of the final product, but more importantly, reduce the cost of the device since more devices can be processed on a single wafer. HBTs demonstrate the highest power density with adequate breakdown characteristics of any wireless PA devices today. Ion-implanted MESFETs and LDMOS parts exhibit very low power densities compared to HBTs.

As in the case of material considerations, the technical characteristics of devices must be weighed against the cost and maturity requirements for wireless product development. Although cost is a primary driver for many radio applications today, it may be difficult to determine fabrication costs of emerging device technologies. For most parts utilized in high volume commercial applications, the cost of fabrication is not closely related to the purchase price for the part. The market supply and demand forces determine purchase price — the price is set as high as the market will bear. If the fabrication costs can be brought below the market price with a comfortable margin, the part will continue to be manufactured. If not, the part will be discontinued. Large semiconductor manufacturing companies may choose to subsidize new technologies for several years before deciding to discontinue them. Although difficult to quantify, cost is critical to the long-term success of a device. Silicon transistors have historically come with a cost advantage over III-V devices. The high volume experience of silicon fabrication facilities and greater maturity of the processes contribute significantly to the low cost advantage.

Examination of Table 15.1 and the discussion of the preceding paragraphs illustrate the difficulty in choosing the optimum device for low voltage/low power applications. Although important, the low

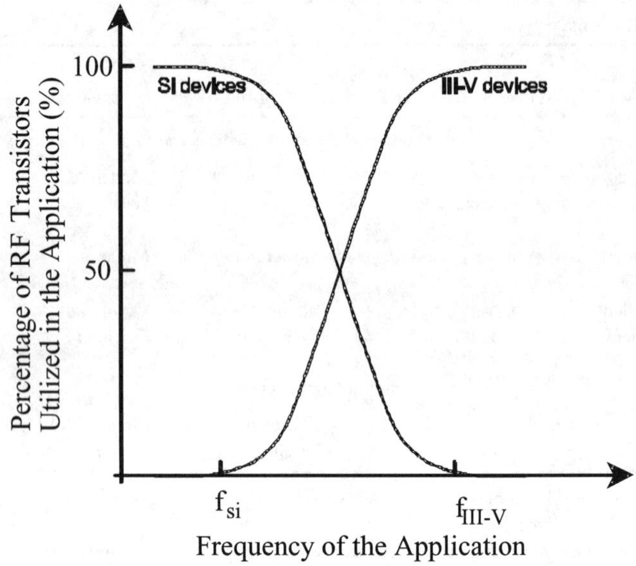

FIGURE 15.2 An illustration of the relationship between frequency of the application and device technology choice.

voltage characteristics of the part are not the only characteristics that must be considered. No transistor is optimal for all requirements and each device choice carries implied compromises for the final product.

Although not included Table 15.1, the frequency of the application can also be a critical performance characteristic in the selection of device technology. Figure 15.2 illustrates the relationship between frequency of the application and device technology choice. Because of the fundamental material characteristics illustrated in Fig. 15.1, silicon technologies will always have lower theoretical maximum operation frequencies than III-V technologies. The higher the frequency of the application, the more likely the optimum device choice will be a III-V transistor over a silicon transistor. Above a certain frequency (labeled $f_{III\text{-}V}$ in Fig. 15.2), III-V devices dominate the transistors of choice, with silicon playing no significant role in the RF portion of the product. In contrast, below a certain frequency (labeled f_{Si} in Fig. 15.2) the cost and maturity advantages of silicon provide little opportunity for III-V devices to compete. In the transition spectrum between f_{Si} and $f_{III\text{-}V}$, the device technology is not an either/or choice, but rather silicon and III-V devices coexist. Although silicon devices are capable of operating above frequency f_{Si}, this operation is often gained at the expense of current drain. As frequency is increased above f_{Si} in the transition spectrum, efficiency advantages of gallium arsenide and other III-V devices provide competitive opportunities for these parts. The critical frequencies, f_{Si} and $f_{III\text{-}V}$ are not static frequency values. Rather, they are continually being moved upward by the advances of silicon technologies — primarily by decreasing critical device dimensions.

The process of choosing device technology is compounded further if higher levels of integration are implemented. Table 15.2 is a companion to Table 15.1. Again, several semiconductor device technologies are ranked in terms of desirable device characteristics. In contrast to Table 15.1, Table 15.2 considers characteristics of interest for integrated radio front-end applications. Although some characteristics are common to both tables, many are not. The evaluation of which technology is superior is likely to change when Table 15.2 characteristics are used to make a device technology choice instead of those listed in Table 15.1. The final column of Table 15.2 replaces LDMOS characteristics with those of CMOS. This change in device technologies that are considered for the application is appropriate and indicative of the problem of choosing an optimum device technology. LDMOS is not a likely device technology for most integrated radio applications while CMOS is not a viable power amplifier technology for most portable wireless products.

TABLE 15.2 Semiconductor Devices are Evaluated against Several Important Device Characteristics for Radio Integration Applications

	MESFET/HEMT	III-V HBT	Si BJT	Si:Ge HBT	CMOS
		Low Voltage Related Characteristics			
Single-polarity supply	No[a]	Yes	Yes	Yes	Yes
Turn-on voltage control	Moderate	Very good	Excellent	Excellent	Good
		General Characteristics			
Cost	Moderate to high	Moderate to high	Low to moderate	Moderate to high	Low
Maturity	Good	Good	Excellent	Moderate	Excellent
RF integration capability[b]	Very good	Very good	Excellent for f<3 GHz	Excellent for f<3 GHz	Excellent for f<2 GHz
Noise figure	Excellent	Very good	Good	Very good	Moderate
Phase noise	Poor	Good	Excellent	Excellent	Good

[a] e-mode MESFETs and HEMTs do make use of single-polarity supply voltage, but at a cost of more difficult manufacturing and with a more limited dynamic range.

[b] Comparing Silicon RF integration capability to III-V RF integration capability is difficult. Silicon process maturity allows for far higher levels of integration (in terms of numbers of transistors), but becomes of limited value above a few GHz because of the parasitics of the Si substrate.

15.5 Circuit Design

RF circuit performance is not generally improved by the reduction of voltage or current. Virtually all types of circuits exhibit degradation in performance with reduced DC power. As an example, when either current or voltage is reduced, a typical RF amplifier circuit will exhibit degradation of gain, linearity, and dynamic range.

If RF power levels are to be maintained, power amplifier load lines must be altered as supply voltage is reduced. Figure 15.3 illustrates the issue of load line and reduced voltage. The load line labeled "a" represents an optimum, idealized, high-efficiency load line for 6 volt power supply operation. The slope of the load line is determined primarily by the matching characteristics between the device and the load. As voltage is reduced from 6 volts to 3 volts, preservation of the slope of the load line would imply a reduction in power by ~3 dB (load line labeled "b"). Power is estimated to first order as the area underneath the load line and contained by the current-voltage plot. If power is to be maintained, the load line for the 3 volt supply case must be made steeper. This is indicated by load line "c." With a steeper load line, the average current along the load line (which represents DC current drain) is increased, while the intersection of the load line and the I-V curve is raised in voltage. When RF power levels are held constant, moving from load line "a" to load line "c" will tend to have the undesirable effect of reducing the amplifier efficiency.

To compensate for this problem, low voltage devices are designed to be larger than their higher voltage counterparts. Increased periphery devices provide higher peak currents required to maintain power level with decreased voltage. Larger devices, however, exhibit lower impedance that must be matched. Thus, low voltage power amplifier parts have high transformation ratios. Typical 1-4 Watt power amplifier parts for commercial portable wireless products exhibit output impedance of less than a few ohms. High Q matching elements provide some advantage in achieving required transformations but come at high cost and reduced integration.

As designers struggle to maintain power, gain, and efficiency with diminishing supply voltage, linearity must also be maintained. This requirement brings increasing importance to the use of linearization circuitry, feedforward, and pre-distortion schemes. Active bias control is also a valuable circuit design technique for many applications. Bias schemes that cause the transistor bias to rise with increased incident RF power help to provide needed maximum power while maintaining efficiency when lower power levels are needed.

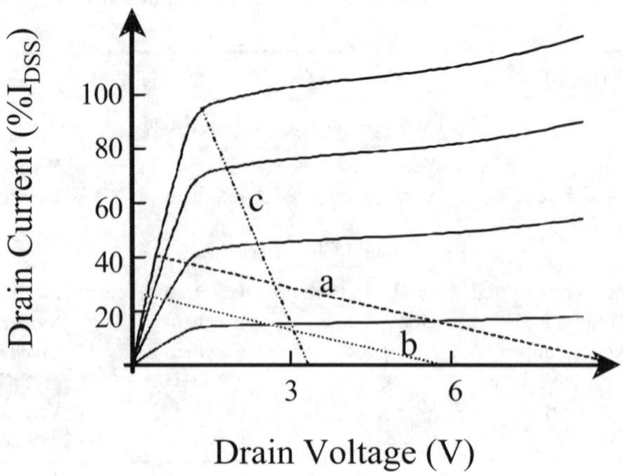

FIGURE 15.3 An illustration of the effect that lowering supply voltage has on power amplifier load line. Load line *a* represents an optimum high efficiency load line for 6-volt supply voltage. Load line *b* results from reducing the supply voltage to 3 volts without modifying the load line slope. The result of using load line *b* is a reduction in output power by more than 3 dB. Load line *c* represents a 3-volt load line where output power level is maintained. The result of using load line *c* is reduced efficiency from the load line *a* case.

Even for low power RF applications, bias control is critical when limited power supply voltage is available. The problem is compounded by device-to-device variability in current-voltage characteristics. Device stacking opportunities are limited and current control bias schemes, as opposed to voltage control schemes, are often preferred.

15.6 Radio and System Architecture

The ongoing communication technology revolution continues to demand greater bit transfer rates and higher bandwidths from wireless products. These demands lead to greater required linearity and higher frequencies of operation from the RF portion of these products. Broader bandwidth, higher frequency operation contributes to increases in battery drain creating a greater challenge to reduce operating voltage and power consumption.

Providing access to significantly increased information content also forces requirements for greater functionality causing a need for greater circuit complexity. Increasing functionality of the product is viewed as a potential method for establishing competitive advantage in an otherwise regulated environment. Built-in address books and more sophisticated displays are just two of the features appearing on new products. Since each new function requires power, this trend also works against the low power consumption imperative.

Other radio architecture trends, such as the movement toward digital modulation, can ease the low voltage/low power design issues. The improved signal-to-noise advantages offered by digital modulation results in reduced output power requirements. Digital modulation schemes can also provide the potential for pulsed (versus CW) operation, reducing power consumption even more.

Some proposed and emerging wireless systems will have a dramatic effect on the war on power consumption, which will change the requirements (and the technologies of choice) entirely.

Micro-cell telephone systems and small area wireless personal area networks will dramatically reduce the battery requirements for these wireless systems. A smaller geographic cell size reduces the required RF output power of both subscriber and base station amplifiers. The required subscriber transmission power is reduced since the total distance that information must be transmitted is reduced. The base station power is also reduced for the same reason and because the total number of users that will be within the smaller cell is reduced. These reductions in required transmission power translate into dramatically reduced power consumption requirements. The power consumption of the RF portion of the radio becomes insignificant compared to the rest of the electronics. The device and material selection process can also become inconsequential. Although from a consumer point of view, this solution offers a significant advantage with no clear downside, the micro-cell infrastructure installation represents a significant investment for the service provider.

In contrast, satellite cellular systems present significant challenges due to the high instantaneous output power requirements of the PA and low minimum noise figure of the front end LNA.

Although the choice of semiconductor material, type of transistor, and circuit topology all affect low voltage design, the radio architecture and system concept can dominate important low voltage performance specifications such as required battery size and battery lifetime.

15.7 Limits to Reductions in Voltage

There are some implications to the reduction of battery size and weight and that will impose serious constraints on the minimum useful operating voltage for portable RF products.

As discussed previously in this section, a primary driver for reduced voltage is the successful evolution of low voltage digital IC technologies. The advantage of running digital and RF circuitry from a single power supply is clear. Issues that limit DC power requirements for RF circuits, however, are fundamentally different than those that limit DC power requirements for associated digital circuits. Digital circuitry is required to store and analyze information that is encoded in a binary manner. This can be accomplished theoretically by the presence or absence of a small charge (even a single electron). Although practical considerations make a single electron memory improbable, and movement of even one electron into and out of storage still requires energy, it is clear that binary data can be manipulated with extremely small amounts of energy. The digital system requirements do not impose arbitrary power requirements on the strength of the digital signal.

In contrast to the digital case, minimum RF power requirements are implied by the radio system design. The RF signal must be transmitted and received across a specified distance or range of distances. Because power is lost in the radiation process, RF circuits must be able to handle power levels that are determined by the propagation media and transmitter-to-receiver separation. If 1 W of transmission power is required for the remote receiver to successfully detect the signal, then reductions in RF power below this level will cause the radio to fail. Unlike the digital situation where any discernable bit is as good as any other bit, the minimum RF transmission power must be maintained. For RF circuitry, increased efficiency and/or increased current must accompany a reduction in voltage.

Most portable units operate at efficiency levels within 2 to 10 percent of the theoretical limits. The implication of near theoretical efficiency is that voltage reductions necessarily involve increased current requirements. If voltage is halved, current must be doubled.

As battery current requirements are increased, the internal resistance becomes a limiting factor in the total power the battery is capable of delivering. For most applications, batteries have been considered ideal voltage sources, but as low voltage electronics trends continue, this assumption will begin to break down.

Figure 15.4 presents the maximum power capability for a battery as a function of nominal voltage and internal resistance. Maximum power for Fig. 15.4 is defined quite optimistically as the nominal voltage times the short circuit current. Although internal resistance is a function of the chemistry as well as the number and size of the battery cells, the values used in Fig. 15.4 are typical for nickel-metal-hydride and Li-ion batteries in common use for portable wireless products today. Further, the trends plotted in the figure will hold for all batteries, regardless of battery chemistry or size. It is clear from the figure that as battery voltages are reduced below approximately 2 volts, the maximum power available from them is reduced to levels on the order of that required from a portable cellular PA alone. When efficiency and other circuit requirements are accounted for, battery power levels are inadequate to support the product.

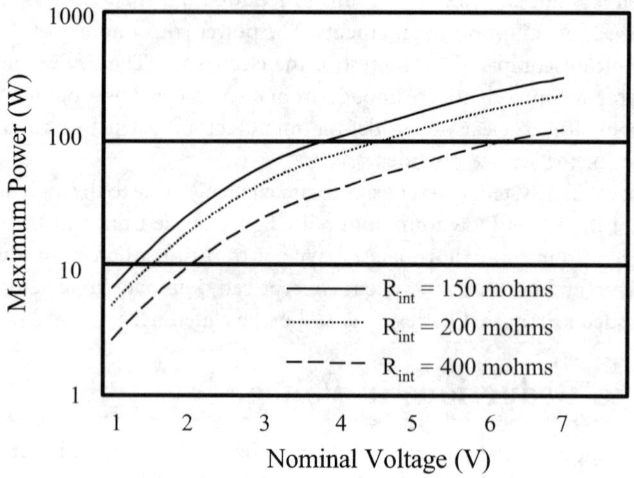

FIGURE 15.4 The maximum power capability for a battery as a function of nominal voltage and internal resistance. Maximum power for this figure is defined as the nominal voltage times the short circuit current.

15.8 Summary

Reductions in voltage and power consumption are important to the development of portable wireless products that are smaller, lighter, and have long battery lifetimes. Low voltage supply requirements are also being driven by a need to make RF ICs compatible with digital ICs. Careful consideration and improvement in the choice of semiconductor material, device type, and circuit topology can help to achieve reductions in voltage. System design and choice of radio architecture also has impact on low voltage/low power electronics. These latter issues have potentially the largest impact on low voltage requirements. The ultimate limit to decreasing battery voltage, however, will be determined by transmitter power requirements and achievable internal resistance of batteries.

The discussion in this section has focused on commercial wireless products. These same trends, however, also describe military, satellite, or avionics communications products — both today and over the past several decades. All of these systems are being driven by the demand to move increasing amounts of information between points via a wireless system.

References

M. Golio, Low Voltage Electronics for Portable Wireless Applications: An Industrial Perspective, 1998 IEEE MTT-S International Microwave Symposium Digest, Baltimore, MD, IEEE Press, June 7–12, 1998.

M. Golio, Low Voltage Microwave Electronics, Santa Clara Valley IEEE MTT Short Course, Santa Clara, CA, Santa Clara Chapter IEEE, MTT, April 4, 1998.

M. Golio, Active Devices for Commercial Subscriber Power Amplifier Products, 1996 URSI Conference, Lille, France, 28 August–5 September 1996.

D. Halchin, M. Golio, C. Farley, Issues with Measurements, Design, and Modeling of Devices for Wireless Applications, Wireless Conference, Sedona, AZ, September 30, 1996.

E. Pettenpaul, New Horizons for the Microwave Business in Europe, *Proceedings of the European Microwave Conference*, Madrid, Spain, 1–5, 1993.

Y. Kitahara, Personal Mobile Communications in Microwave Band, *Proceedings of the European Microwave Conference*, Madrid, Spain, 94–95, 1993.

R. C. Dixon, Portable Communications Needs and No-Nos, *GaAs IC Symposium Digest*, 29–32, 1993.

P. Briere, E. Perea, O. de Saint Leger, System and Technology Aspects of Future Digital Cellular Phones in Europe, *GaAs IC Symposium Digest*, Monterey, CA, 33–36, 1991.

E. Pettenpaul, L. Scharf, K.J. Schopf, Enhanced GaAs Device Concepts for the New Digital Mobile Communication Systems, *Proceedings of the European Microwave Conference*, Madrid, Spain, 132–134, 1993.

G. P. Donzelli, GaAs versus Silicon Technology, *Proceedings of the European Microwave Conference*, Madrid, Spain, 110–112, 1993.

C. Kermarrec, G. Dawe, T. Tewksbury, B. Meyerson, D. Harame, M. Gilbert, SiGe Technology: Applications to Wireless Digital Communications, *IEEE Microwave and Millimeter-Wave Monolithic Circuits Symposium Digest*, San Diego, CA, 1–4, 1994.

M. Hirayama, K. Honjo, M. Obara, N. Yokoyama, HBT IC Technologies and Applications in Japan, *GaAs IC Symposium Digest*, 3–6, 1991.

D. Whipple, The CDMA Standard, *Applied Microwave & Wireless*, 6, 24–37, Spring 1994.

M. Sullivan, Wireless LANs, Status and Future, *Applied Microwave & Wireless*, 6, 54–61, Spring 1994.

Low Voltage Silicon Bipolar Transistors, *Microwave Journal*, 37, 126–128, June 1994.

T. Yamaguchi, T. Archer, R. Johnston, J. Lee, Process and Device Optimization of an Analog Complementary Bipolar IC Technology with 5.5-Ghz fT PNP Transistors, *IEEE Trans. Electron. Devices*, 41, 1019–1026, June 1994.

D. Vook, T. Kamins, G. Burton, P. Vande Voorde, H. Wang, R. Coen, J. Lin, D. Pettengill, P. Yu, S. Rosner, J. Turner, S. Laderman, H-S. Fu, A. Wang, Double-Diffused Graded SiGe-Base Bipolar Transistors, *IEEE Trans. Electron. Devices*, 41, 1013–1018, June 1994.

V. Nair, R. Vaitkus, D. Scheitlin, J. Kline, H. Swanson, Low Current GaAs Integrated Down Converter for Portable Communication Applications, *GaAs IC Symposium Digest*, IEEE Press, 41–44, 1993.

V. Nair, Low Current Enhancement Mode MMICs for Portable Communication Applications, *GaAs IC Symposium Digest*, IEEE Press, 67–70, 1989.

K. Kanazawa, M. Kazumura, S. Nambu, G. Kano, I. Teramoto, A GaAs Double-Balanced Dual-Gate FET Mixer IC for UHF Receiver Front-End Applications, *IEEE Trans. Microwave Theory & Techn.*, MTT-33, 1548–1553, December 1985.

O. Ishikawa, Y. Ota, M. Maeda, A. Tezuka, H. Sakai, T. Katoh, J. Itoh, Y. Mori, M. Sagawa, M. Inada, Advanced Technologies of Low-Power GaAs ICs and Power Modules for Cellular Telephones, *GaAs IC Symposium Digest*, IEEE Press, 131–134, 1992.

P. Philippe, M. Pertus, A 2 Ghz Enhancement Mode GaAs Down Converter IC for Satellite TV Tuner, *Proceedings of the European Microwave Conference*, Madrid, Spain, 61–64, 1993.

H. Heaney, F. McGrath, P. O'Sullivan, C. Kermarrec, Ultra Low Power Low Noise Amplifiers for Wireless Communications, *GaAs IC Symposium Digest*, IEEE Press, 49–51, 1993.

J.O. Plouchart, H. Wang, M. Riet, HBT Monolithic Integrated Phase Locked Oscillator for DCS 1800 Mobile Communications, *Proceedings of the European Microwave Conference*, Madrid, Spain, 834–836, 1993.

M. Nakatsugawa, Y. Yamaguchi, M. Muraguchi, An L-Band Ultra Low Power Consumption Monolithic Low Noise Amplifier, *GaAs IC Symposium Digest*, IEEE Press, 45–48, 1993.

K. R. Cioffi, Monolithic L-Band Amplifiers Operating at Milliwatt and Sub-Milliwatt DC Power Consumption, *IEEE Microwave and Millimeter-Wave Monolithic Circuits Symposium Digest*, Albuquerque, NM, 9–12, 1992.

H. Takeuchi, M. Muraoka, T. Hatakeyama, A. Matsuoka, M. Honjo, S. Miyazaki, K. Tanaka, T. Nakata, A Si Wide-Band MMIC Amplifier Family for L-S Band Consumer Product Applications, *IEEE Microwave Theory and Techniques Symposium Digest*, 1283–1284, 1991.

Y. Imai, M. Tokumitsu, A. Minakawa, Design and Performance of Low-Current GaAs MMICs for L-Band Front-End Applications, *IEEE Trans. Microwave Theory & Techn.*, 39, 209–214, February 1991.

Y. Imai, M. Tokumitsu, A. Minakawa, T. Sugeta, M. Aikawa, Very Low-Current and Small-Size GaAs MMICs for L-Band Front-End Applications, *GaAs IC Symposium Digest*, IEEE Press, 71–74, 1989.

W. Baumberger, A Single Chip Rejecting Downconverter for the 2.44 Ghz Band, *GaAs IC Symposium Digest*, 37–40, 1993.

B. Khabbaz, A. Douglas, J. DeAngelis, L. Hongsmatip, V. Pelliccia, W. Fahey, G. Dawe, A High Performance 2.4 Ghz Transceiver Chip-Set for High Volume Commercial Applications, *IEEE Microwave and Millimeter-Wave Monolithic Circuits Symposium Digest,* San Diego, CA, 11–14, 1994.

M. Wang, M. Carriere, P. O'Sullivan, B. Maoz, A Single-Chip MMIC Transceiver for 2.4 Ghz Spread Spectrum Communication, *IEEE Microwave and Millimeter-Wave Monolithic Circuits Symposium Digest,* San Diego, CA, 19–22, 1994.

J. Moniz, B. Maoz, Improving the Dynamic Range of Si MMIC Gilbert Cell Mixers for Homodyne Receivers, *IEEE Microwave and Millimeter-Wave Monolithic Circuits Symposium Digest,* San Diego, CA, 103–106, 1994.

R. Herman, A. Chao, J. Deichsel, C. Mason, A. McKay, Highly Integrated Mixed-Signal L-Band Configurable Receiver Array, *IEEE Microwave and Millimeter-Wave Monolithic Circuits Symposium Digest,* San Diego, CA, 23–26, 1994.

T. Apel, E. Creviston, S. Ludvik, L. Quist, B. Tuch, A GaAs MMIC Transceiver for 2.45 Ghz Wireless Commercial Products, *IEEE Microwave and Millimeter-Wave Monolithic Circuits Symposium Digest,* San Diego, CA, 15–18, 1994.

R. Meyer, W. Mack, A 1-Ghz BiCMOS RF Front-End IC, *IEEE J. of Solid-State Circuits,* 29, 350–355, March 1994.

K. Sakuno, M. Akagi, H. Sato, M. Miyauchi, M. Hasegawa, T. Yoshimasu, S. Hara, A 3.5W HBT MMIC Power Amplifier Module for Mobile Communications, *IEEE Microwave and Millimeter-Wave Monolithic Circuits Symposium Digest,* San Diego, CA, 63–66, 1994.

Y. Ota, M. Yanagihara, T. Yokoyama, C. Azuma, M. Maeda, O. Ishikawa, Highly Efficient, Very Compact GaAs Power Module for Cellular Telephone, *IEEE Microwave Theory and Techniques Symposium Digest,* Albuquerque, NM, 1571–1520, 1992.

M. Maeda, M. Nishijima, H. Takegara, C. Adachi, H. Fujimoto, Y. Ota, O. Ishikawa, A 3.5 V, 1.3 W GaAs Power Multi Chip IC for Cellular Phone, *GaAs IC Symposium Digest,* Atlanta, GA, 53–56, 1993.

A. Herrera, E. Artal, E. Puechberty, D. Masliah, High Efficiency, Highly Compact L-Band Power Amplifier for DECT Applications, *Proceedings of the European Microwave Conference,* Madrid, Spain, 155–157, 1993.

S. Makioka, K. Tateoka, M. Yuri, N. Yoshikawa, K. Kanazawa, A High Efficiency GaAs MCM Power Amplifier for 1.9 Ghz Digital Cordless Telephones, *IEEE Microwave and Millimeter-Wave Monolithic Circuits Symposium Digest,* San Diego, CA, 51–54, 1994.

T. Kunihisa, T. Yooyama, H. Fujimoto, K. Ishida, H. Takehara, O. Ishikawa, High Efficiency, Low Adjacent Channel Leakage GaAs Power MMIC for Digital Cordless Telephone, *IEEE Microwave and Millimeter-Wave Monolithic Circuits Symposium Digest,* San Diego, CA, 55–58, 1994.

T. Yoshimasu, N. Tanaba, S. Hara, An HBT MMIC Linear Power Amplifier for 1.9 Ghz Personal Communications, *IEEE Microwave and Millimeter-Wave Monolithic Circuits Symposium Digest,* San Diego, CA, 59–62, 1994.

P. O'Sullivan, G. St. Onge, E. Heaney, F. McGrath, C. Kermarrec, High Performance Integrated PA, T/R Switch for 1.9 Ghz Personal Communications Handsets, *GaAs IC Symposium Digest,* Albuquerque, NM, 33–35, 1993.

S. Dietsche, C. Duvanaud, G. Pataut, J. Obregon, Design of High Power-Added Efficiency FET Amplifiers Operating with Very Low Drain Bias Voltages for Use in Mobile Telephones at 1.7 Ghz, *Proceedings of the European Microwave Conference,* Madrid, Spain, 252–254, 1993.

J. Griffiths, 3 Volt MMIC Power Amplifier for Wireless Phones, *Applied Microwave & Wireless,* 6, 38–53, Spring 1994.

K. Inosako, K Matsunaga, Y. Okamoto, M. Kuzuhara, Highly Efficient Double-Doped Heterojunction FET's for Battery-Operated Portable Power Applications, *IEEE Electron. Device Letters,* 15, 248–250, July 1994.

M. Golio, Large Signal Microwave Device Modeling Challenges for High Volume Commercial Communications Applications, *Workshop Proceedings of the European Microwave Conference: Advanced Microwave Devices, Characterisation and Modelling,* Madrid, Spain, 30–33, 1993.

R. Darling, Subthreshold Conduction in Uniformly Doped Epitaxial GaAs MESFETs, *IEEE Trans. Electron Devices,* 36, 1264–1273, July 1989.

M. Golio, Characterization, Parameter Extraction and Modeling for High Frequency Applications, *Proceedings of the European Microwave Conference*, Madrid, Spain, pp. 69-72, 1993.

J. Pedro, J. Perez, A Novel Non-Linear GaAs FET Model for Intermodulation Analysis in General Purpose Harmonic-Balance Simulators, *Proceedings of the European Microwave Conference*, 714–716, 1993.

T. Yoshimasu, K. Sakuno, N. Matsumoto, E. Suematsu, T. Tsukao, T. Tomita, Low Current GaAs MMIC Family with a Miniaturized Band-Stop Filter for Ku-Band Broadcast Satellite Applications, *GaAs IC Symposium Digest*, 147–150, 1991.

N. Shiga, T. Sekiguchi, S. Nakajima, K. Otobe, N. Kuwata, K. Matsuzaki, H. Hayashi, MMIC Family for DBS Downconverter with Pulse-Doped GaAs MESFETs, *GaAs IC Symposium Digest*, 139–142, 1991.

C. Caux, P. Gamand, M. Pertus, A Cost Effective True European DBS Low Noise Converter, *GaAs IC Symposium Digest*, 143–145, 1991.

K. W. Kobayashi, R. Esfandiari, M. Hafizi, D. Streit, A. Oki, M. Kim, GaAs HBT Wideband and Low Power Consumption Amplifiers to 24 Ghz, *IEEE Microwave and Millimeter-Wave Monolithic Circuits Symposium Digest*, 85–88, 1991.

J. M. Golio, M. Miller, J. Staudinger, W.B. Beckwith, E.N. Arnold, *Microwave MESFETs and HEMTs*, Norwood, MA: Artech House, 1991.

R. A. Pucel, D. Masse, R. Bera, Performance of GaAs FET Mixers at X-band, *IEEE Trans. Microwave Theory & Techn.*, MTT-24, 351–360, June 1976.

S. Maas, B. Nelson, D. Tait, Intermodulation in Heterojunction Bipolar Transistors, *IEEE Microwave Theory and Techniques Symposium Digest*, IEEE Press, 91–93, 1991.

M. Kim, A. Oki, J. Camou, P. Chow, B. Nelson, D. Smith, J. Canyon, C. Yang, R. Dixit, B. Allen, 12–40 Ghz Low Harmonic Distortion and Phase Noise Performance of GaAs Heterojunction Bipolar Transistors, *GaAs IC Symposium Digest*, 117–120, 1988.

W. Liu, T. Kim, P. Ikalainen, A. Khatibzadeh, High Linearity Power X-Band GaInP/GaAs Heterojunction Bipolar Transistor, *IEEE Electron. Device Letters*, 15, 190–192, June 1994.

16

Technology Computer Aided Design

Peter A. Blakey
Northern Arizona University

16.1 Introduction

Computer-aided design (CAD) is used in response to complexity in engineering problems. The complexity may be associated with the manipulation and storage of large amounts of relatively simple information; or it may be associated with the interactions of complicated nonlinear physical phenomena. Both types of complexity are encountered in the field of microelectronics. Electronics CAD (ECAD) is used to design new products using an *existing* semiconductor technology, and technology CAD (TCAD) is used to accelerate the development of *new* semiconductor technologies. ECAD is able to manage large amounts of relatively simple information, such as the number, size, and locations of the polygons used to define lithographic masks. TCAD predicts the way in which nonlinear process and device physics would impact the performance of proposed technologies. The subject matter of this chapter is TCAD and its application to the development of RF, microwave, and millimeter-wave semiconductor technologies.

16.2 An Overview of TCAD

The core tools of technology CAD (TCAD) are process simulators and device simulators. Process simulators predict the structures that result from applying a sequence of processing steps to an initial structure. Device simulators predict the electrical characteristics of specified structures. Process and device simulators are often used in combination to predict the impact of process variations on electrical behavior. The basic flow of information is indicated in Fig. 16.1.

Process and device simulators use numerical techniques to solve mathematical equations that describe the underlying physics of an experiment. They are able to predict the result of experiments, without requiring the experiments to be run. Predictive simulation is useful whenever it is time consuming, expensive, dangerous, or otherwise difficult to run real experiments. In the case of microelectronics development, it is both time consuming and expensive to run real experiments in a semiconductor fab. As a result, it can be cost effective to substitute simulated experiments for some of the real experiments that would otherwise be required by purely empirical development procedures.

0-8493-1562-3/02/$0.00+$1.50

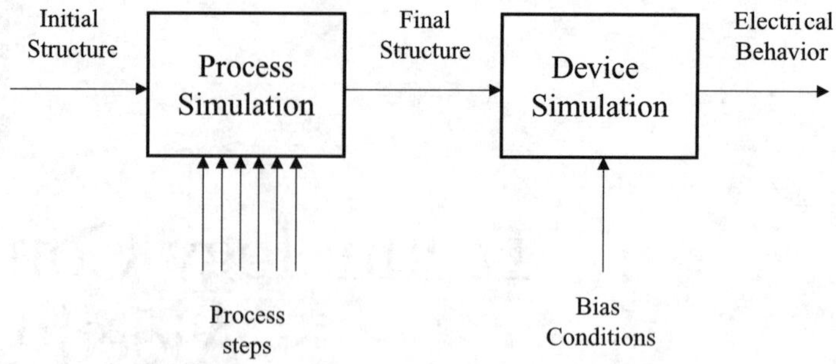

FIGURE 16.1 The basic flow of TCAD information.

Process simulators incorporate physical models for phenomena such as diffusion, oxidation, ion implantation, deposition, and etching. Device simulators solve equations that express charge continuity and the dependency of the electrostatic potential on the distribution of charge. In order to solve the charge continuity equations, device simulators need models for charge transport and for the generation and recombination of charge carriers. Implementing process and device simulators is a challenging task that requires knowledge of physics, numerical methods, visualization techniques, user interfaces, and software engineering. Research groups in universities and in industry were the first to develop this type of software. However, these groups were not well suited to the task of maintaining the software and providing long-term support to users. General-purpose process and device simulation software is now available from several commercial vendors who provide maintenance and support. Most microelectronics companies use commercial software as the foundation of their TCAD capabilities.

Commercial TCAD systems also include tool integration utilities that make it convenient to use the core tools in combination, and task integration environments that automate large-scale simulation-based experimentation. The tool integration utilities include run time environments, visualization tools, structure editors, layout editors and interfaces, and optimization capabilities. Task integration usually supports a "Virtual Wafer Fab" paradigm that makes it convenient to define and run simulated split lot experiments. After a simulated experiment is defined, the system automatically generates all of the associated jobs, farms out the jobs across a computer network, and collects the results. It may also provide modeling capabilities that make it easy to analyze the simulated results.

Predictive, physics-based simulation is different from data modeling. Simulation predicts the data that an experiment would yield if it were run. Modeling provides a compact description of existing data that may come from either real or simulated experiments. Data models are used to interpolate, but cannot be used to extrapolate. Expectations regarding "accuracy" are quite different for data models and for physics-based simulators. The agreement between modeled and experimental data is normally very good. The agreement between simulated data and experimental data is subject to several sources of uncertainty that can lead to significant discrepancies between simulated data and experimental data. These sources of uncertainty are discussed in a later section.

Process and device simulators were implemented initially to assist the development of digital Si VLSI technologies. In recent years the capabilities of TCAD software have been extended to cover a range of different material systems, and to provide convenient RF-oriented analysis capabilities. The widespread use of epitaxially grown materials and low thermal budget processing means that process simulation for III-V hetero-structure technologies can often be reduced to the definition of material layers and doping densities, with geometric specification of deposition and etch steps. This avoids difficult problems associated with process simulation of silicon technologies, which must account for moving boundaries during oxidation, complicated diffusion mechanisms, and so forth. Device simulation for advanced material systems often requires advanced models for charge transport, and it is usually necessary to account for the impact of deep level traps. The appropriate models have been incorporated into some of the commercially available device simulators.

In principle, device simulators can predict any electrical behavior that can be measured. Simulators targeted toward the development of digital silicon VLSI technologies are primarily concerned with the calculation of DC I-V characteristics, but they also provide basic capabilities for calculating small-signal AC behavior and large-signal transient behavior. Mixed-mode device-and-circuit simulation capabilities are also available. Mixed-mode simulation is basically SPICE-like circuit simulation that uses numerical device simulation to model the behavior of one or more of the active devices in a circuit. Other types of analysis are feasible, but are not available in commercial simulators. One example is the prediction of noise behavior. Another example is harmonic balance analysis of device operation that is implemented at the level of the underlying physics, rather than at the circuit level. Computational load-pull systems seek to mirror the functionality of experimental load-pull systems. These systems predict RF output power, power-added efficiency, gain, linearity, and tunability, and present the results as a function of load (or source) impedance plotted on a Smith chart. Such systems have been implemented in industry, but are not yet offered by commercial vendors.

16.3 Benefits of TCAD

Simulated experiments can replace a significant fraction of the real experiments that are otherwise required by traditional, empirical development procedures. Simulated experiments are almost always performed more rapidly and more cheaply than real experiments. The use of TCAD can therefore yield significant reductions in both cycle times and development costs. Reduced development times lead in turn to a faster time to market, and this normally translates into higher profits. The magnitude of the financial benefits depends both on the skill with which TCAD is deployed and on the characteristics of the market for a given technology. TCAD is most useful when the market is high volume, and both performance and cost driven. Until recently, this meant that TCAD was applied primarily to the development of silicon VLSI technologies. As a result of the wireless revolution, TCAD is now being used to accelerate the development of analog, RF, microwave, and millimeter-wave technologies.

The high costs of semiconductor fabs, material, and personnel mean that reducing development costs by even a few percent is enough to justify significant investments in TCAD capabilities. When TCAD is used effectively it can yield direct savings in development times and development costs of around 50%, and can result in enhanced profits (due to faster time to market) that are much greater than the direct cost savings. Savings of this magnitude can, in principle, justify investments of tens or even hundreds of millions of dollars in TCAD capabilities. The investments that companies make in TCAD have traditionally been much smaller than this.

Simulation can predict all of the quantities that are measured routinely. It can also predict the values of quantities that are either very difficult or impossible to measure. Examples of such quantities are the doping profiles, carrier profiles, and electrostatic potential distributions within a semiconductor device. By examining these quantities, engineers gain insight into the subtleties of device operation. This insight can guide the direction of future development, and it can lead to the development of simple analytic theories that capture the essence of device operation. Although the enhanced physical insight is very useful, it is difficult to quantify the value in monetary terms.

Process and device simulators encapsulate and integrate the knowledge of experts in a way that makes detailed knowledge and experience accessible on demand to less expert process and device engineers. For example, it is possible for a device engineer to account for the impact of complicated transport phenomena on device performance, without having any knowledge of the associated physics. This function of encapsulation and integration helps to ameliorate both the shortage of experienced technical personnel and the comparatively narrow technical span of most experts. Once again, it is difficult to quantify the value in monetary terms.

16.4 Limitations of TCAD

For certain widely used types of simulation, the underlying physics is well established and, for practical purposes, "exact." For example, electromagnetic simulation is based on Maxwell's equations, thermal simulation is based on the heat flow equation, and circuit simulation is based on Kirchoff's Current and Voltage Laws. The underlying physics of process and device simulation is less well defined. For each physical phenomenon (e.g., ion implantation, or electron transport) there is a hierarchy of models of varying complexity. All of these models introduce approximations. In general, the more complicated models provide a more complete description, but are more difficult to implement and involve longer (often much longer) computation times. The need to select appropriate accuracy-efficiency trade-offs can make it difficult for nonexperts to utilize TCAD effectively.

After a physical model has been selected, it is necessary to define values for the parameters associated with the model. These parameters must usually be defined over a range of different conditions (e.g., for different temperatures, doping concentrations, or electric fields.) The values of the parameters may be obtained from experiments or from theory. In either case, there is often considerable uncertainty in the accuracy of the model parameters.

Process and device simulators solve systems of coupled, nonlinear equations that do not have general analytic solutions. The equations are therefore solved using numerical techniques that are implemented in computer software. Numerical techniques provide an approximate solution that is defined on a discrete "mesh" of points. The choice of mesh and numerical techniques has a major impact on the accuracy and efficiency with which solutions are obtained. Using a finer mesh provides solutions that have higher accuracy but are calculated more slowly. Numerical techniques for solving nonlinear equations employ sequences of approximations that are expected to converge to a solution. In some cases the sequence of approximations fails to converge. Because of these issues, users of TCAD need some expertise in defining meshes and numerical techniques that provide acceptable trade-offs between accuracy, efficiency, and stability.

Discrepancies between measured and simulated results arise because of difficulties in specifying equivalent experiments. For example, a real diffusion step may specify a thermal ramp in terms of the times and temperatures that are dialed in as input to a controller. Process simulation will normally use the same times and temperatures. However, in the real experiment the temperature of the wafer as a function of time is different from the nominally specified ramp. The temperature as a function of time may also vary from wafer to wafer within a lot, and from point to point on a given wafer. Another example is provided by device simulation. The lattice temperature used in device simulation is often assumed to be 300 K, but the real lattice temperature is a function of both the true ambient temperature and any self-heating that occurs. Other sources of uncertainty are systematic measurement and repeatability.

In order to apply TCAD effectively, users need technical skills that encompass physics, numerical techniques, and the application domain. They need to understand and employ effective application methodologies, and they need good judgment to assess the validity and usefulness of TCAD predictions that are impacted by several sources of uncertainty. They also need to be able to "sell" their insights to process and device engineers who do not understand the role of simulation and are skeptical of its value, and they need to be comfortable with having a secondary support role within a technology development organization.

16.5 The Role of Calibration

By calibrating simulated results to experimental results, it is possible to ameliorate some of the issues that were outlined in the previous section. The existence of high-quality calibration is a prerequisite for using TCAD for certain applications. Achieving a good calibration involves a considerable amount of effort, and poor calibration can exacerbate the problems outlined previously. In view of the importance of calibration to the effective use of TCAD, a review of the associated issues is presented in this section.

Calibration attempts to reduce some of the errors associated with simulation. The measure of the errors is taken as the differences between simulated and measured results. Calibration seeks to reduce these errors by adjusting the values of the model parameters used by the simulators. In essence, it is assumed that the various sources of errors can be partially compensated for by using appropriate "effective" values of model parameters. A good calibration must use a sufficiently complete set of output variables, i.e., quantities for which the predicted and measured results are to be brought into better agreement. It must also use an appropriate set of input variables, i.e., model parameters whose values are adjusted to improve the agreement between predicted and measured values. A good calibration must not allow the adjusted values of the model parameters to assume implausible values.

Calibration is often performed using ad hoc procedures that lead to unsatisfactory results. A very common error is to adjust too few input variables, or the wrong input variables. A related error is to allow input variables to assume values that are implausible or nonphysical. Another common problem is that some key output variables are neglected, and the associated errors diverge as a result of the calibration attempt. Calibration should be set up as a clearly defined optimization problem and the techniques used for calibration should draw on known methods for nonlinear optimization involving multiple constrained input variables. Black-box optimization can be performed, using small-change parameter sensitivity matrices calculated using TCAD. However, it is usually more efficient to use a sequential, step-by-step calibration procedure in which only a subset of the input and output variables are varied at each step. For example, in the case of a very simple calibration of a GaAs MESFET simulation, the threshold voltage could be matched by adjusting the channel charge. The low-field mobility could then be adjusted to fit the on-resistance in the linear region, and the saturated carrier velocity could be adjusted to fit the saturated current.

Performing a good calibration takes a lot of effort, but certain applications of TCAD only become viable after a good calibration has been achieved. Regardless of the amount of effort expended on calibration, there will still be residual errors. This is because calibration does not remove approximations in the underlying physics; it simply allows a portion of their impact to be absorbed into the values of the model parameters. No set of model parameters can correct for an underlying physical model that is inadequate. The converse of this observation is that the degree to which a calibration is successful provides insight into the adequacy of the underlying physics.

16.6 Applications of TCAD

The block diagram shown in Fig. 16.2 provides a high-level view of the activities associated with traditional, empirical development of semiconductor technologies. The development process involves a make-and-measure cycle that is repeated until acceptable results are obtained. The process recipe is then transferred to a fab; and layout rules and electrical models for active and passive components are transferred to the circuit design community. Circuits are designed, masks are made, and the circuits are fabricated.

The block diagram shown in Fig. 16.3 provides a high-level view of the activities that are associated with more modern simulation-oriented development strategies. Depending on the ways in which TCAD is used, modern development methodologies may be viewed as being "TCAD influenced," "TCAD driven," or "system driven via TCAD." TCAD-influenced development uses process and device simulation to accelerate the make-and-measure phase of technology development. TCAD-driven development makes more extensive use of process and device simulation in the areas of modeling, characterization, and manufacturing. System-driven development supplements TCAD-driven development by adding explicit bidirectional links to system requirements and basic architectural choices. Much of modern development is TCAD influenced. Some companies use TCAD-driven methodologies, and a few companies are implementing system-driven methodologies.

It is useful to order individual TCAD activities by the amount of calibration required to achieve useful results. One such list, ordered by increasing calibration requirements, is as follows:

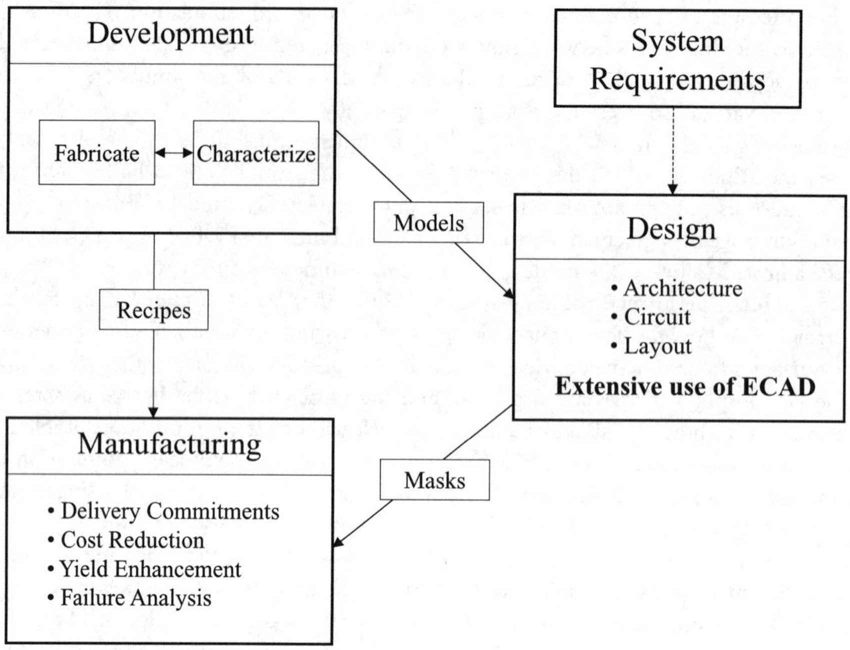

FIGURE 16.2 A high-level view of the traditional development methodology.

- Concept evaluation
- Technology discrimination
- Technology development
- Simulation-based model development
- Simulation-assisted characterization
- Statistical process design
- Technology transfer
- Process control
- Yield enhancement and failure analysis

Prior to the availability of TCAD, there was considerable reliance on simplified analytic models. Gross oversimplification of these models would lead to unrealistic performance projections, e.g., to the prediction of useful transistor performance at THz frequencies. TCAD is now used to perform rapid evaluations of proposed new concepts. Accounting for Poisson's equation, one or more carrier continuity equations, the coupling between these equations, and the need for device contacts ensures more realistic estimates of device performance. Projections of wildly overoptimistic device performance have been substantially eradicated by the widespread availability of TCAD. Concept evaluation usually requires little calibration effort, since all that is required is a go/no-go assessment.

At the initial stages of technology development, it is sometimes necessary to select between two or more competing approaches. TCAD is well suited to this purpose. Calibration is usually not very critical because the errors that are introduced by simulation tend to affect both technologies approximately equally, and are unlikely to cause a simulation-based ranking to be different from an experimentally determined ranking.

Once an overall approach has been selected, TCAD is used to explore different sub-options, and to assist process optimization. This is presently the single most important application of TCAD. Simulation-based design and optimization can and should be done systematically, but in practice it is often done

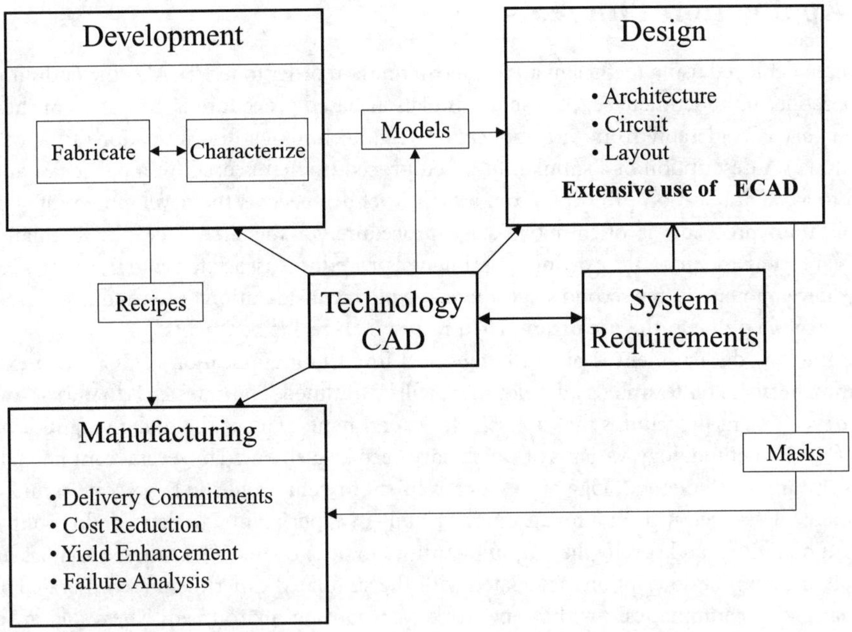

FIGURE 16.3 A high-level of modern development methodologies.

using inefficient ad hoc techniques. Application methodologies associated with simulation-based technology development are discussed in the next section. As the results of real experiments become available they should be fed back to refine the simulator calibrations, in order to improve the usefulness of simulation at each stage of design and optimization. A general goal is for the calibration achieved for one generation of a technology to provide a solid starting point for the simulation-based design and optimization of the next generation of the technology.

TCAD can also be used to assist modeling and characterization activities. In the area of device modeling, TCAD can be used as a reference standard that enables careful testing of assumptions made in the course of developing device models, and supports systematic assessments of the global accuracy of a device model. TCAD is especially useful for assessing the accuracy of large-signal models that are constructed from DC and small-signal AC data. The large-signal behavior predicted by the model is compared to the large-signal behavior predicted directly using TCAD. The development and characterization of charge-based models is also facilitated by the use of TCAD, since the calculated charge distributions within a device can easily be integrated and partitioned in accordance with charge-based modeling approaches. TCAD can also be used to assist in device characterization. For example, it is quite straightforward to predict the values of bias- and temperature-dependent source and drain access resistances. Simulation can also assist in the interpretation of pulsed I-V measurements that are used to characterize thermal effects and low-frequency dispersive effects associated with deep-level traps.

When there is sufficiently high confidence in the quality of an underling calibration, the uses of TCAD can be expanded further. Calculation of process sensitivities provides a foundation for statistical process design, design centering, and yield modeling. Information concerning process sensitivity is also a useful input for process control. Simulation of the impact of large-change variations (e.g., the impact on electrical behavior of skipping a process step) can provide "signature" information that assists failure analysis. The ideal situation is when a comprehensive set of simulated data, together with an associated set of input data, forms an integral part of the technology transfer package that is supplied by a development organization to a manufacturing organization.

16.7 Application Protocols

TCAD is often deployed using inefficient ad hoc techniques. In order to use TCAD efficiently, it is helpful to develop, refine, and document sets of useful simulation-based procedures. The details of these procedures often vary considerably from one application (e.g., technology discrimination) to another (e.g., failure analysis.) A description of a simulation-based procedure that is used for a particular application is referred to as an application protocol. From a high-level perspective, the development of simulation-based application protocols is often a two-stage procedure. In the first stage, the simulation-based procedures are set up to mirror the existing experimental procedures associated with traditional, empirical technology development. In the second stage, there is explicit consideration of the additional possibilities that can be exploited due to the use of simulated rather than real experiments.

Since technology development is presently the most important application of TCAD, an example of an application protocol for technology development will be outlined. The first step is to mirror traditional, empirical development procedures that use split-lot experiments. This is the underlying idea of Virtual Wafer Fab (VWF) methodology, variants of which have been implemented by several commercial vendors of TCAD software. VWF methodology makes it convenient to define and run large-scale simulated split-lot experiments. Users select from a broad range of built-in experimental designs, define split variables among input quantities, and specify the output quantities to be extracted. Once these variables have been specified, all of the input descriptions associated with the simulated experiment are generated automatically, and farmed out automatically within a networked computing environment. The results are collected automatically, stored in a database, and are conveniently viewed and analyzed using built-in tools.

The next step is to define ways in which a simulation-based approach can go beyond traditional development procedures. When using TCAD it is possible to use sophisticated "black box" optimization techniques that require the ability to calculate accurate derivatives in a multidimensional search space. Due to long cycle times, high costs, and noisy experimental data, this approach is not feasible in the real world. However, it is quite suitable for use in the simulation world, where results can be obtained quickly and reproducibly. The use of black-box optimization focuses explicit attention on the need to define an objective function, and also on the need to define all of the constraints on the input variables. Split-lot experimentation also involves objective functions and constraints, but they are usually defined and implemented less formally.

A focus on objective functions and constraints leads to a useful high-level view of technology development in general and the role of simulation in particular. Technology development can be viewed as a search, within a multidimensional design space, for the point corresponding to an optimum design. Various techniques are available for narrowing down the regions of the design space that need to be searched. These techniques include analytic theory and scaling arguments, numerical simulation, and real experiments. Each of these techniques has its advantages and limitations. Analytic theory and scaling arguments are quick and easy to apply, and can eliminate large regions of the design space, but their absolute accuracy is poor. Simulation can provide much better accuracy, and is relatively fast and inexpensive, but is subject to model errors and numerical error. Experiments provide "real" data, but are expensive, time consuming, and subject to experimental error. Analytic theory, simulation, and experiments can be viewed as a hierarchy of search techniques that can be applied sequentially and iteratively to maximize the efficiency of the search process.

A simplified view of the search for an optimum design in a two-dimensional space is presented in Fig. 16.4. Analytic theories and scaling arguments are used to establish the boundaries of the region of the design space that needs to be searched. Simulation is then used to predict the optimum point in the design space. Due to modeling and numerical errors, this predicted optimum (indicated by the closed circle) is not exactly coincident with the true optimum point. The region around the predicted optimum is therefore searched experimentally, to determine the location of the true optimum (the open circle). According to this view, the goal of simulation is to predict an optimum design with sufficient accuracy that the true optimum can be found using only a small number of real experiments. Thus, simulation

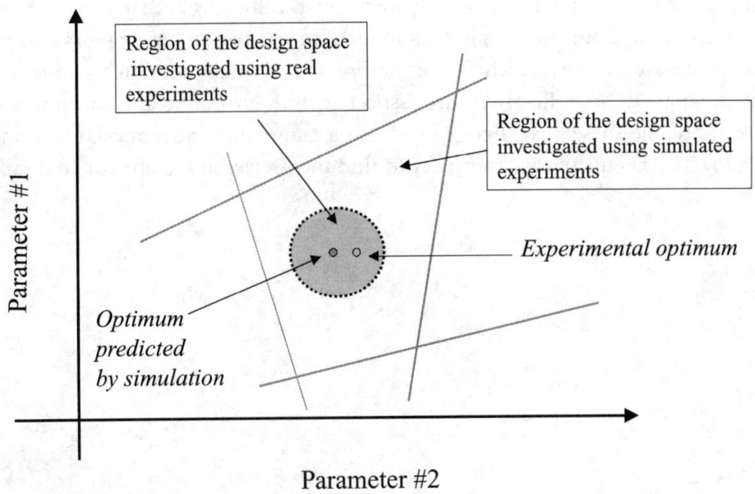

FIGURE 16.4 A simplified view of the search process.

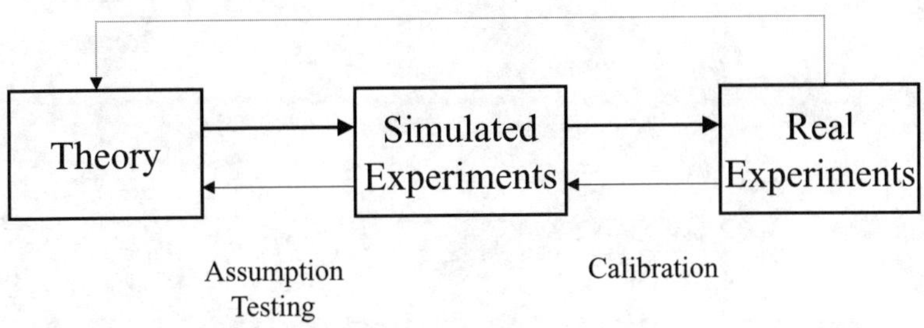

FIGURE 16.5 A three-level hierarchical approach to technology development.

does not eliminate the need for experiments, but it is realistic to expect a reduction of 50 to 80% in the number of real experiments that would be required if simulation is not used.

Figure 16.5 provides another view that indicates how theory, simulated experiments, and real experiments are linked in modern development procedures. The forward arrows indicate the basic sequence of development activities as outlined in the previous paragraph. The reverse arrows indicate that information is fed back to allow simulator calibration and assumption testing. Subsets of this view correspond to other development methodologies. For example, eliminating simulation results in a "traditional" development methodology, in which analytic theory and scaling arguments are used to bracket a design space that is then searched experimentally. Eliminating theory results in a black-box-simulation-reliant development methodology that will usually be inefficient in its search of an unconstrained design space.

Similar approaches can be used to develop application protocols for the other uses of TCAD.

16.8 Conclusions

Technology CAD provides benefits associated with faster development times, lower development costs, enhanced physical insight, and the encapsulation and integration of expert knowledge. These benefits are especially important in the context of high-performance, high-volume, consumer-oriented applications. Technology CAD is very well established as an integral part of the procedures used to develop silicon VLSI technologies, and is assuming an important role in the development of semiconductor technologies for RF, microwave, and millimeter-wave applications.

Although the benefits of using TCAD can be very great, achieving the benefits is not easy. Some of the challenges, such as the definition of physical models, model parameters, calibration procedures, and application protocols, are primarily technical in nature. Other challenges, such as a shortage of trained manpower able to apply TCAD effectively, are associated with economic and social factors. Companies that are able to overcome these challenges can obtain a competitive advantage. Conversely, companies that do not deploy TCAD effectively can expect to find themselves at a competitive disadvantage.

17

Nonlinear Transistor Modeling for Circuit Simulation

Walter R. Curtice
W.R. Curtice Consulting

17.1 Modeling in General

By definition, a transistor model is a simplified representation of the physical entity, constructed to enable analysis to be made in a relatively simple manner. It follows that models, although useful, may be wrong or inaccurate for some application. Designers must learn the useful range of application for each model.

It is interesting to note that all transistors are fundamentally nonlinear. That is, under any bias condition, one can always measure harmonic output power or intermodulation products at any input RF power level, as long as the power is above the noise threshold of the measurement equipment. In that sense, the nonlinear model is more physical than the linear model.

The purpose of this work is give a tutorial presentation of nonlinear transistor modeling. After reviewing the types of models, we will concentrate on equivalent circuit models, of the type used in SPICE.[1] Recent improvements in models will be described and the modeling of temperature effects and the effects of traps will be discussed. Finally, parameter extraction and model verification is described.

0-8493-1562-X/03/$0.00+$1.50

17.1.1 Two-Dimensional Models

The models constructed for describing the nonlinear behavior of transistors fall into several distinctly different categories, as depicted in Table 17.1. The most complex is the "physics-based" model. Here electron and hole transport is described by fundamental transport and current continuity relationships and the physical geometry may be described in one-, two-, or even three-dimensional space. The electric field is found by solution of Poisson's equation consistent with the distribution of charge and boundary conditions. Such a model may use macro-physics, such as drift-diffusion equations,[2-4] or more detailed descriptions, such as a particle-mesh model with scattering implemented using Monte-Carlo methods.[5,6] Two- and three-dimensional models must be used if geometrical effects are to be included. The matter of how much detail to put into the model is often decided by the time it takes for the available computer to run a useful simulation using the model. In fact, as computers have increased their speed, modelers have increased the complexity of the model simulated.

TABLE 17.1 Types of Large-Signal Transistor Models

I. Physical or "Physics-Based" Device Models
II. Measurement-Based Models
1 - Anaytical Models, such as SPICE Models
2 - Black Box Models
Table-Based Models
Artificial Neural Network (ANN) Models

The lengthy execution time required for Monte-Carlo analysis can be reduced by using electron temperature[7] as a measure of electron energy. Electron temperature is determined by the standard deviation of the energy distribution function and is well defined in the case of the displaced Maxwellian distribution function. Electron and hole transport coefficients are developed as a function of electron temperature, and nonequilibrium effects, such as velocity overshoot in GaAs, may be simulated in a more efficient manner. However, the solution of Poisson's equation still require appreciable computational time.

BLAZE[8] is a good example of a commercial, physically-based device simulator that uses electron temperature models. BLAZE is efficient enough to model interaction of a device with simple circuits.

The physics-based model would be constructed with all known parameters and simulations of current control for DC and transient or RF operation, then compared with measured data. Using the data, some transport coefficients or physical parameters would be fine-tuned for best agreement between the model and the data. This is the process of calibration of the model.[9] After calibration, simulations can be trusted to be of good accuracy as long as the model is not asked to produce effects that are not part of its construction. That is, if trapping effects[10] have not been incorporated into the model, the model will disagree with data when such effects are important. With the physics-based model, as with all others, a range of validity must be established.

Present physics-based models still require too much computational time to be used to any extent in circuit design work. Optimization of a circuit design will involve invoking the device model frequently enough to be impractical with physics-based models. These models can be used if only one or two nonlinear transistors are used in a specific circuit, but typically, the circuit designer has a larger number of nonlinear devices.

Several quasi two-dimensional models[11,12] have been developed that execute more efficiently. Initial results look good, but accuracy may depend upon the simplifications made in the development of the code and will vary with the application.

17.1.2 Measurement-Based Models

The next general category is that of "measurements-based" models. These are empirical models either constructed using analytical equations and called "analytical models" or based upon a lookup table developed from the measured data. The latter are called "table-based" models. Multidimensional spline functions are used to fit the data in some of these models[13] and only the coefficients need be stored.

In the case of analytical models, the coefficients of the equations serve as fitting parameters to permit the equations to approximate the measured data. Functions are usually chosen with functional behavior similar to measured data so that the number of fitting parameters is reduced.

The advantages of analytical models are: computational efficiency, automatic data smoothing, accommodation of device statistics, physical insight, and the ability to be modified in a systematic manner. Disadvantages are: restriction of behavior often due to use of over-simplified expressions, difficulty in parameter extraction, and guaranteed nonphysical behavior in some operating condition. The nonphysical behavior is often associated with the use of a function, such as a polynomial, to fit data over a specific range of voltages and subsequent application of the model to voltages outside this range. The function may not behave well outside the fitting range. The best example of analytical models is the set of transistor models used in the various forms of the SPICE program. A major advantage of analytical models is that all the microwave nonlinear simulators provide some sort of user-defined model interface for analytical model insertion.

Table-based models have some properties of black-box models. The equations used result from fitting to the data, using splines, or other such functions. These models can therefore "learn" the behavior of the nonlinear device and are ideal for applications where the functional form of the behavior is unknown. Table-based models are efficient but do not provide the user with any insight, since there is a minimal "circuit model." They have difficulty incorporating dispersive effects, such as "parasitic gating" due to traps (see the section on Modeling the Effects Due to Traps) and do not accommodate self-heating effects.

The model cannot be accurately extrapolated into regions where data was not taken, and the models are often limited in their application due to the particular coding used by their author. This means that the model cannot be tailored by users other than the author. Customization of models is important to improve the "performance" of a model. The first table-based model that has been widely used is the Root model.[13]

17.1.3 Physical Parameter Models

One may argue that there is a class of models between physics-based and analytic models, namely, physical-parameter models. A good example would be the Gummel-Poon model.[14] Here, analytical equations are used but the fitting parameters or equation coefficients have physical significance. For example, NF, the ideality factor of the emitter-base junction, is one model parameter. This model is an analytical model but more useful device information may be gleaned from the value of the coefficients. This is often the case and the model is widely used for various forms of bipolar devices.

A useful physical parameter model for the AlGaAs/InGaAs/GaAs PHEMT has been published and verified by Daniel and Tayrani.[15] No information has been given on the range of validity of the analytical model, and it is expected that such a simple model will have inaccuracy when two-dimensional effects or nonequilibrium effects are important. It is interesting that the HEMT structure has less two-dimensional effects than the MESFET because of the sheet current layer produced in the HEMT.

There are many analytical models for which the coefficient has the name and dimensions of a physical parameter but the coefficient is only a fitting parameter and is not strongly related to the physical parameter. Khatibzadeh and Trew[16] have presented one commonly called the Trew model and Ladbrooke[17] has presented a second model commonly called the Ladbrooke model. Ladbrooke's model is an extension of the much earlier Lehovec and Zuleeg[18] model and it is more empirical than physical, as Bandler has shown.[19] Such models must be tested to see how strong the relationship is between the model parameter and the physical parameter. It is a matter of the degree of correlation between the two quantities. It is dangerous to attach too much physical significance to these coefficients. One should verify the relationship before doing so.

There is a unique case where physical parameters have been installed in a previously developed analytical model. The Statz-Pucel (analytical) GaAs MESFET model[20] has been converted to a physical parameter model by D'Agostino et al.[21]

17.1.4 Neural Network Modeling

Rather recently, a new approach has been developed for the modeling of nonlinear devices and networks. It utilizes artificial neural networks, or ANN. ANN models are similar to table-based or black-box models in that there is no assumption of particular analytical functions. As with table-based models, the ANN model "learns" the relationship between current and voltage from the data, and model currents are efficiently calculated after application of voltages. ANN analysis can treat linear or nonlinear operation of devices or complex circuits. ANN models have many of the advantages and deficiencies of table-based models. An excellent special issue of *RF and Microwave Computer-Aided Engineering*[22] has been devoted to this modeling method. Unfortunately, discussion of this approach is beyond the scope of this work.

17.2 Scope of This Work

The purpose of this work is to present a tutorial on the modeling of the nonlinear behavior of transistors. The scope is limited to nonlinear models useful for the development of circuit designs. It would not be possible to cover all the important material on other types of models, such as physical models, in this article. This paper will deal primarily with nonlinear analytical models for MESFETs, (P)HEMTs and HBTs. The emphasis is on GaAs device models, although many of these models are also used for transistors fabricated in InP, silicon, and other materials. The RF LDMOS power device is also discussed because it can be treated as a three-terminal device, much like a MESFET.

We address the concerns of analog and digital circuit designers who must choose between a wide variety of nonlinear models for transistors. Of particular concern here is the MMIC designer who must select the proper model to use for his GaAs microwave transistor. Even with very complex models presently supplied in circuit simulators, some specific behaviors are not modeled and new model features are required. We will spend much time on SPICE models and SPICE-type models. We will inspect some of the recent models that incorporate important device effects, omitted in previous models. We will discuss the modeling of gate charge as a function of local and remote voltages, the modeling of self-heating effects, the modeling of trapping effects, and model verification.

Unfortunately, the references presented will only be representative of prior work because there is a wealth of papers in each area of transistor modeling. I apologize to any author not included, as there are now and have been many people working in this area. I do recommend some modeling tutorial articles, previously published. Trew[23] and Snowden[24] and Dortu et al.[25] have presented excellent reviews of SPICE type transistor models and are recommended reading.

17.3 Equivalent Circuit Models

We concern ourselves here with equivalent circuit models because they are formulated to be efficiently exercised in a circuit simulator, and thus are efficient for circuit design and optimization. This is because the simulator is accustomed to dealing with resistors, capacitors, inductors, and voltage or current-controlled sources.

One problem is that "de facto" standard models have evolved in the industry, and often these models are inadequate to describe the device behavior. This is even true for small-signal equivalent-circuit models. Still, all circuit simulators utilize standard model topologies for small-signal and large-signal MESFET and PHEMT models. These models represent a minimum number of elements and are efficient for evaluation of transistor characteristics. Figures 17.1 and 17.2 show the conventional topologies for small-signal and large-signal simulation, respectively. It is conventional to separate the extrinsic parameters from the intrinsic device parameters, as shown in Fig. 17.1. The intrinsic parameters are assumed to contain all the bias-dependent behavior and the extrinsic parameters are assumed to be of constant values.

Curtice and Camisa[26] and Viakus[27] have discussed the trade-offs that exist between simple models with a small number of parameters, and complex models with a large number of parameters. A primary

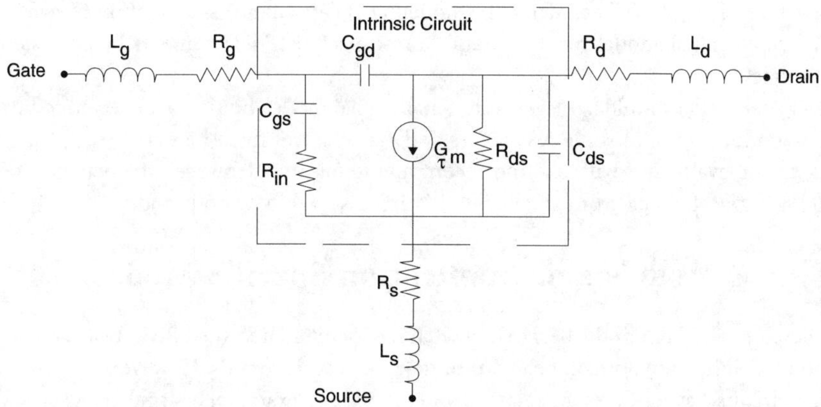

FIGURE 17.1 The conventional small-signal model for a MESFET, showing intrinsic and extrinsic elements.

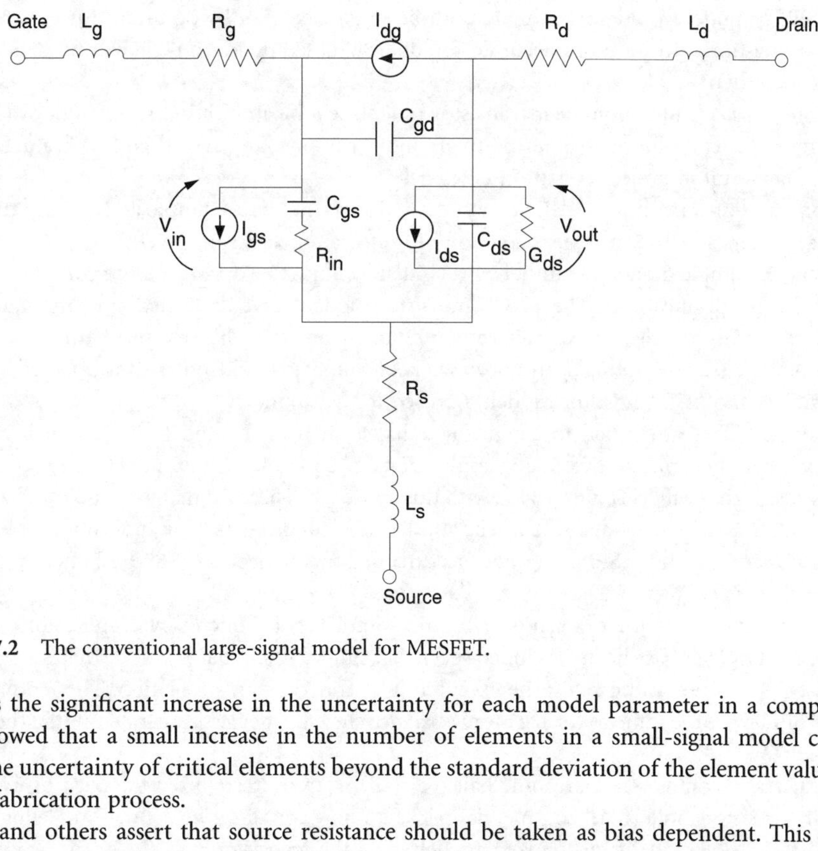

FIGURE 17.2 The conventional large-signal model for MESFET.

concern is the significant increase in the uncertainty for each model parameter in a complex model. Viakus showed that a small increase in the number of elements in a small-signal model could easily increase the uncertainty of critical elements beyond the standard deviation of the element value resulting from the fabrication process.

Byun[28] and others assert that source resistance should be taken as bias dependent. This decision is actually a choice made by the modeler. If source and drain resistance are taken as constant, then the reference planes defining these resistances are taken as being close to the metal ohmic contacts and not too close to the Schottky contact. That is, no region that may become depleted of change is included. All bias-dependent behavior is then lumped into the intrinsic elements. This is the convention followed by most modelers. It results in a simpler model, with fewer parameters.

Unfortunately, many of the published nonlinear device models have inconsistencies with the conventional small-signal model. Some of the inconsistencies may go unnoticed but can cause design errors. A good example that will be described later is the modeling of capacitance as a function of two independent voltages. The small-signal model must contain "transcapacitance" elements to be consistent with the large-signal model.

The large-signal model should agree with the small-signal model, but is not expected to be as efficient. Because transconductance and some resistances and capacitances must be evaluated from functions of voltages, numerical evaluation will take more computational time. However, the behavior of the large-signal model will gracefully go from small-signal to large-signal in a good model.

17.4 SPICE Models and Application-Specific Models

SPICE was developed to help in the design of switching circuits. Thus, the SPICE transistor models were developed to model the time domain behavior of devices in such circuits. However, transistors operating in RF analog circuits have a different locus of operation. For example, a simple class A amplifier will have locus of operation around its quiescent bias point. This should be compared with a logic circuit where the transistors go from a biased-off condition (high voltage, low current) to a strongly turned on condition (low voltage, high current). If the same transistor is used in these two applications, one would expect the SPICE model would approximate both behaviors, but not be optimum for either. In fact, if the model is fine-tuned to be more accurate for one of these applications, it will by default, be less accurate for the other.

For these reasons, accurate nonlinear transistor models will be application specific. In order to make the model more general, the model can be made more complex and more model parameters will be added. This may result in poorer execution efficiency.

Clearly one goal of transistor modeling should be to keep the model simple and to keep the number of model parameters small so that the extraction of these parameters is more efficient. My experience is that designers like simple models for initial work and are willing to work with more complicated models for difficult design specifications. The SPICE transistor models serve the function of the initial, simple models. These models are also universally known by name and have history and familiarity associated with them. A designer attempting to use a new GaAs foundry would not be intimidated by obscure nonlinear models if SPICE transistor models are used in that foundry.

Many modelers have attempted to extend or enhance the SPICE models so that their accuracy is improved, particularly in microwave analog applications. Usually, the default model is the original SPICE model and the designer will feel comfortable with this approach. There are many examples. The Gummel-Poon BJT model has been extended by Samelis and Pavlidis[29] and others[30] for application to heterojunction bipolar devices. The JFET SPICE model was extended by Curtice[31] in 1980 for better application to GaAs MESFET logic circuits.

Because of the increasing use of harmonic-balance simulators for microwave applications, many new equivalent circuit models have been developed specifically for these simulators. Nevertheless, these models are "SPICE-type" models, and can also be executed in a time domain simulation. The requirements of a model for SPICE are the same as for harmonic balance since the device is operated in the time domain in both simulators.

The producers of commercial harmonic balance (HB) software recognized the need of users to customize their transistor models. All commercial packages now contain user-defined modeling interfaces that permit the installation of customized models into the transistor model library. The process of installing or customizing a model in SPICE is much more difficult and not available to the average user. However, the ease of installation of models into HB software has produced a rash of new models for many transistor types.

Table 17.2 shows the typical array of SPICE equivalent circuit models available as part of a commercial simulator software package. The models are categorized, in general, as diode models, GaAs MESFET or (P)HEMT models, MOS models, and bipolar device models. The list is not complete for any specific product but representative of the models available. The models listed in Table 17.2 are in most commercial simulator products whether a version of SPICE or a harmonic balance simulator.

TABLE 17.2 SPICE Models

GaAs MESFET/HEMT	MOS Models
Curtice (Cubic and quadratic)	BSIM 1,2,3 3v3
STATZ (Raytheon)	UC Berkeley 2 and 3
JFET (N & P)	HSPICE
TOM (TriQuint's Own Model)	MOSFET (various levels)
Materka	
Diode Models	**BJT Models**
P/N diode	Gummel-Poon
PIN diode	METRAM
	VBIC

17.5 Improved Transistor Models for Circuit Simulation

Early SPICE models have shown a number of deficiencies. One problem is that the models developed before 1980 were developed for silicon devices and they do not reflect the behavior of GaAs devices. Most SPICE models need to be customized to be accurate enough for present design requirements. With regard to GaAs MESFET and PHEMT modeling, the strong dependency of gate-source capacitance upon drain-source voltage as well as gate-source voltage is not modeled in the early SPICE models. None of the standard SPICE models accommodate self-heating effects. These effects are more important in GaAs applications due to the poorer thermal conductivity of GaAs compared to silicon. Some GaAs transistors exhibit important dispersion effects in transconductance as well as in drain admittance. All the GaAs models in Table 17.2 were added during the 1980s. These models represented major improvements; however, deficiencies remain. These deficiencies are summarized below for SPICE large-signal models:

- Insufficient accuracy for GaAs applications.
- Poor modeling of nonlinear capacitance.
- Poor modeling of self-heating effects.
- No modeling of dispersion of transconductance.
- Model parameter extraction not defined.
- Poor modeling of nonlinear effects dependent upon higher order derivatives.

The improved large-signal models of the 1990s exhibit some common features. It is quite popular to utilize analytical functions that have an infinite number of derivatives. For example, in the modeling of PHEMTs, Angelov et al.[32] have relied heavily on the hyperbolic tangent function for current because its derivative with respect to gate-source voltage is a bell-shaped curve, much like transconductance in PHEMTs. All further derivatives also exit. The Parker[33] model also utilizes higher order continuity in the drain current description and its derivatives.

Some SPICE models do have continuous derivatives but may not be accurate. The differences between the COBRA[34] model and the previous Materka[35] model are more evident when the derivatives of current (first through third) are compared. Since the COBRA model has derivatives closer to the data, Cojocaru and Brazil[34] show that the model predicts intermodulation products more accurately.

Many SPICE models use a simple expression for junction capacitance in MESFETs and PHEMTs. However, capacitance values extracted from data show that the gate-source and gate-drain capacitance depends strongly on the remote voltage as well as the local voltage (or capacitance terminal voltage). The Statz (Raytheon), the TOM, and the EEFET3 SPICE models[36] all have detailed equations for the gate, drain, and source charge as a function of local and remote voltages. Extraction of coefficients for these expressions is not simple and this will be discussed in the next section.

17.6 Modeling Gate Charge as a Function of Local and Remote Voltages in MESFETS and PHEMTS

Small-signal modeling of GaAs and InP MESFETs and PHEMT show that both C_{gd}, the gate-drain capacitance, and C_{gs}, the gate-source capacitance, vary with change of both V_{gs}, the gate-source voltage, and V_{ds}, the drain-source voltage. Thus, these capacitances are dependent upon the local, or terminal voltage, and a remote voltage. The dependency upon the local voltage is expected for capacitances, but the dependency upon the remote voltage leads to a term called "transcapacitance." The modeling of these capacitances can lead to nonphysical effects if not handled properly, as Calvo et al.[37] have shown.

Harmonic balance simulators work with charge functions whose derivatives are capacitive terms. The conventional approach is to find some charge function for total gate charge, such as Q_g (V_{gs}, V_{ds}). Then,

$$C_{11} = \text{Partial Derivative of } Q_g \text{ with respect to } V_{gs}$$

$$C_{12} = \text{Partial Derivative of } Q_g \text{ with respect to } V_{ds}$$

For consistency with small-signal models:

$$C_{11} = C_{gs} + C_{gd}$$

$$C_{12} = -C_{gd}$$

The problem remaining is to partition the total gate charge Q_g into charge associated with the gate-source region, Q_{gs}, and charge associated with the gate-drain region, Q_{gd}. Then, for large signal modeling, the gate node has charge Q_g, the source node has charge $-Q_{gs}$, and the drain node has charge $-Q_{gd}$, and

$$Q_g = Q_{gs} + Q_{gd}.$$

One scheme for partitioning the charge is used in the EEFET model and described in the HP-EEsof Manual.[36] One advantage of this approach is that the model becomes symmetrical, meaning that drain and source may be interchanged and the expressions are still valid.

A simpler approach is presented by Jansen et al.[38] Jansen assumes that all of the gate charge is associated with the gate source region and:

$$Q_{gd} = 0$$

$$Q_g = Q_{gs}.$$

The topology for this approach is different than the conventional model. There is no drain-gate capacitance element. Instead, the transcapacitance term accounts for the conventional drain-to-gate capacitive effects. That is, a change in V_{ds} produces current in the gate source region through the transcapacitance term. Figure 17.3 shows the new topology for the intrinsic circuit.

The procedure for modeling the capacitive effects is the same for both approaches and is the following:

1. Measure small-signal values of C_{gs} and C_{gd} as a function of V_{gs} and V_{ds}
2. Choose a Q_g function and optimize the coefficients of the Q_g expression for best fit of

$$C_{11} = C_{gs} + C_{gd}$$

$$C_{12} = -C_{gd}$$

A good example of the fitting functions and the type of fit obtained is given by Mallavarpu, Teeter, and Snow[39] for a PHEMT.

In summary, large-signal capacitive effects are modeled by constructing a total gate charge function, $Q_g(V_{gs}, V_{ds})$, whose partial derivative approximates the measured capacitance functions. If device sym-

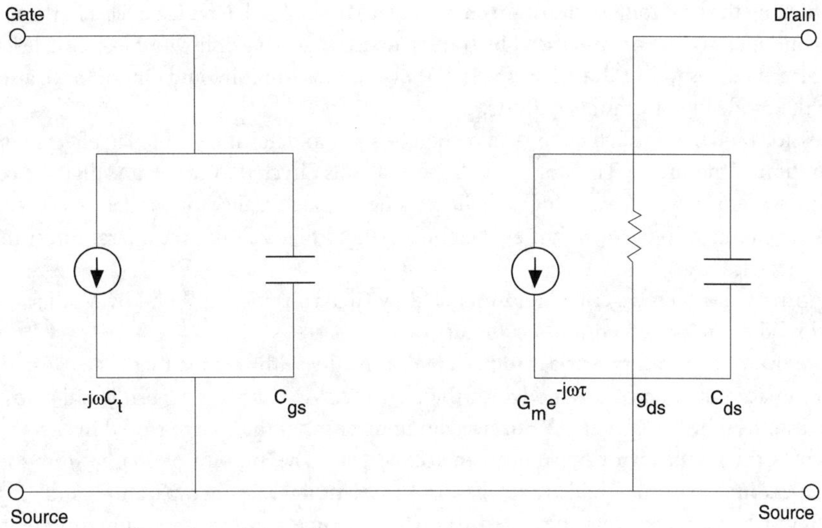

FIGURE 17.3 Jansen's topology for the intrinsic circuit.

metry is important, the EEFET charge partitioning scheme may be used. For amplifier applications, the Jansen model is simplest to code and implement because there is no charge partitioning expression. However, the Jansen model uses a topology that is not conventional.

17.7 Modeling the Effects Due to Traps

Electron and hole traps exist in GaAs materials and cause numerous effects during operation of a GaAs MESFET or PHEMT transistor. The following is a brief listing of these effects:

- Dispersion in transconductance and output admittance
- Backgating
- Parasitic bipolar effects
- "Kinks" in the I/V relationship
- Surface gating
- Gate and drain lag effects during switching
- Light sensitivity
- Substrate current or lack of current pinchoff

These effects have been studied and circuit-level models developed to simulate the effects. In many cases, the details of the behavior have been made clear using two-dimensional simulation modeling. For example, Li and Dutton[10] used PISCES-IIB to show that the common EL2 trap causes dispersion in the output conductance of a GaAs MESFET up to several hundred Hz.

The circuit-level modeling of dispersion is described by Cojocaru and Brazil.[34] They extend the previous conventional modeling of dispersion of the output conductance to include dispersion of the transconductance. The circuit is very simple. A second voltage-control current source in parallel with a resistance is capacitively coupled to the internal drain-source terminals. This enables the model's transconductance and drain-source conductance to be tailed for high frequencies using these new elements.

Horio and Usarni[40] use two-dimensional simulation to show that a small amount of avalanche breakdown in the presence of traps causes excess hole charge in the substrate that produces "kinks" in the low-frequency I/V data. Since the traps cannot be easily eliminated, the kinks may be removed by removing the conditions initiating avalanche breakdown.

Upon switching the gate voltage, the drain current of a MESFET will have lag effects in the microsecond and millisecond regions that are produced by traps. Curtice et al.[41] have shown the circuit-level modeling of such gate lag effects as well as drain lag effects. Others, such as Kunihiro and Ohno[42] have also presented circuits for the modeling of drain lag effects.

The transistor model with such circuits may then be used to determine if the lag effects interfere with proper operation of the circuit. The work of Curtice et al. was directed toward GaAs digital circuits where the switching waveform must be of high quality. Using the new transistor model, one may determine not only if the circuit will perform, but also circuit changes that will permit operation in the presence of strong lag effects.

Light sensitivity has been described and modeled by Chakrubarti et al.[43] and by Madjar et al.[44] Some circuit-level models are presented in their discussions.

In the microwave application arena, the principal difficulty with traps is that they cause difficulty in determining an accurate microwave model for the transistor. An excellent experimental study of surface gating effects is given by Teyssier et al.[45] Surface gating means that the charge stored in surface states and traps influences the I/V behavior by acting as a second gate. The amount of charge stored in the traps will vary, depending upon the applied voltages, the ambient light, the temperature, and trapping time constants. Teyssier et al. show that measured trap capture time constants are quite different than trap emission time constants. They show how they are able to accurately characterize the I/V behavior for RF operation by using 150 ns bias pulse width. They also describe how they characterize the thermal behavior using longer pulse width.

Many others have published data showing surface gating effects (see, for example, Platzker et al.[46]). The approach a modeler should use is to first determine how important such trapping effects are in the transistor operation. In many transistor designs, the active region is shielded sufficiently from surface charge so that negligible surface gating occurs. In that case, low-frequency I/V data and the resulting transconductance may be very predictive of microwave-frequency behavior. If the trapping effects are found to be of importance, then short-pulsed characterization is required and low-frequency I/V data will not suffice.

In any case, device characterization must include the behavior changes due to self-heating and ambient temperature effects. The modeling of heating effects will be discussed next.

17.8 Modeling Temperature Effects and Self-Heating

Anholt and Swirhun[47] and others have documented the changes of GaAs MESFETs and HEMTs at elevated temperatures. However, the modeling of a device over a temperature range is often accomplished using temperature coefficients. With regard to drain current and DC transconductance, the effects of elevated temperature is quite different at large channel current as compared to operation near pinch off. At large channel current, the electron mobility decrease with temperature increase is most important, whereas at low current, the decrease in the pinch-off voltage with temperature is most important. This produces the interesting effect that transconductance decreases with temperature at large current but increases with temperature at very low currents. This behavior is best modeled using a temperature analog circuit that will be described later.

Figure 17.4 shows the behavior of DC drain current for a 0.25-μm PHEMT at four different ambient temperatures. Heating effects are obvious in Fig. 17.4 where current (and transconductance) near positive V_{gs} is reduced and current (and transconductance) near pinch off is increased.

The effects of temperature upon drain current in MESFETs and PHEMTs may be considered second order but they are of first order in bipolar devices. The reason is due to the current exponential dependence upon temperature in a bipolar device. The gate current effects due to temperature in MESFETs and PHEMTs are of first order for the same reason. First order and even some second order changes with temperature may be modeled using linear temperature coefficients if the changes are reasonably linear.

Many devices are operated such that their self-heating effects are quite important. This can occur, for example, in a power amplifier design where the standby biasing current is low, but the advent of input

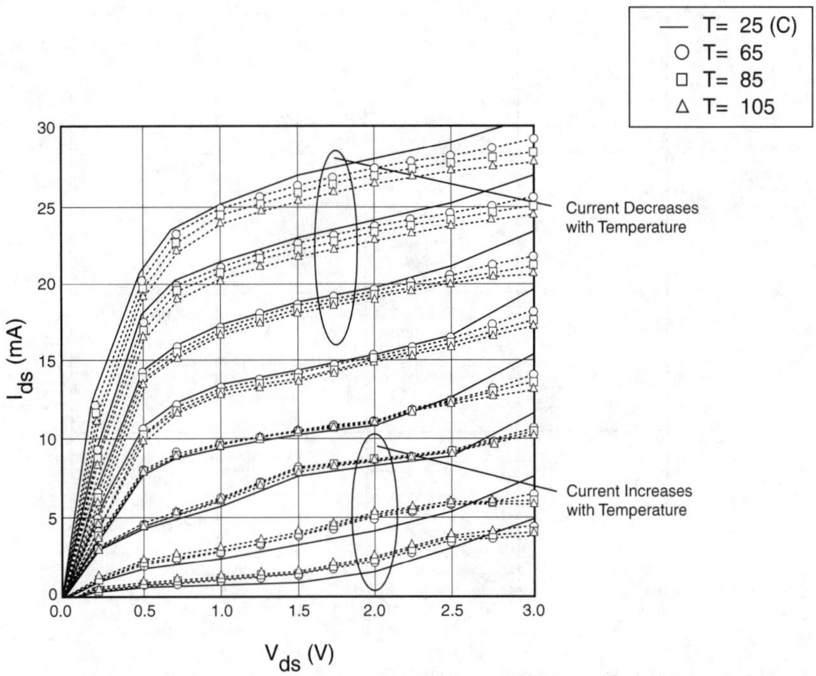

FIGURE 17.4 The behavior of DC drain current for a 0.25-μm PHEMT at four different temperatures.

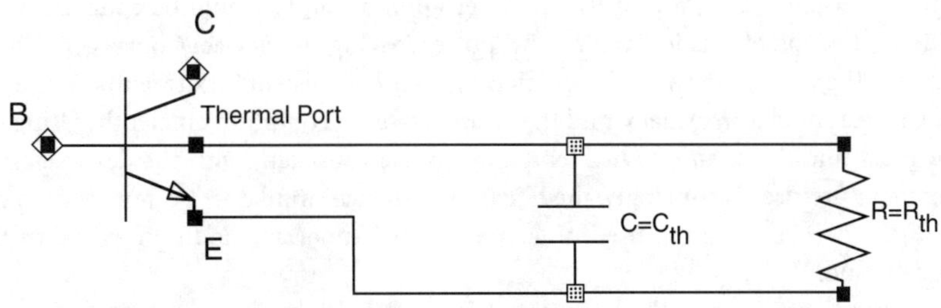

FIGURE 17.5 The HBT transistor model with the thermal analog circuit.

RF power turns on the drain current and output RF power. The ambient temperature may not change but the device will operate with more elevated temperature with the application of the RF power. The use of temperature coefficients may not be accurate for such simulations.

More accurate modeling of self-heating effects in transistors circuits simulators has been done for about ten years using a thermal analog circuit, first used in SPICE applications. It has been found to work well for bipolar simulation as well as for MESFET and PHEMT simulations. The early studies were reported by Grossman and Oki,[48] F. Q. Ye,[49] and others.[50-52] The main differences in these early studies relate to the description of temperature effects in the transistor and not to the CAD model used for simulation in SPICE.

Figure 17.5 shows an HBT transistor model with the additional thermal analog circuit. The device can be of any type but must have well-defined descriptions of the model coefficients as a function of temperature. The analog circuit consists of a current source, a resistor, and a capacitor. R_{th} is the value of thermal resistance and the R_{th}*C_{th} time constant is the thermal time constant of the device. The current source to the thermal circuit, I_{th}, is equal in magnitude to the instantaneous internal dissipated power to the device. For DC biasing, I_{th} to the thermal circuit would be equal to the total biasing power to the device and the temperature rise would be numerically equal to I_{th} times R_{th}. For RF or transient conditions,

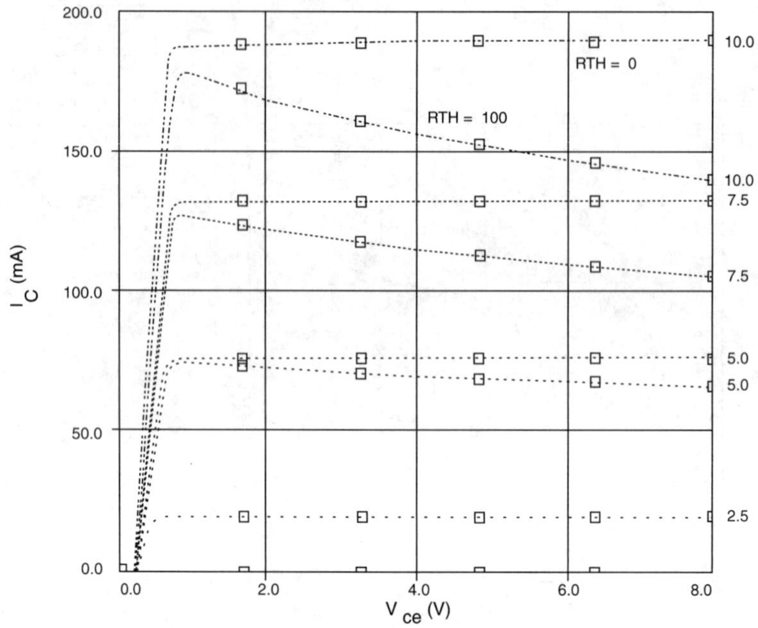

FIGURE 17.6 The collector I-V relationship for an HBT with and without self-heating effects.

the average temperature rise would be that evaluated over some period of time including the effects of the thermal time constant. Thus, for an RF amplifier application, I_{th} would be equal to the DC biasing power plus the RF heating effects less the net RF power leaving the device. Convergence in a simulator is not assured since the value of the model coefficients must be consistent for the temperature of the device.

Harmonic balance simulators usually find the steady-state RF condition efficiently. Using this method, the simulator must find the solution with temperature rise consistent with the device parameters producing the temperature rise. Surprisingly, the harmonic balance simulators do not seem to be much less efficient when the thermal circuit is used, unless a thermal runaway condition exists. In that case, no solution will be found.

Figure 17.6 shows the collector I-V relationship for an HBT exhibiting self-heating effects. The current curves without heating effects are flat in the saturation region. Heating of the lattice reduces the electron mobility, and thus reduces the collector current.

17.9 Enhancing the Gummel-Poon Model for Use with GaAs and InP HBTs

The Gummel-Poon model, or the GP model,[14] is a complex, physical parameter model with 55 parameters, and is widely used. It was developed early for SPICE, and all colleges and universities teach their electrical engineering students to use this model. Although the GP model has many parameters, the current expressions are relatively simple. In addition, the current parameters are more closely tied to material parameters rather than manufacturing tolerances, so that there is less variation in current control characteristics than with MESFETs and PHEMTs. The standard bipolar device has less two-dimensional effects than do MESFETs and PHEMTs.

Much effort has been expended to improve the accuracy of compact BJT circuit models for silicon devices. Fossum[53] has reviewed the effort to 1989 and it continues to this day.

Whether the bipolar device is all silicon or a heterojunction device made with SiGe on Si, AlGaAs on GaAs, GaInP on GaAs, or InP on InGaAs, the bipolar action with current gain is the same physical process. So one expects some similarities in the analysis and modeling of the device. However, the heterojunction with the wide bandgap emitter causes the details of the analysis and model to have significant differences from a homojunction device.

The standard SPICE GP model has a number of major deficiencies that must be addressed before it can be used to accurately model a GaAs-based HBT in a large-signal microwave application. First of all, the SPICE code has silicon bandgap parameters hard coded into it and this must be changed to produce the correct temperature effects upon the bandgap. Next, collector-to-base avalanche breakdown must be added because it is important to most GaAs applications. The GP model uses PTF, a phase function, to accommodate the time delay associated with transconductance. It is more convenient for microwave engineers to use the time delay term TAU, as used in MESFET models.

The parameter "Early Voltage" is not as important in GaAs modeling, as it is usually very large. This is because the base doping can be made an order of magnitude larger than for silicon devices, because of the wide bandgap emitter. The large base doping reduces the importance of collector biasing upon the base region, and thus upon the collector current.

There may be dispersion in the collector admittance in HBTs, so some RF conductive element may be needed between collector and emitter.

The manner in which F_t, the frequency for unity current gain, changes with voltage and current is quite different in GaAs devices than with silicon devices. Therefore, a new functional form is need here. Most of the behavior of F_t with respect to collector voltage is related to the electron velocity-electric field (v-E) curve for the material. In the case of silicon devices, F_t generally increases with collector voltage and saturates until heating effects cause a decrease. Figure 17.7 shows such behavior for a SiGe HBT. The I/V characteristic of the device is given in Fig. 17.8.

In the case of GaAs HBTs, F_t peaks at a low voltage and monotonically decrease with further increase of collector voltage. Figure 17.9 shows such data and the devices I/V characteristics are given in Fig. 17.10. This difference in behavior reflects the striking differences between silicon and GaAs v-E curves in the high field region.

There are a multitude of new CAD models formulated to model the behavior of GaAs and InP HBTs. Two examples are the VBIC model[54] and the MEXTRAM model[55] and these are installed on a number

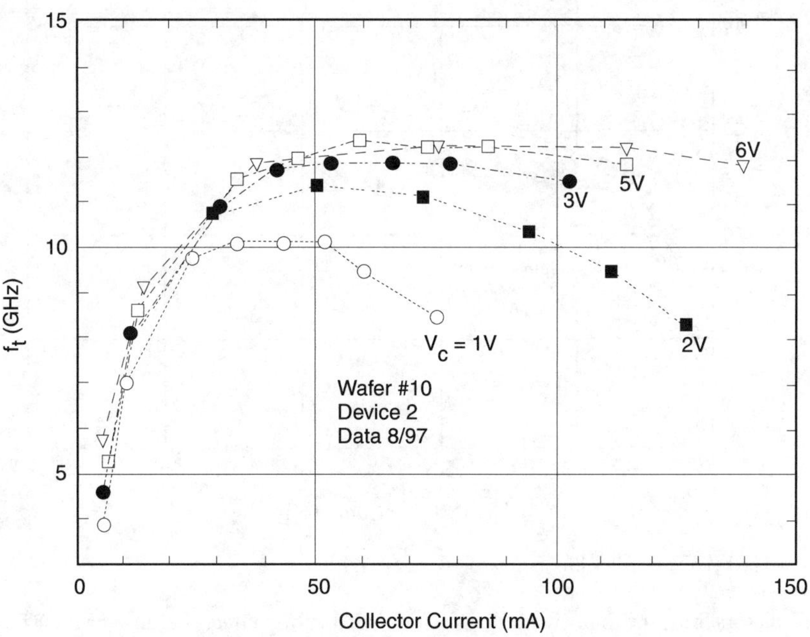

FIGURE 17.7 The behavior of F_t with biasing for a SiGe HBT.

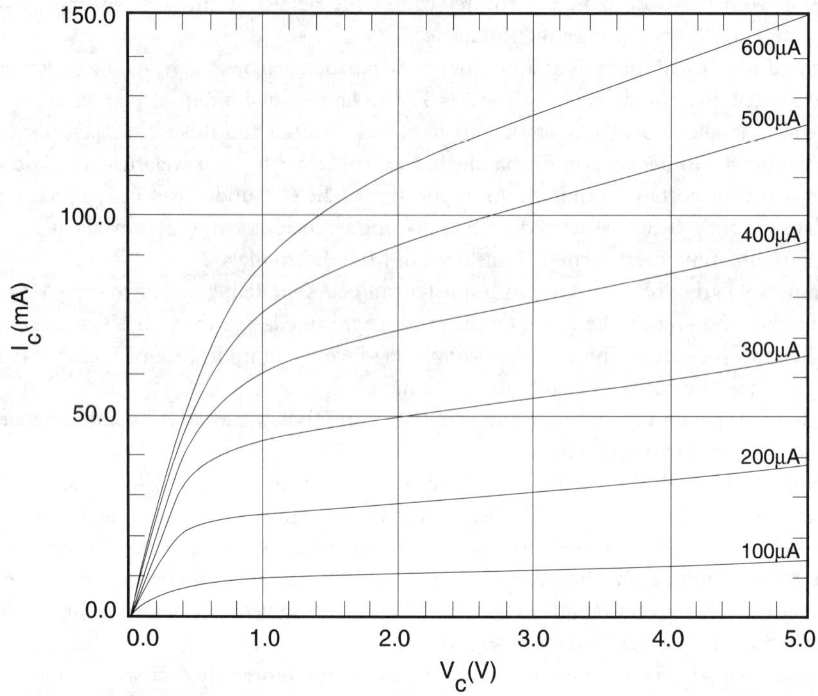

FIGURE 17.8 The I-V relationship for the device of Fig. 17.7.

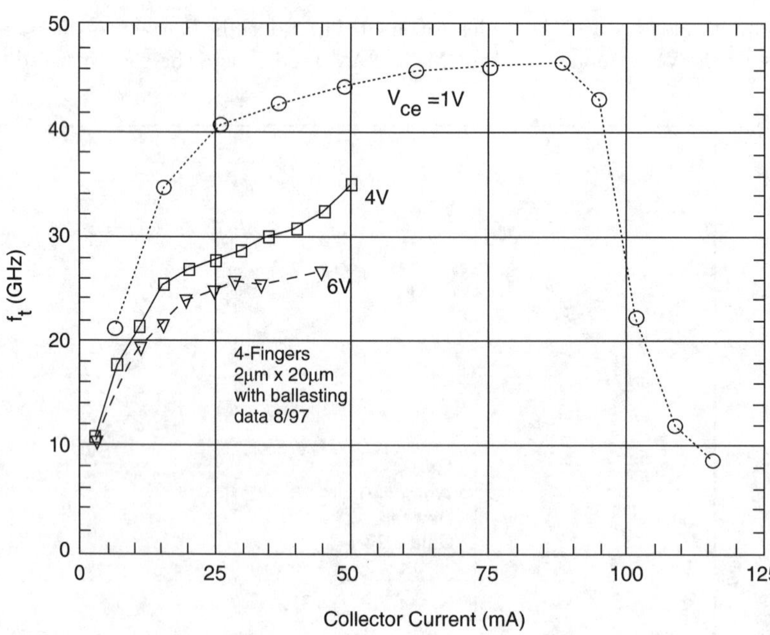

FIGURE 17.9 The behavior of F_t with biasing for GaAs HBT.

of circuit simulators. Most of the new models include self-heating effects because of their importance to device operation and accurate modeling. Because of the significantly poorer thermal conductivity in GaAs compared to silicon and because of the higher power density for best operation in GaAs, self-heating effects are usually important to the operation of the GaAs-based HBTs.

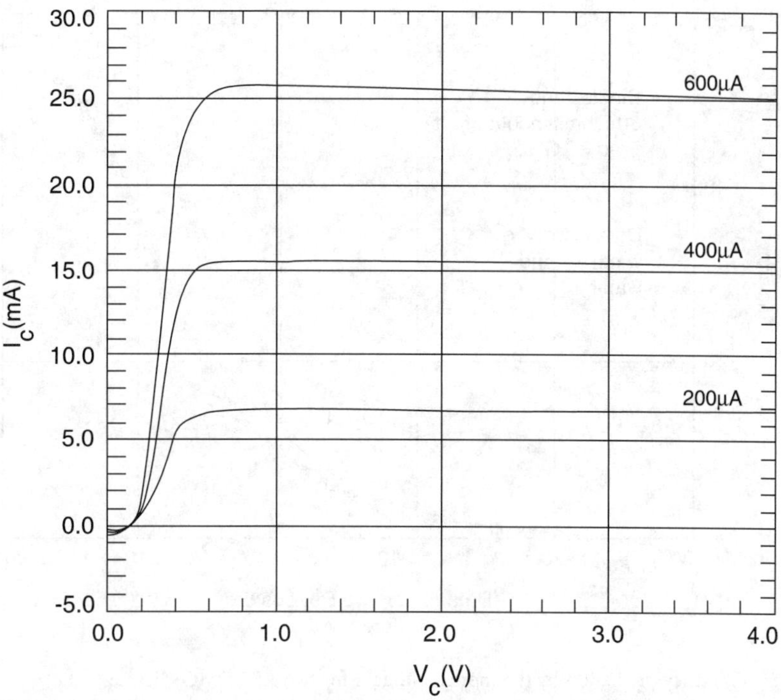

FIGURE 17.10 The I-V relationship for the device of Fig. 17.9.

17.10 Modeling the RF LDMOS Power Transistor

There are numerous silicon MOS models available for DC and RF modeling of silicon transistors. Because the silicon LDMOS transistor has become important for cost-effective consumer applications, many companies have developed nonlinear models specifically for this device. The device incorporates a p-type sinker diffusion to ground the source to the substrate, and thus can be treated as a three-terminal device. This makes it possible to construct a much simpler model, one very similar to the SPICE models developed for the GaAs MESFET.

Perugupalli et al.[56] have used a SPICE circuit network incorporating the standard NMOS SPICE element. Motorola uses the Root model developed for GaAs MESFETs for characterizing the device. However, self-heating effects, important for power applications, cannot be incorporated into this model. The Ho, Green, and Culbertson model[57] based upon the SPICE BSIM3v3 model suffers from the same problem.

Miller, Dinh, and Shumate[58] developed analytical current equations for the device, which led to the development of a new, simpler SPICE model by Curtice, Pla, Bridges, Liang, and Shumate.[59] The model includes self-heating effects, is accurate for both small and large-signal simulations, and operates in transient or harmonic balance simulators. Figure 17.11 shows the model predicts power-added efficiency in excellent agreement with the data for an RF power sweep. A similar model based upon the same equations has been developed and verified by Heo et al.[60]

17.11 Parameter Extraction for Analytical Models

The extraction of parameters for a device model has become less laborious since the advent of new extraction and equipment control programs, such as IC-CAP by Hewlett Parkard and UTMOST by Sylvaco International. These programs provide data acquisition and parameter extraction. Such software

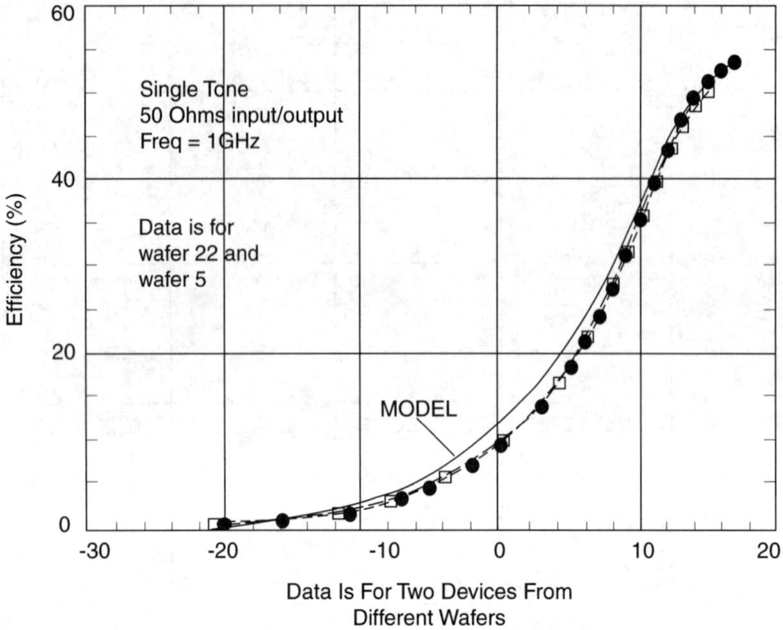

FIGURE 17.11 Efficiency prediction by the model and data for two RF LDMOS devices.

first enables the engineer to collect I/V and RF data in a systematic fashion on each device tested. This provides consistency between I/V and RF data. The parameter extraction routines then permit the extraction to specific, standard (SPICE) models, such as the Gunnel-Poon, or to new models for which equations may be user defined. Optimizers are used to provide the best fit between the data and the model. Testing can be with pulsed biasing or DC. Heating effects can be separately studied using thermal chucks during testing. Teyssier et al.[45] have discussed the merits of long- and short-pulse testing.

It is usually necessary to use devices of small sizes for characterization and then scale the model to devices actually used in the circuit design. Most SPICE models provide scaling with device area. However, the scaling laws for devices should be verified. It is usually possible to scale MESFETs and HEMTs accurately for a larger number of fingers of the same size. Golio[61] shows the scaling rules if the finger width is different. Because the biasing power may not be uniform on large devices and because the interelectrode capacitance does not scale simply, more complicated scale rules may be found for relatively large devices at high frequencies, i.e., above 5 GHz.

17.12 The Vector Nonlinear Network Analyzer

A large-signal, waveform measurement system has been used by many researchers to measure device characteristics dynamically. The equipment provides time domain voltage and current waveforms during RF excitation of the transistor. The equipment is often called the Vectorial Nonlinear Network Analyzer, or VNNA.

Demmler and Tasker[62] have shown that it is possible to accurately determine the drain current relationship to gate voltage for RF excitation at 2 GHz of a MODFET. The characteristic time delay is found by adding delay until "looping" is minimized. Furthermore, the drain-source I-V relationship can also be evaluated at 2 GHz. There is some difference from that obtained from the DC data. Thus the VNNA provides large-signal transfer characteristics from which more accurate model extraction can be done. Wei et al.[63] have also utilized this techniques to provide the data used for device model parameter extraction for a GaAs HBT.

17.13 Model Verification

The usual approach to verification of a large-signal model is to compare measured device performance with simulations under the same conditions. Initial verification should be comparison of a power sweep of the transistor at the application frequency and with no matching at the input and the output. For this test, we know that all harmonics at the input and output see 50-ohm impedance.

One should measure not only the output power at the fundamental, but also the power at second and third harmonics. If the model is not fully optimized, it will usually agree well with fundamental output power, gain, and efficiency, but not agree with the harmonic power production. The usual cause is due to poor modeling of the I/V relationship; however, in some cases, the nonlinear capacitive modeling may be the problem. After this problem is fixed, testing of third- and fifth-order IMD (intermodulation distortion) should be made, again in a 50-ohm system. If the harmonics now agree, the third-order IMD should agree and some further work may be required to get agreement for the fifth-order IMD.

In the previous test, it is important that the large-signal model be reasonably accurate at small-signal levels. It need not be as accurate as the best small-signal model for most applications.

Further verification work would involve power sweeps under tuned conditions. That is, the transistor may be tuned for best efficiency and the tuner impedance measured. It is important to measure the tuner impedance for fundamental, second, and third harmonics and to use these values in the simulation. The effects of the second harmonic voltage at either input[64] or output can be extremely important.

Testing of the load-pull characteristics should be made and compared with the model's behavior. Here, again, one has to be careful about the effects of harmonics. There are load-pull systems that operate separately on fundamentals and harmonics.

Further tests that may be important to the application may be testing with various ambient temperatures, noise testing, switch-on testing, and others. The specific application of the transistor will dictate the importance of the agreement for each test as well as the RF frequencies, power, and modulation to use for testing.

17.14 Foundry Models and Statistics

GaAs chip foundries provide design manuals that utilize small-signal as well as large-signal models. These are developed from measurements and statistical analysis of the data. However, these are guidelines for the designer, and often the best procedure is to obtain foundry test devices and develop more accurate models based upon new data. Device uniformity has improved greatly and yield prediction is becoming more accurate.

In addition, software programs such as IC-CAP and others provide statistical analysis of models extracted from test wafers. Both corner models and standard statistical patterns are available.

There is the fundamental problem that most foundries continue to tune their processes. It is often the case that the process has been changed and previous statistics are no longer valid. However, the design engineer is better off with approximate guidelines as to statistical patterns or corner models, than none at all.

17.15 Future Nonlinear Transistor Models

One can expect that with the ever-increasing speed of computers, circuit simulators will be able to utilize more physics-based models. This will aid in determining the effect of device design parameters upon chip yield and performance.

Improvements will be made in nonlinear model extraction software. The extraction parameters will be much less dependent upon the expertise of the tester. There will be improved collection schemes for transistor model statistics.

Finally, one expects that the nonlinear models will be made to be more easily tailored for adaptation to specific device behaviors. One often would like to start with a template for one of the standard nonlinear models and then tailor its behavior. Future simulators should make this procedure simpler than present procedures.

References

1. Nagle, L. W., SPICE 2: A computer program to simulate semiconductor circuits, Electronics Research Laboratory, College of Engineering, University of California, Berkeley, Memo, ERL-M520, 1975.
2. Pinto, M. R., Conor, R. S., and Dutton, R. W., PIECES2 - Poisson and continuity equation solver, Stanford Electronics Laboratory, Technical Report, Stanford University, 1984.
3. Wada, T. and Frey, J., Physical basis of short-channel MESFET operation, *IEEE Trans. on Electron Devices*, ED-26, 476, 1979.
4. Curtice, W. R., Analysis of the properties of three-terminal transferred electron logic gates, *IEEE Trans. on Electron. Devices*, ED-24, 1553, 1977.
5. Warriner, R. A., Computer simulation of gallium arsenide field-effect transistors using Monte-Carlo methods, *Solid-State Electron Devices*, 1, 105, 1977.
6. Moglestue, C., A self-consistent Monte Carlo particle model to analyze semiconductor microcomponents of any geometry, *IEEE Trans. on CAD*, CAD-5, 326, 1986.
7. Curtice, W. R. and Yun, Y-H, A temperature model for the GaAs MESFET, *IEEE Trans. on Electron Devices*, ED-28, 954, 1981.
8. *ATLAS User's Manual*, Silvaco International, Santa Clara, CA, Version 4.0, 1995.
9. Curtice, W. R., Direct comparison of the electron-temperature model with the particle-mesh (Monte-Carlo) model for the GaAs MESFET, *IEEE Trans. on Electron Devices*, ED-29, 1942, Dec. 1982.
10. Li, Q. and Dutton, R. W., Numerical small-signal AC modeling of deep-level-trap related frequency-dependent output conductance and capacitance for GaAs MESFETs on semi-insulating substrates, *IEEE Trans. on Electron Devices*, 38, 1285, 1991.
11. Snowden, D. M. and Pantoia, R. R., Quasi-two-dimensional MESFET simulation for CAD, *IEEE Trans. on Electron Devices*, 36, 1989.
12. Morton, C. G., Atherton, J. S., Snowden, C. M., Pollard, R. D., and Howes. M. J., A large-signal physical HEMT model, 1996 *International Micrwave Symposium Digest*, 1759, 1996.
13. Root, D. E. et al., Technology independent large-signal non quasi-static FET models by direct construction from automatically characterized device data, 21*st European Microwave Conference Proceedings*, 927, 1991.
14. Gummel and Poon, An integral charge-control relationship for bipolar transistors, *Bell System Tech. Journal*, 49, 115, 1970.
15. Daniel, T. T. and Tayrani, R., Fast bias dependent device models for CAD of MMICs, *Microwave Journal*, 74, 1995.
16. Khatibzadeh, M. A. and Trew, R. J., A large-signal analytical model for the GaAs MESFET, *IEEE Trans. on Microwave Theory and Tech.*, 36, 231, 1988.
17. Ladbrooke, P. H., *MMIC Design: GaAs FETs and HEMTs*, Artech House, Inc., Boston, 1989, chap. 6.
18. Lehovec, K. and Zuleeg, R., Voltage-current characteristics of GaAs JFETs in the hot electron range, *Solid State Electron.*, 13, 1415, 1970.
19. Bandler, J. W. et al., Statistical modeling of GaAs MESFETs, 1991 *IEEE MTT-S International Microwave Symposium Digest*, 1, 87, 1991.
20. Statz, H, Newman, P., Smith, I. W., Pucel, R. A., and Haus, H. A., GaAs FET device and circuit simulation in SPICE, *IEEE Trans. Electron Devices*, 34, 160, 1987.
21. D'Agostino, S. et al., Analytic physics-based expressions for the empirical parameters of the Statz-Pucel MESFET model, *IEEE Trans. on MTT*, MTT-40, 1576, 1992.
22. *International Journal of Microwave and Millimeter-Wave CAE*, 9, No. 3, 1999.

23. Trew, R. J., MESFET models for microwave CAD applications, *Internation Journal of Microwave and Millimeter-Wave CAE*, 1, 143, 1991.
24. Snowden, C. M., Nonlinear modeling of power FETs and HBTs, *International Journal of Microwave and Millimeter-Wave CAE*, 6, 219, 1996.
25. Dortu, J-M, Muller, J-E, Pirola, M., and Ghione, G. Accurate large-signal GaAs MESFET and HEMT modeling for power MMIC amplifier design, *International Journal of Microwave and Millimeter-Wave CAE*, 5, 195, 1995.
26. W. R. Curtice and R. L. Camisa, Self-consistent GaAs FET models for amplifier design and device diagnostics, *IEEE Trans. on Microwave Theory and Tech.*, MTT-32, 1573, 1984.
27. R. L. Vaitkus, Uncertainty in the Values of GaAS MESFET Equivalent Circuit Elements Extracted from Measured Two-Port Scattering Parameters, Presented at 1983 IEEE Cornell Conference on High Speed Semiconductor Devices and Circuits, Cornell University, Ithaca, NY, 1983.
28. Byun, Y. H., Shur, M. S., Peczalski, A., and Schuermeyer, F. L., Gate voltage dependence of source and drain resistances, *IEEE Trans. on Electron Devices*, 35, 1241, 1998.
29. Samelis, A. and Pavlidis, D., Modeling HBT self-heating, *Applied Microwave & Wireless*, Summer Issue, 56, 1995.
30. Teeter, D. A. and Curtice, W. R., Comparison of hybrid pi and tee HBT circuit topologies and their relationship to large-signal modeling, 1997 *IEEE MTT-S International Microwave Symposium Digest*, 2, 375, 1997.
31. Curtice, W. R., A MESFET model for use in the design of GaAs integrated circuits, *IEEE Trans. on Microwave Theory and Techniques*, 23, 448, 1980.
32. Angelov, I., Zirath, H., and Rorsman, N., New empirical nonlinear model for HEMT and MESFET and devices, *IEEE Trans. on Microwave Theory and Techniques*, 40, 2258, 1992.
33. Qu, G. and Parker, A. E., Continuous HEMT model for SPICE, *IEE Electronic Letters*, 32, 1321, 1996.
34. Cojocaru, V. I. and Brazil, T. J., A scalable general-purpose model for microwave FETs including the DC/AC dispersion effects, *IEEE Trans. on Microwave Theory and Techniques*, 12, 2248, 1997.
35. Materka, A. and Kacprzak, T., Computer calculation of large-signal GaAs FET amplifier characteristics, *IEEE Trans. on Microwave Theory and Tech.*, 33, 129, 1985.
36. Circuit Network Items, Series IV, Hewlett Packard, HP Part. No. E4605-90038, 1161, 1995.
37. Calvo, M. V., Snider, A. D., and Dunleavy, L. P., Resolving Capacitor Discrepancies Between Large and Small Signal FET Models, 1995 IEEE MTT-S International Microwave Symposium, 1251, 1995.
38. Jansen, P. et al., Consistent small-signal and large-signal extraction techniques for heterojunction FET's, *IEEE Transaction on Microwave Theory and Tech.*, 43, 1, 87, 1995.
39. Mallavarpu, R., Teeter. D., and Snow, M., The importance of gate charge formulation in large-signal PHEMT modeling, *GaAs IC Symposium Technical Digest*, 87, 1998.
40. Horio, K. and Usarni, K., Analysis of kink-related backgating effect in GaAs MESFETs, *IEEE Electron Devices Letters*, 537, 16, 1995.
41. Curtice, W. R., Bennett, J. H., Suda, D., and Syrett, B. A., Modeling of current lag effects in GaAs IC's, 1998 *IEEE MTT-S International Microwave Symposium Digest*, 2, 603, 1998.
42. K. Kunihiro and Y. Ohno, An equivalent circuit model for deep trap induced drain current transient behavior in HJFETs, *1994 GaAs IC Symposium Digest*, 267, 1994.
43. Chakrabarti, P., Shrestha, S. K., Srivastava, A., and Skxena, D., Switching characteristics of an optically controlled GaAs-MESFET, *IEEE Trans. on Microwave Theory and Techniques*, 42, 365, 1994.
44. Madjar, K., Paolella, A., and Herczfeld, P. R., Modeling the optical switching of MESFET's considering the external and internal photovoltaic effects, *IEEE Trans. on Microwave Theory and Tech.*, 42, 62, 1994.
45. Teyssier, J-P, Bouysse, P., Ouarch, A., Barataud, D., Peyretaillade, T., and Quere, R., 40-GHz/150-ns versatile pulsed measurement system for microwave transistor isothermal characterization, *IEEE Trans. on Microwave Theory and Tech.*, 46, 2043, 1998.
46. A. Platzker et al., Characterization of GaAs devices by a versatile pulsed I-V measurement system, *1990 IEEE MTT Symposium Digest*, 1137.

47. Anlholt and Swirhun, Experimental characterization of the temperature dependence of GaAs FET equivalent circuits, *IEEE Trans. on Electron Devices*, 39, 2029, 1992.
48. Grossman, P. C. and Oki, A., A large signal DC model for GaAs/GaAlAs heterojunction bipolar transistors, *Proc. IEEE BCTM*, 258, 1989.
49. Ye, F. Q., A BJT model with self-heating for WATAND computer simuation, M. S. Thesis, Youngstown State University, Youngstown, OH, 1990.
50. McAndrew, C. C., A complete and consistent electrical/thermal HBT model, *Proc. IEEE BCTM*, 200, 1992.
51. Corcoran, J., Poulton, K. and Knudsen, K., GaAs HBTs: an analog circuit design perspective, *Proc. IEEE BCTM*, 245, 1991.
52. Fox, R. M. and Lee, S-G, Predictive modeling of thermal effects in BJTs, *Proc. IEEE BCTM*, 89, 1991.
53. Fossum, J. G., Modeling issues for advanced bipolar device/circuit simulation, *Proc.* 1989 *IEEE BCTM*, 234, 1989.
54. McAndrew, C. C., Seitchik, J., Bowers, D., Dunn, M., Foisy, M., Getreu, I., Moinian, S., Parker, J., van Wijnen, P., and Wagner, L., VBIC95: An improved vertical IC bipolar transistor model, *Proc.* 1995 *BCTM*, 170, 1995.
55. de Graaff, H. C. and Kloosterman, W. J., New formulation of the current and charge relations in bipolar transistor modeling for CACD purposes, *IEEE Trans. on Electron Devices*, ED-32, 2415, 1986.
56. Perugupalli, P., Trivedi, M., Shenai, K., and Leong, S. K., Modeling and characterization of 80v LDMOSFET for RF communications, *1997 IEEE BCTM*, 92, 1997.
57. Ho, M. C., Green, K., Culbertson, R., Yang, J. Y., Ladwig, D., and Ehnis, P., A physical large signal Si model for RF circuit design, *1997 MTT-S International Microwave Symposium Digest*, 1997.
58. Miller, M., Dinh, T., and Shumate, E., A new empirical large signal model for silicon RF LDMOS FET's, *1997 IEEE MTT Symposium on Technologies for Wireless Applications Digest*, Vancouver, Canada, 19, 1997.
59. Curtice, W. R., Pla, J. A., Bridges, D., Liang, T., and Shumate, E., A new dynamic electro-thermal model for silicon RF LDMOS FET's, *1999 IEEE MTT-S International Microwave Symposium Digest*, 1999.
60. Heo, D., Chen, E., Gebara, E., Yoo, S., Lasker, J., and Anderson, T., Temperature dependent MOSFET RF large signal model incorporating self-heating effects, *1999 MTT-S International Microwave Symposium Digest*, 1999.
61. Golio, J. M., *Microwave MESFETs and HEMTs*, Artech House, Boston, 1991.
62. Demmler, M. and Tasker, P. J., A vector corrected on-wafer large-signal waveform system for novel characterization and onlinear modeling techniques for transistors, Presented at the workshop on New Direction in Nonlinear RF and Microwave Characterization, 1996 International Microwave Symposium, 1996.
63. Wei, C.-J., Lan, Y. E., Hwang, J. C. M., Ho, W.-J., and Higgins, J. A. Waveform-based modeling and characterization of microwave power heterojunction transistors, *IEEE Trans. on Microwave Theory and Techniques*, 43, 2898, 1995.
64. Watanabe, S, Takatuka, S., Takagi, K., Kukoda, H., and Oda, Y., Simulation and experimental results of source harmonic tuning on linearity of power GaAs FET, 1996 MTT-S International Microwave Symposium, 1996.

Index

A

B

C

D

E

F

G

H

I

J

K

L

M

T

U

V

W

Y

Z